Volker Achim Weberruß

Universality in Statistical Physics and Synergetics

Volker Achim Weberruß

Universality in Statistical Physics and Synergetics

A Comprehensive Approach to Modern Theoretical Physics

with 38 Figures

Set by Vieweg, Braunschweig
Printed on acid-free paper

ISBN 978-3-528-06513-3   ISBN 978-3-663-13894-5 (eBook)
DOI 10.1007/978-3-663-13894-5

Preface

This book is based on my research work I did between 1986 and 1992 at the *Institut für Theoretische Physik und Synergetik, Universität Stuttgart*. It might be of interest to all students and scientists who are interested in modern mathematical physics. It deals with one of the essential topics of modern physical research, namely the problem of universal aspects in statistical physics, and in doing so multi-component systems, like thermodynamic systems, laser systems and high-dimensional quantum systems, are considered. Synergetic aspects of multi-component systems will be discussed. Remarks to biological systems will be made, in particular remarks to the problem of EEG analysis. Non-linear aspects will be discussed. The problem of a macroscopic access to multi-component systems was an essential part of my research work. In this context it has to be emphasized that in this book the concept of hyper-surface equations will be introduced, which is a totally analytical strategy to determine distribution function parameters in an exact way by using only given measurement quantities. This concept can be used to determine both distribution functions of macroscopic systems and path integrals of quantum systems. In this context the property *self-similarity* will play a crucial role. A special extreme principle, the maximum information entropy principle, can be taken as a basis to introduce this concept. This principle will be discussed. In order to determine the dynamic behavior of multi-component systems, special evolution equations are needed. Such evolution equations will be discussed. In particular, equations of the *Langevin* type, the *Fokker-Planck* type and the *Schrödinger* type will be considered, and it will be shown that an elementary access to both equations of the *Fokker-Planck* and the *Schrödinger* type is possible. A special chapter will consider quantum system theory, in particular the concept of *Feynman* path integrals. In this context a general concept to calculate *Feynman* kernels will be introduced. An additional possibility to describe the behavior of multi-component systems is given by information theory. The problem of description of such systems with the tools of information theory will also play a crucial role in this book. It will be discussed that the concept of information theory allows a universal access to arbitrary multi-component systems. Additionally, basics of a relativistic system theory will be considered. In this context a connection of quantum systems and cosmological systems will be worked out. Then it will be shown that a universal access to all physical systems discussed in this book is possible by introducing the notion *Riemann universe*.

Due to the fact that physical systems of the micro-world, of the direct observable surrounding and cosmological systems are considered, and due to the fact that universal properties will be discussed, the structure of this book can be called *comprehensive* or *holistic*.

I am grateful to Prof. Dr. Dr. h. c. mult. Hermann Haken for his support. I wish to thank Dr. Wolfgang Schwarz for lektorship. Furthermore, I wish to thank my parents Ingeborg and Rolf Weberruß for their personal support. Last but not least I wish to thank Dorothee Klink for various corrections of my modest English.

Volker Achim Weberruß
Im Lehenbach 18
D-7065 Winterbach
Dezember 1992

Contents

1 Introduction

One of the most astonishing aspects in nature is the occurrence of *universal behavior*, i. e. behavior which occurs in many classes of physical systems. An example is represented by the phenomenon of period-doubling, an intermediate process which very often connects highly regular periodic motions with chaotic motions in oscillatory systems like dropping watertaps, technical oscillators or the human heart. For the human heart this means that a beating heart changes its state in such a way that first an alternating series of strong and weak heartbeats and then a totally irregular motion occurs. Such dramatic changes of states are normally called *phase transitions*. Phase transitions of several kinds occur when specific *control parameters* cross critical values. Such phase transitions are observable in the *thermodynamic equilibrium* as well as *far from thermodynamic equilibrium*. Examples of equilibrium phase transitions are phase transitions of matter like the transition of water from the liquid into the gas state. Examples of non-equilibrium phase transitions are the laser transitions. For example, special laser transitions separate non-lasing from lasing states. Other non-equilibrium phase transitions are biological transitions like the above described heart transition or the change of the movement state of walking animals like the transition of horses' gait from trot to gallop. A widely used classification scheme for phase transitions is the scheme of *Ehrenfest*, i. e. the separation of phase transitions into *phase transitions of first* and *second order*. First order transitions are transitions of matter, i. e. melting, boiling or condensing. Other first order transitions are special paraelectric transitions. Second order transitions are laser transitions, transitions from the paramagnetic to the ferromagnetic state, special ferroelectric transitions and the transition from normal conductivity to supraconductivity. The transition from He 1 to suprafluid He 2 has to be classified as a second order transition. All these phase transitions are defined by special criteria, which shall now not be discussed. These facts thus show that phase transition phenomena of first and second order – in the above sense – are also *universal phenomena*. A physical system which crosses a critical point so that a phase transition occurs fulfills special *phase transition conditions*. Such conditions very often describe universal rules. An example is represented by a macroscopic laser condition, which also occurs in a statistical theory of magnetism. Many of these phase transitions are governed by a physical principle called the *slaving principle*. A consequence of this principle is that only a few specific quantities determine the behavior of a complex physical system near a critical point. Such quantities are called *order parameters*. They govern the whole physical system near a critical point. An example is given by the first laser threshold, which separates a lasing from a non-lasing state. The order parameters then are the real amplitudes of the relevant electric field modes. Other examples can be found in fluid dynamics and especially in the animated nature.

Thus, slaving is also a *universal principle*.

This physical behavior can be expressed in a universal language, the language of mathematics. Universal behavior and far-reaching mathematical expressions correspond to each other. For example, the statistical behavior of thermodynamic systems and laser systems near critical points can be described with one *exponential distribution function*. A determination of the distribution function parameters then gives the special system. The time-space evolution of such distribution functions is given by special *differential evolution equations* like a *Fokker-Planck* equation or a *Schrödinger* equation. For example, laser and thermodynamic distribution functions can be found as solutions of a special kind of a stationary *Fokker-Planck* equation. Such evolution equations can be formulated in a far-reaching way, i. e. it is possible to use one basic evolution equation to describe many classes of physical systems, in which case occuring evolution equation parameters need to be determined to get the special physical system, i. e. a compressed formulation of universal behavior can be found on the level of differential evolution equations.

Distribution functions and evolution equations are essential especially when *multi-component systems* are considered, i. e. physical systems which consist of so many (but not necessarily different!) components that two levels exist (the level of the components and the level of the whole system), in which case both levels can be considered without considering the other one. In this context the level of the whole system is a *mean value level*. Examples of such multi-component systems are the well-known thermodynamic systems. One level then is the level of the molecules, and one level is the level of the whole material system, for example, the level of the gas. But not only material systems are multi-component systems in the above sense. Systems which consist of non-material variables are as well multi-component systems in the above sense. Thus, the space of mode amplitudes of a multi-mode laser has to be classified as a multi-component system. A relevant multi-component system of the animated world is the human brain. Such a system consists especially of many neurons which are the relevant components for information processing, i. e. they form a neural network. Other animated systems are sociological systems, i. e. groups of people. Many aspects of such multi-component systems can be treated with the above mentioned formalisms in a universal way.

Another relevant aspect, which has to be mentioned at this place, is the aspect of *self-organization*, i. e. only by changing control parameters new patterns arise, where complicated patterns are possible if non-linear systems are taken as a basis. An artificial example is represented by the laser which shows laser activity when the energy input (described by a so-called pump parameter) rises above a critical value. Then electromagnetic field patterns arise. An example in the natural world is the human brain, which works on a *collective level*. Such *synergetic systems* can be considered as well as *non-synergetic systems* in a universal way.

Naturally, multi-component systems do not show only two levels. Normally, a physical system represents a cascade of different measurement levels. An example is shown in figure 1.1. Figure 1.1 shows a sociological system, i. e. interacting people. Thus, the components of this system are the people themselves. Every

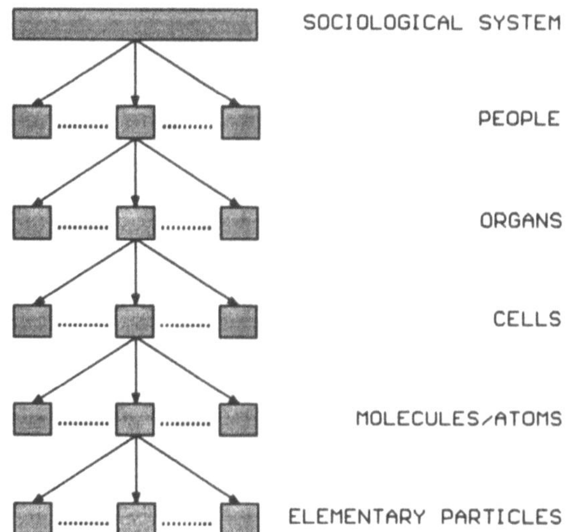

Figure 1.1
A cascade of dependent
systems

human being consists of organs like the the human brain. The human brain it-self consists of neurons (axon+soma+dendrites) and glia cells. Such cells consist of molecules/atoms like the molecule of the genetic code, desoxyribonucleinacid (DNA). And such molecules show a structure themselves, i. e. they consist of elementary particles. All these systems are typical statistical systems.

Two successive levels of such a cascade can be named *microscopic level* and *macroscopic level*, respectively. Thus, it has to be taken into consideration that the terms *macroscopic* and *microscopic* are relative terms. However, if such a notation is used, the microscopic level is the level of the components, and the macroscopic level is the corresponding mean value level. Connections between such system levels can then be considered in a universal way. An example is represented by equations which arise in laser theory as well as in thermodynamics. In both cases these equations describe the connection between the statistical level and the macroscopic level. However, in laser theory this means that the behavior of fluctuating mode amplitudes near a critical point is connected with correlation functions which describe the output intensity of the laser light, and in thermodynamics this means that the statistics of fluctuating magnetism is connected with correlation functions of the measurable magnetism.

One aspect has to be emphasized here. Normally, the starting level to consider a statistical system is a microscopic level. In this book, however, the macroscopic starting level will play a crucial role. The *maximum information entropy principle* will then be used. It allows to determine the form of statistical distribution functions by starting from a macroscopic level, i. e. from a measurement level. To determine the distribution function parameters, an additional mathematical concept is necessary. In this book an analytical concept to determine such parameters will be introduced, the *concept of hyper-surface equations*.

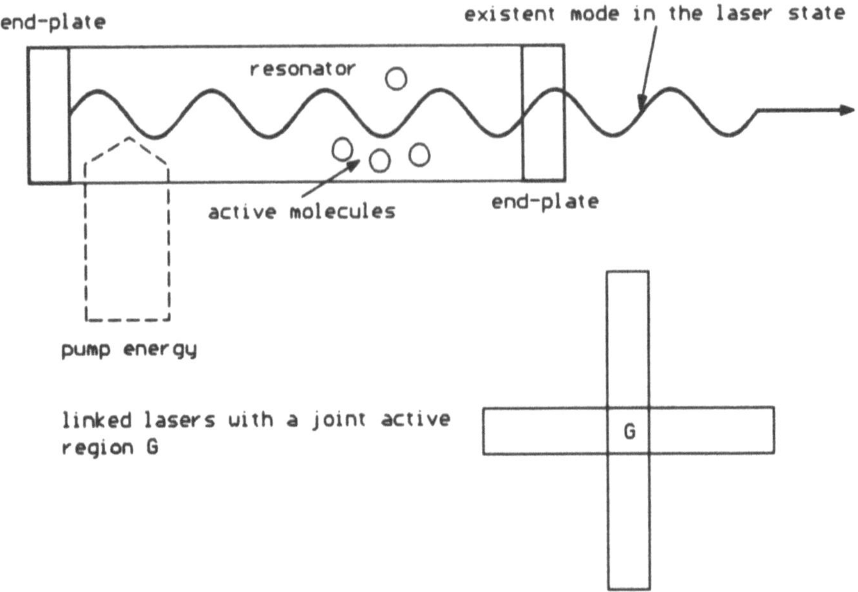

Figure 1.2 A trivial laser system and a coupled two-laser system which is able to process information

Some mathematical concepts occurring here can easily be extended to quantum systems. Then in particular *Feynman path integrals* will be considered. In this context a general procedure to calculate *Feynman* integrals will be introduced. This procedure can be considered as an analytical continuation of a concept which will be used to calculate partition functions. Therefore, it will be shown that concepts evolved in this book have a high degree of universality.

A new insight into the problem of multi-component systems will be gained by using tools of information theory.

In summary it may be said that it has to be emphasized that *universality in statistical physics* is a widespread property. Some points of universality will be discussed in this book, especially the universal aspects of the statistical behavior of multi-component systems and the universal aspects of statistical time-space evolution of such systems. Systems in thermodynamic equilibrium as well as systems far from thermodynamic equilibrium (e. g. laser systems) will be discussed. (An example of a laser system is shown in figure 1.2. Laser systems which contain more lasers are able to process information. An example is given in figure 1.2, too. An example of a thermodynamic system is given in figure 1.3. This figure shows a spin system near a phase transition point.) Some remarks about biological systems (which can also be considered as physical systems) like the human brain will be made. Additionally, high-dimensional quantum systems will be considered.

The starting point of this book will be a short view of elementary multi-component

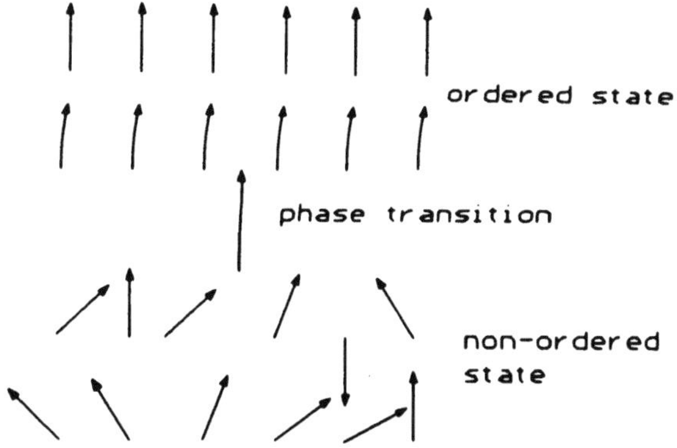

Figure 1.3
A trivial thermodynamic
system: A spin system near
a phase transition point

systems, namely thermodynamic systems. This chapter (chapter 2) shall be a motivating chapter. Then the problem of a macroscopic determination of distribution functions arises as well as the necessity to introduce a far-reaching mathematical concept to gain phase transition conditions. It can be made clear that one needs quantities such as *entropy* to describe the evolution of multi-component systems and to classify multi-component systems. In the next three chapters (chapters 3, 4, 5) a general theory will then be evolved. First, some aspects of system theory will be considered (chapter 3). In particular the generalized form of the term entropy, the term *information*, will be introduced. An extremal principle, the maximum information entropy principle, will be discussed. An analytical concept of system analysis will then be introduced (chapter 4). The hyper-surface equations will then be calculated. The differential aspects will be discussed afterwards (chapter 5). In this context it will be shown that an elementary access to both equations of a *Fokker-Planck* and a *Schrödinger* type is possible. During these considerations normally thermodynamic systems are the examples used. The fact that the evolved formalisms have a universal meaning will be shown in chapter 6, where laser systems will be considered. Chapter 7 is dedicated to the extension of the introduced concepts to the field of quantum systems. Then the universality of the evolved concepts will become apparent. In chapter 8 the term *information* will be discussed in more detail. Then examples of thermodynamics, laser theory and quantum system theory will be used. Apart from that, connections to the problems *genetic code, texture analysis* and *human behavior* will be presented. In chapter 9 basics of a relativistic system theory will be presented, and it will be shown that both physical systems considered in this book and cosmological systems can be submitted under one universal notion, the notion *Riemann universe*. Furthermore, in this context it will be discussed that evolution equations of quantum system theory and theory of gravitation are *self-consistency equations* which describe the *self-consistency of matter*.

2 Thermodynamic Systems

Systems which are directly observable by human observers are solid, liquid or gaseous objects, in which case a more detailed observation shows that the inner level is a molecular one. On the macroscopic level – the object level – quantities like *temperature* and *heat* are definable. If one pays attention to the problem of the influence of temperature and/or heat, such a system is called a *thermodynamic system*. Thus, thermodynamic systems are elementary multi-component systems, where a microscopic level is given by the molecule system, and a macroscopic level is given by the mean value level, which is the more easily measurable object level. The quantities of the microscopic level are impulses and positions of the molecules, and macroscopic quantities are pressure, volume and temperature (in particular the absolute temperature). Derived quantities are quantities such as inner energy, free energy, entropy, etc.. Very often thermodynamic systems are equilibrium systems. Such systems I would like to consider in this chapter. However, it already has to be remarked that only systems far from thermodynamic equilibrium are able to produce more complicated patterns. Examples, which will be discussed later, are laser systems. If such systems are driven by pump energy, they produce relatively complicated electric field patterns, i. e. they reach a lasing state. The systems of the animated nature are also systems far from thermodynamic equilibrium, which becomes very obvious when the reader considers the many gradients which hold the human body active. (The absence of gradients of physical variables describes the thermodynamic equilibrium!) In this chapter, I would like to discuss linear and non-linear states of thermodynamic systems. In particular, non-linear states are interesting, because then a wealth of phenomena, like phase transitions, is possible. (Far from thermodynamic equilibrium non-linear mathematical expressions are necessary!) This chapter shall be a motivating chapter, because I would like to show the basic problems in the theory of multi-component systems in a relatively easy way, i. e. with the help of relatively easily understandable thermodynamic systems in thermodynamic equilibrium. But what does this mean in detail? Firstly, it means such systems show pathways to describe multi-component systems on a microscopic as well as on a macroscopic level, and, in this context, they point out how to construct relevant quantities on both levels. Secondly, it means they show pathways to expand the treatment to other systems, far from thermodynamic equilibrium, too. Thirdly, it means they point out the necessity of a concept to determine specific parameters by macroscopic quantities, i. e. they give rise to introduce the concept of hyper-surface equations. Considering all these facts, I would like to begin with the consideration of universal structures in the theory of multi-component systems.

First, the macroscopic level will be considered, in particular the terms *absolute temperature* and *heat* as well as the laws of thermodynamics will be considered.

The macroscopic entropy concept will be introduced in this context. Afterwards the microscopic level will be considered, in particular the statistical (microscopic) entropy concept will be discussed. In this context the maximum entropy principle will be taken into account. To show the importance of non-linear phenomena, the *Landau* theory of phase transitions shall be considered. Then non-linear distribution functions of a special kind occur, which play a crucial role in the following, namely exponential functions of a special order.

2.1 The Macroscopic Level: The Fundamental Laws of Thermodynamics

In this context the macroscopic level is the level of the system, i. e. the level which can be described with the quantities pressure, volume, temperature, etc.. In order to describe a thermodynamic system on this level of consideration, the laws of thermodynamics are essential. While the 0., 1. and 2. law describe empirical experiences about variables of state (namely temperature, inner energy and entropy), the 3. law describes the behavior of a thermodynamic system near the zero point of the absolute temperature T. Such macroscopic laws represent very efficient decriptions of high-dimensional systems, i. e. without considering the behavior of each component of such multi-component systems, it is possible to decribe the essential behavior of the whole system. Such a kind of description is well-known for thermodynamic systems. Thus, the reader may think it is not of interest to think about such description methods. However, it has to be mentioned here that such laws are not known for other highly complex systems like the neural network of the human brain. Due to the fact that one of the goals of this book is to discuss possibilities of microscopic and macroscopic description methods (as well as their connections), also of systems beyond the thermodynamic region, it makes sense to start the considerations by using a relatively easy and well-known example. First, the 0. law shall be discussed.

2.1.1 The 0. Law of Thermodynamics: The Existence of Temperature

For us observers it is a well-known fact that in a directly observable physical system the macroscopic quantity *temperature* is needed to describe the system. The existence of the quantity *temperature* is postulated by the 0. law of thermodynamics, i. e. the 0. law is represented by:

> On the directly observable macroscopic level there exists a variable of state, the temperature.

In order to introduce this quantity in an exact way, it is possible to analyse only the macroscopic behavior of special physical systems like the behavior of heat-engines.

The absolute temperature, which will only be used in this book, can be defined by a process of a special heat-engine, the *Carnot* machine (or by a gas thermometer if the gas behaves like an ideal gas). As the absolute temperature T is a variable of state, i. e. a quantitiy which does not depend on the way the state has been reached, the equation

$$\oint dT = 0 \tag{2.1}$$

holds. This definition then requires the equation

$$dT = \sum_i K_i d\alpha_i \,, \quad K_i = \frac{\partial T}{\partial \alpha_i} \,, \tag{2.2}$$

i. e. dT is a total differential. The quantities α_i are specific parameters or variables like the quantitiy *pressure* or the quantity *volume*. Such an absolute temperature describes a multi-component system on a macroscopic level. It already has to be mentioned here that the macroscopic quantity T occurs within various statistical distribution functions like the distribution function of *Boltzmann* or functions which occur in *Landau's* theory of phase transitions. In this context the macroscopic quantity T determines the statistical state (for example, a paraelectric or a ferroelectric state). It has to be mentioned as well that the absolute temperature is defined for thermodynamic equilibrium, i. e. for states which do not show any gradients. Therefore, the quantity *temperature* makes sense only in or at most near thethermodynamic equilibrium if a customary definition is taken as a basis. (Near the thermodynamic equilibrium small space elements exist which are in an equilibrium state! In order to extend the term *temperature* to states far from thermodynamic equilibrium, other definitions than the customary ones are necessary!) As later laser systems shall be considered, too, it has to be mentioned here that a population inversion, which arises in the lasing state, can be described with an inverse *Boltzmann* factor. Then a negative temperature can be introduced, which forces an inverse *Boltzmann* factor. However, such an inverse *Boltzmann* factor is only possible if the energy states are limited (otherwise a divergent function would occur!). Indeed this is the case if laser systems are considered.

The temperature of a physical system is directly connected with the inner energy of a thermodynamic system. The thermodynamic law of conservation of energy shall now be discussed.

2.1.2 The 1. Law of Thermodynamics: The Macroscopic Energy Principle

In this book the 1. law of thermodynamics shall be discribed by the following statement:

> In every thermodynamic system a macroscopic variable of state, the inner energy U, exists, which represents the total energy of the system. The inner energy of a system changes by exchange of heat or work with the surrounding.

Mathematically, the 1. law of thermodynamics can be expressed in the differential form

$$dU = dQ + dA \ . \tag{2.3}$$

(2.3) represents the thermodynamic law of conservation of energy, i. e. a macroscopic law. As the reader can see, a change of the inner energy can be reached by influx (removal) of heat (dQ) and/or of work (dA). However, instead of U, Q and A there are no variables of state. The change of heat fulfills the equation

$$dU = c_{V,p} \, \Delta m \, dT \ , \tag{2.4}$$

where Δm is the observed mass, $c_{V,p} \, \Delta m$ is the heat capacity, and where $c_{V,p}$ is the specific heat. V, p are indices which show that volume V or pressure p, respectively, have to be constant. In order to describe to possibilities of evolution of a physical system, an additional law is necessary, namely the 2. law of thermodynamics. This law shall now be considered.

2.1.3 The 2. Law of Thermodynamics: The Law of Entropy

Many processes which are energetically possible have never been observed. An example for this phenomenon is represented by a solid body lying on the ground. Energetically it might be possible that heat energy will be exchanged between the solid body and the ground so that the solid body soars up like a bird. In fact, this has never been observed. A law which describes the possible evolution processes is the 2. law of thermodynamics. This law can be described by the following statement:

> In a thermodynamic system a variable of state exists, namely the thermodynamic entropy, which can never decrease within a closed system. In a closed system which is in thermodynamic equilibrium the entropy has a maximum.

Mathematically expressed, it means the validity of the equation

$$\int_{Z_1}^{Z_2} dS = S(Z_2) - S(Z_1) \overset{\text{closed system}}{\geq} 0 \ , \tag{2.5}$$

where Z_1 and Z_2 describe two different states of the considered system. The change of the entropy S can be calculated by using an imagined reversible process, because the equation

$$dS = dQ_{\text{rev}}/T \tag{2.6}$$

holds. (*Reversibility* means that a process in the inverse direction is possible, in which case no change of the environment is observable.) dQ_{rev} describes the reversible exchanged heat. (2.6) can be used as a macroscopic definition of the entropy. A relevant fact in this context is that the entropy shows a special property, namely the property *additivity*, i. e. a relevant fact is the validity of the equation

$$S_{1,2} = S_1 + S_2 + \Delta S \, , \tag{2.7}$$

where ΔS is an additional term, which describes the interaction between the two systems 1 and 2. $S_{1,2}$ is the total entropy of the interacting systems.

In the following the 3. law of thermodynamics shall be considered. This law makes a statement about the behavior of the entropy near the zero point of the absolute temperature.

2.1.4 The 3. Law of Thermodynamics: The Limit Value Condition at the Zero Point of the Absolute Temperature

In this book the 3. law of thermodynamics shall be described by the following statement:

> The entropy of a physical system which approaches the zero point of the absolute temperature takes on a constant value, which does not depend on specific variables. This constant can be identified with the value 0.

Mathematically expressed this means that the equation

$$\lim_{T \to 0} \left(\frac{\partial S}{\partial \alpha_i} \right) = 0 \tag{2.8}$$

holds, with the quantities α_i being specific variables such as volume or magnetic field strength.

These thermodynamic laws describe a macroscopic level of thermodynamic multi-component systems. In particular, it has to be mentioned that this thermodynamic example shows that physical systems do not behave on a macroscopic level like a high-dimensional system so that relatively easy rules can be found to describe such multi-component systems. Nowadays many macroscopic quantities and rules introduced in the thermodynamic context can be extended to other systems if a general form is used, and very often a universal structure can be found. The generalized form of the term *entropy*, namely the term *information*, will be considered later. Then laser systems and quantum systems will be used as examples. However, it should be mentioned here that it also makes sense to use such formalisms to describe biological systems like the human brain. The term *information* can then be used to classify a human EEG. Furthermore, it has to be mentioned here that it seems to be possible to extend all concepts intoduced in thermodynamics to other physical systems if general forms are used. However, the problem of generalization and universality shall be considered later. Now the correspondending level, namely the microscopic (statistical) level of the discussed thermodynamic systems shall be considered. This is very logical, because the problem of generalization of the term *entropy* can be resolved by using a statistical formulation of the entropy. Moreover, other problems can be discussed in this context, which occur in a modified way also in other physical systems like laser systems, so that the following considerations represent a reasonable starting point.

2.2 The Microscopic Level: Molecular Statistics

In this context the microscopic level is the level of the molecules. Then a statistical treatment makes sense and shall now be discussed. Firstly, in a very short way, the energetical aspects shall be considered. Secondly, the statistical definition of the entropy shall be dealt with. Then occuring thoughts allow an easy extension to other physical systems.

2.2.1 The Molecular Energy

The inner energy U represents the total energy of the molecular system, i. e. U consists of translation energy, interaction energy and the inner-molecular vibrational and rotational energy. To describe the inner energy on a mean value level, the formula

$$\overline{U}_{\text{molecule}} = \frac{f}{2}\, k\, T \tag{2.9}$$

can be used. (2.9) describes the mean energy of one molecule at the temperature T. k is *Boltzmann's* constant and f represents the number of degrees of freedom, which consists of translational, rotational and vibrational degrees of freedom. (The vibrational ones have to be counted twice!) (2.9) represents the so-called *principle of equipartition of energy*, bearing in mind that in this context low absolute temperatures cause so-called *frozen degrees of freedom*, i. e. only special degrees of freedom are observable. For example, for a gas which consists of hydrogen molecules H_2 the number of degrees of freedom is 7 only for high temperatures (3 translational coordinates, 2 angle coordinates, 1 vibrational coordinate which describes the space between the two atoms and has to be counted twice to describe as well the potential energy). It has to be borne in mind that also high pressures cause a failure of (2.9). However, these energetic considerations are not very important in this book. The following considerations about the thermodynamic entropy are much more important.

2.2.2　The Statistical Definition of Entropy

The above introduced entropy is a variable of state which was defined on a macroscopic level with the law of entropy restricting the possibilities of evolution of a thermodynamic system. Such a restriction is given by the inequality sign. Such an entropy refers to systems in thermodynamic equilibrium. Now the question arises what *restriction of the evolution of a thermodynamic system* means on a microscopic level. Well, the answer is easy to give: Position coordinates and energy states of the molecules change, where the evolution of position-energy coordinates is restricted. Then the question arises which entropy definition is needed. This question shall now be answered.

Deterministic Mechanics, μ-Space and Γ-Space

In order to describe a mechanical multi-component system it is possible to use the *Hamiltonian* canonical equations, i. e. to use equations of the type

$$\dot{p}_k = -\frac{\partial H(\boldsymbol{q},\boldsymbol{p})}{\partial q_k} \; , \;\; \dot{q}_k = \frac{\partial H(\boldsymbol{q},\boldsymbol{p})}{\partial p_k} \; . \tag{2.10}$$

(2.10) represents a multi-dimensional system of mechanical evolution equations, which contain the *Hamiltonian* function H and the generalized position coordinates q_k as well as the generalized impulse coordinates p_k. The vectors $\boldsymbol{q}, \boldsymbol{p}$ represent all single coordinates. Such a *Hamiltonian* function is connected with the *Lagrangian* function L by the equation

$$H(\boldsymbol{q},\boldsymbol{p}) = \sum_k p_k \dot{q}_k - L(\boldsymbol{q},\dot{\boldsymbol{q}}) \; . \tag{2.11}$$

((2.11) describes a special case of a *Legendre* transformation.) The evolution of a space element in the position-impulse space is regulated by the *Hamiltonian* equations. Both deformations and changes of the position of a space element are possible. During the movement through the phase space the volume of the space element does not change, only the form of the element can vary. This fact is given by the theorem of *Liouville*, which can be expressed by the equation

$$\frac{d\Delta\Omega}{dt} = 0 \; , \tag{2.12}$$

where $\Delta\Omega$ describes a space element of the phase space, and t is the time coordinate. However, (2.12) describes the behavior of canonical systems in the phase space. Dissipative systems have to be described by an inequality, i. e. in this case the volume of a space element will not be preserved.

This high-dimensional space is called a Γ-*space*, i. e. a Γ-space is the space of position and impulse coordinates of interacting molecules. However, very often an ensemble of molecules can be a non-interacting system. An example is represented by the ideal gas, which can be realized at higher temperatures. In this case a low-dimensional space, the μ-*space*, can be taken as a basis. A μ-space is the phase space of the 3 position and the 3 impulse coordinates. A phase space of the first or the second kind has to be taken as a basis to deal with mechanical statistical systems. In the following the statistical entropy has to be dealt with. First, the μ-space shall be considered.

The Statistical Entropy of μ-Space Systems

A statistical entropy definition which corresponds to the introduced macroscopic definition is *Boltzmann's* entropy. The validity of this entropy is restricted to μ-space systems. A short derivation of this entropy shall now be given.

An observation of a mechanical system, which consists of N elements, will show a distribution of the states of the elements. In a phase space it will then turn out that n_i elements depend on the space element $\Delta\Omega_i$. (Such a space element contains both the energy coordinates and the position coordinates.) A macroscopic state, i. e. a special distribution without labeled elements, can be realized in many ways, in which case the number W of the realization possibilities is given by the equation

$$W = \frac{N!}{n_1!n_2!\ldots n_i!\ldots n_n!} \, . \tag{2.13}$$

Then the equation

$$W = \prod_k W_k \tag{2.14}$$

holds, i. e. the number of the realization possibilities of an ensemble can be written as a product of the realization numbers W_k of the different ensembles. (2.13) describes the states of all molecules on a statistical level. Every observable state is then correlated with a number W so that the entropy needs to be correlated with W. An obvious thought then is that an observable macroscopic state in thermodynamic equilibrium is a state which is labeled by the largest possible number W. But as the entropy is an additive quantity – this was above mentioned – it is not possible to use W itself as a measure for the thermodynamic entropy (W is a multiplicative quantity, see (2.14)). Therefore, it makes sense to use the relation

$$S_\mu = k \ln W \tag{2.15}$$

to describe the thermodynamic entropy, with k being any constant. As a μ-space system is taken as a basis, an index μ has been chosen. (2.15) is an expression which is consistent with the requirement *additivity*. Thus, the question arises whether (2.15) describes the thermodynamic entropy. Indeed (2.15) represents the thermodynamic entropy if the constant is taken as *Boltzmann's* constant, because (2.15) can then be used in a self-consistent way in the theory of thermodynamics. For example, it is possible to gain *Boltzmann's* distribution function of thermodynamic equilibrium by using this statistical entropy. To show this it shall be assumed that the population number of the various space elements is a large one. *Stirling's* formula

$$N! \stackrel{\text{large } N}{\simeq} \left(\frac{N}{e}\right)^N \ , \quad n_i! \stackrel{\text{large } n_i}{\simeq} \left(\frac{n_i}{e}\right)^{n_i} \tag{2.16}$$

can then be used to reformulate (2.15). By using the standardization condition

$$N = \sum_i n_i \tag{2.17}$$

the expression (2.15) can be written in the form

$$S_\mu = -k\,N \sum_i \frac{n_i}{N} \ln \frac{n_i}{N} \ . \tag{2.18}$$

n_i/N represents a probabilitiy w_i so that instead of (2.18) the relation

$$S_\mu = -k\,N \sum_i w_i \ln w_i \ , \tag{2.19}$$

can be used. (2.19) can be taken as a basis to gain *Boltzmann's* distribution function of thermodynamic equilibrium. This fact shall now be considered.

The Maximum Entropy Principle (MEP)

(2.19) needs to have a maximum if equilibrium states are considered – this was above mentioned. In an equilibrium state a thermodynamic system can be described by *Boltzmann's* distribution function of thermodynamic equilibrium, which gives the probability to find a molecule with a special energy. Thus, it has to be possible to gain this function by maximizing the entropy. Then specific constraints have to be considered. Mathematically expressed this means that the equation

$$\delta S_\mu = 0 \tag{2.20}$$

has to be used, where additional constraints have to be considered. The first constraint is the standardization relation

$$\sum_i w_i = 1 \ \text{or} \ \delta \sum_i w_i = 0 \ , \tag{2.21}$$

and the second constraint is given by

$$\sum_i w_i E_i = \overline{U} = \text{const.} \ \text{or} \ \delta \sum_i w_i E_i = 0 \ , \tag{2.22}$$

i. e. the mean energy is a fixed quantity. By using the method of *Lagrangian* multipliers the variational expression

$$\delta \sum_i w_i C_i = 0 \ , \tag{2.23}$$

occurs, where the central part of this expression is given by the equation

$$C_i = kN \left(\ln w_i + \lambda_{\mathrm{norm}} + \lambda_{\mathrm{energy}} E_i \right) . \tag{2.24}$$

The parameters λ_{norm} and $\lambda_{\mathrm{energy}}$ are so-called *Lagrangian multipliers*, which allow to include the constraints. The solution of the variational problem (2.23) is given by the exponential function

$$w_i = Z_{\mathrm{thermo}}^{-1} \exp \left(-\lambda_{\mathrm{energy}} E_i \right) , \tag{2.25}$$

wherein Z_{thermo} is the partition function, which guarantees the standardization and can be written in the form

$$\begin{aligned} Z_{\mathrm{thermo}} &= \exp \left[- \left(1 + \lambda_{\mathrm{norm}} \right) \right] \\ &= \sum_i \exp \left(-\lambda_{\mathrm{energy}} E_i \right) . \end{aligned} \tag{2.26}$$

The choice

$$\lambda_{\mathrm{energy}} = \frac{1}{kT} \tag{2.27}$$

generates *Boltzmann's* distribution function of thermodynamic equilibrium. This example shows that the expression (2.19) is very logical. The question arises whether it is possible to deduce an entropy expression which contains interactions, too. This can be done by the method of *Gibbs* and shall now be described.

The Statistical Entropy of Γ-Space Systems

General considerations about the behavior of trajectories in the phase space allow to find a general form of *Boltzmann's* distribution function, i. e. a form which contains interactions. Then it is possible to derive a general form of the thermodynamic entropy. These considerations shall now be done.

The probabilities w_1 and w_2 to find a physical system within phase space elements $\Delta\Omega_1$ and $\Delta\Omega_2$ at times t_1 and t_2 are given by

$$\Delta w_1 = \rho(\boldsymbol{q}_1, \boldsymbol{p}_1) \, \Delta\Omega_1 , \quad \Delta w_2 = \rho(\boldsymbol{q}_2, \boldsymbol{p}_2) \, \Delta\Omega_2 , \tag{2.28}$$

in which case the functions $\rho(\boldsymbol{q}_1, \boldsymbol{p}_1)$ and $\rho(\boldsymbol{q}_2, \boldsymbol{p}_2)$ are probability densities, which depend on the position variables q_k and the impulse coordinates p_k. Both probabilities w_i are equal, because always the same phase space points are considered. Due to the fact that *Liouville's* theorem holds, both probability densities are equal. Thus, the equation

$$\rho(\boldsymbol{q}_1, \boldsymbol{p}_1) = \rho(\boldsymbol{q}_2, \boldsymbol{p}_2) \tag{2.29}$$

holds. In thermodynamic equilibrium a time-independent density has to be used. This shall be done. Further the function $\rho(\boldsymbol{q}, \boldsymbol{p})$ is an integral of motion. If the total energy $H(\boldsymbol{q}, \boldsymbol{p})$ can be used as the only integral of motion, the relation

$$\rho(\boldsymbol{q}, \boldsymbol{p}) = \rho(H) \tag{2.30}$$

is then possible. Furthermore, it has to be considered that two coupled thermodynamic systems which are in thermodynamic equilibrium after decoupling are also in thermodynamic equilibrium. As the absolute temperature does not change very much, interaction terms in the *Hamiltonian* function H can be neglected. Then the equation

$$\rho(H) = \rho(H_1 + H_2) = \rho(H_1)\rho(H_2) \tag{2.31}$$

holds. As changes of H, H_1 or H_2 cause the same change within (2.31), the differential equation

$$\partial_H \rho(H) = \rho(H_1)\, \partial_{H_2}\rho(H_2) = \rho(H_2)\, \partial_{H_1}\rho(H_1) \tag{2.32}$$

has to be used, which is equal to

$$\frac{\partial_H \rho(H)}{\rho(H_1)\rho(H_2)} = \frac{\partial_{H_1}\rho(H_1)}{\rho(H_1)} = \frac{\partial_{H_2}\rho(H_2)}{\rho(H_2)} = \text{constant}, \tag{2.33}$$

where the solution of (2.33) is given by

$$\rho(H) \sim \exp(-\lambda_{\text{energy}} H)\,. \tag{2.34}$$

The choice (2.27) then generates a distribution function, which is consistent with other thermodynamic relations, in which case the proportional constant can be found by using a suitable standardization condition. Then the distribution function

$$\rho_{\text{thermo}}(H) = Z_{\text{thermo}}^{-1} \exp\left(-\frac{H}{kT}\right)\,. \tag{2.35}$$

holds, where the partition function is given by

$$Z_{\text{thermo}} = \int \exp\left(-\frac{H}{kT}\right)\, d\Omega\,. \tag{2.36}$$

(2.35) is *Boltzmann's* distribution function if interaction is possible, i. e. if a Γ-space system is considered. By using the partition function as a basis all macroscopic thermodynamic relations can be constructed, i. e. a complete formalism to describe thermodynamic behavior is possible by using the relevant partition function. In the following, this shall be shown. Then the entropy belonging to such a distribution function can easily be found.

If the partition function is taken as a basis, all relevant macroscopic relations of thermodynamics can be found by using suitable operators (such as differential operators or algebraic operators). Then very often the natural logarithm of the partition function is necessary. As this logarithm is not defined for quantities which have physical dimensions, instead of the partition function itself,

$$Z_d = d\, Z_{\text{thermo}} \tag{2.37}$$

shall be used, with the factor d guaranteeing a dimensionless partition function. This factor will be defined later. Now some relevant thermodynamic relations, formulated with the partition function formalism, shall be considered.

By using (2.37) *Boltzmann's* distribution function (2.35) can be written in the form

$$\rho_{\text{thermo}}(H) = -kT\frac{\delta \ln Z_d}{\delta H}. \tag{2.38}$$

This can be calculated by executing the variational differentiation.

The inner energy can also be calculated in this way. The inner energy is a mean value of H, thus the equation

$$U = \frac{1}{Z_{\text{thermo}}} \int \exp\left(-\frac{H}{kT}\right) H\, d\Omega = -\frac{\partial \ln Z_d}{\partial(1/kT)} \tag{2.39}$$

holds. This can be shown by calculating the differential coefficient.

By using both the definition of entropy (2.6) and the 2. law of thermodynamic the variational expression

$$\delta S = \frac{\delta Q_{\text{rev}}}{T} = \frac{\delta U - \delta A}{T} = \frac{\delta U - \int \rho(H)\,\delta H\, d\Omega}{T} \tag{2.40}$$

can be found. By using the partition function (2.37) the formulation

$$\delta S = k\,\delta\left(\frac{U}{kT} + \ln Z_d\right) \tag{2.41}$$

is possible. Thus, the equation

$$S = k\left(\frac{U}{kT} + \ln Z_d\right) \tag{2.42}$$

is obvious. (2.42) represents a macroscopic expression to describe the thermodynamic entropy by using the relevant partition function. (It might be better to name such an expressions a *quasi-macroscopic* expression, because the microscopic quantity *partition function* is included. However, this shall not be done.)

Other thermodynamic relations can be found in the same way. It has to be remarked that such a formalism is also possible to describe other physical systems such as laser systems. Then, however, other partition functions are necessary (see chapter 6), and the meaning of the expressions is another one. It also has to be remarked that a generalized form of such a partition function can be introduced, which contains the statistics of a wealth of different systems. This function will be called a *statistical basic function* and will be introduced in chapter 7. Such a statistical basic function includes quantum systems, too.

In order to find a statistical entropy definition, the expression (2.42) can be taken as a basis. This shall now be shown.

The expression (2.42) can be written in the form

$$S = -k \int_m \rho(H) \ln \left[\frac{\rho(H)}{d} \right] d\Omega \tag{2.43}$$

(insert the distribution function (2.35), use the energy expression (2.39) and the standardization condition), in which case the choice

$$d^{-1} = d\Omega \tag{2.44}$$

guarantees an entropy expression which generates the borderline case (2.19). (Then (2.43) represents the limit of a sum which contains probabilities $w = \rho(H)d\Omega$ just like the entropy expression (2.19). Additionally, it has to be remarked that the element number N of (2.19) is included in (2.43), because in the discrete case within (2.43) a product of N elementary distribution functions arises.) However, (2.43) does not represent an ordinary integral, (2.43) represents a so-called *measurement integral*, i. e. an expression which bases on a limit condition of the form

$$S = -k \lim_{\Delta\Omega \to d\Omega} \sum_{\Omega_i} \rho(H) \ln \left[\rho(H) \Delta\Omega \right] \Delta\Omega \, , \tag{2.45}$$

where $d\Omega$ denotes elements of accuracy, which are given by *Heisenberg's* uncertainty relation. Thus, (2.45) describes a special limit value, namely a limit value which one gets by using finite small elements $d\Omega$, where these elements are the space elements with respect to the best possible measurement, and in which case these elements are given by *Heisenberg's* relation. Such measurement integrals will be used very often later, however, in a more general form. A generalization will be given in chapter 3. Then further details will be considered. Thus, it has to be remarked that a space element $d\Omega$ within a measurement integral is no operation symbol so that algebraic operations are possible. This possibility will be used later. Furthermore, the following has to be remarked: If connections between such an entropy expression and other integral expressions are considered, the other integral expressions (such as the inner energy) also have to be treated as measurement integrals to get a self-consistent formalism. It has to be emphasized here that this treatment is the exact one, i. e. then a formalism arises which describes a measurement in an exact way. If elements of accuracy are taken as a basis, an integral formalism is possible if a convergent sum exists. However, an elementary property of the entropy is the divergence with respect to infinite small elements of accuracy. So an integral formulation of entropy is not possible, only a formulation in the above sense (see (2.45)) is usable. To describe such an entropy also other ways exist, this has to be remarked here. Very often the natural logarithm is written in form of a sum $\ln \rho(H) + \ln \Delta\Omega$ and the then occuring part $\sum_{\Omega_i} \rho(H) \ln \Delta\Omega = \ln \Delta\Omega$ is neglected. Then an ordinary integral instead of a measurement integral can be used. Such an integral expression gives values which make sense, because the entropy is defined with the exeption of a constant. However, the argument of the natural logarithm is not dimensionless so that the entropy itself is not dimensionless. This is in contradiction to the basic requirements. Other problems arise, too. However, these problems shall not be discussed. Therefore, in this book measurement integrals shall be prefered.

By using the abbreviation

$$\ln W = - \int_m \rho(H) \ln \left[\rho(H) \, d\Omega \right] \, d\Omega \tag{2.46}$$

the entropy (2.43) can be written in the form

$$S = k \ln W \; , \tag{2.47}$$

i. e. in *Boltzmann's* form. (2.47) is a general statistical formulation of entropy. (2.47) also holds if interactions are included. By using the 2. law of thermodynamics the entropy describes the evolution possibilities of a thermodynamic system. Furthermore, the entropy is a measure of the width of a considered statistical distribution function. Very often this means that the entropy is a measure of the disorder of a thermodynamic system. Thus, the entropy also is a quantity which describes the state of a thermodynamic system in a compressed way. This quantity can be used to classify different states of thermodynamic systems.

Thus, the problem of microscopic and macroscopic description methods in the theory of multi-component systems was introduced by considering relatively easy examples, namely thermodynamic examples. Typical macroscopic and microscopic methods to describe multi-component systems were considered. In particular, I want to refer the reader to the method of describing a multi-component system by statistical entropy, the method of gaining a statistical distribution function by using the MEP and the method of system description by using the partition function formalism. Therefore, in the next chapter the generalization of the concept of entropy as well as the generalization of the MEP will be considered. The partition function formalism will be considered as well. These generalizations will then play a crucial role in the whole book. In particular, the partition function formalism will be used in chapter 4. Then a general basic equation system of 0. order will be used as a basis to derive the above mentioned hyper-surface equations, in which case this basic equation system is nothing but a system of general equations which is formulated by the partition function formalism. The universal aspects of all these concepts will then be worked out.

During these general considerations non-linear distribution functions will be taken as a basis. Such non-linear functions occur also in thermodynamics, namely in *Landau's* theory of phase transistions. In the following section some short remarks to the problem of phase transitions and non-linear distribution functions in the thermodynamic context will be made.

2.3 Statistics and Phase Transitions

A transition of a physical system from one state to another represents a so-called *phase transition* if characteristic parameters change dramatically. Thermodynamic phase transitions are transitions of matter (for example, melting, boiling and condensing) or transitions of the electric state and magnetic state, respectively. These phase transitions are typical equilibrium phase transitions. A widely used classification scheme is the scheme of *Ehrenfest*, which separates phase transitions of first and second order. Transitions of matter are phase transitions of first order. Such transitions show coexistence regions, where all phases (for example, ice modifications and water) are existent at the same time. Hysteresis effects (for example, overheating) are possible. Transitions from the ferromagnetic to the paramagnetic state are transitions of second order. Then always one well-defined phase is observable. Transitions from the ferro- to the paraelectric state are of first or second order. A characteristic quantity to describe such phase transitions is the *Curie* temperature, which gives the temperature of the actually observable transition point. Instead of this *Curie* temperature normally the temperature T_0 is used, which describes the evolution of a possible new state. On grounds of hysteresis effects T_0 is not identical to the *Curie* temperature. For the simple reason that fluctuations exist a thermodynamic state is a dynamical state, i. e. on a low scale a dynamical behavior of specific quantities such as magnetism or polarization is observable. Thus, methods of statistics are necessary to describe the behavior near a phase transition point. Then non-linear distribution functions are necessary to describe the observable behavior. A reasonable starting point to deal with the problem of statistics near a phase transition point is *Landau's* theory of phase transitions. Some essential parts of this theory shall now be considered.

2.3.1   Remarks on Landau's Theory

In oder to describe a thermodynamic system, special thermodynamic potentials such as the entropy or the free energy can be taken as a basis. Extreme points of such a thermodynamic potential are correlated with special states of a physical system. An example is represented by the entropy, i. e. the maximum of the entropy is correlated with an equilibrium state of a thermodynamic system. Another example is represented by the free energy. Then the minimum of the free energy describes equilibrium states. To describe ferromagnetic or ferroelectric phase transitions of liquid crystals or ionic crystals, the free energy can be taken into account, where the free variables are temperature T and volume V. If an ionic crystal in an electric field is considered, additionally the electric field strength E has to be used. In order to show the essential facts of *Landau's* theory, a ferroelectric phase transition in a solid shall be exemplary considered, in which case it has to be taken into account that in a solid the volume is nearly constant during a ferroelectric phase transition, so that V can be neglected.

A macroscopic polarization p occurs if a transistion from a para- to a ferroelectric state takes place within a solid. If fluctuating effects are neglected, the value of p is equal zero in the paraelectric state, and p has a finite value after transition. Thus, the free energy of the system depends on p, and an expansion with respect to p is possible. Such an expansion is of the form

$$
\begin{aligned}
F(E,T,\boldsymbol{p}) = F_0(T) \;&+\; \sum_{i=1}^{3} \lambda^{(1)}_{i,\text{Landau}} p_i \;+ \\
&+\; \sum_{i,k=1}^{3} \lambda^{(2)}_{i,k,\text{Landau}}\, p_i p_k \;+ \\
&+\; \sum_{i,k,l=1}^{3} \lambda^{(3)}_{i,k,l,\text{Landau}}\, p_i p_k p_l \;+ \\
&+\; \sum_{i,k,l,m=1}^{3} \lambda^{(4)}_{i,k,l,m,\text{Landau}}\, p_i p_k p_l p_m \;+ \\
&+\; \sum_{i,k,l,m,n=1}^{3} \lambda^{(5)}_{i,k,l,m,n,\text{Landau}}\, p_i p_k p_l p_m p_n \;+ \\
&+\; \sum_{i,k,l,m,n,o=1}^{3} \lambda^{(6)}_{i,k,l,m,n,o,\text{Landau}}\, p_i p_k p_l p_m p_n p_o \;+ \\
&+\; \text{terms of higher order} \ .
\end{aligned}
\tag{2.48}
$$

$F_0(T)$ is the total free energy if all elements p_i of the polarization $\boldsymbol{p}$ are equal zero. The parameters which are denoted by the symbol λ are specific parameters, which depend on temperature etc.. As the free energy is unchanged under the action of the symmetry group of the crystal, such an expansion can be reduced. For example, a realistic expression is an expression of the form

$$
F_{p,1}(E,T,p) = F_0(T) + \lambda^{(2)}_{\text{Landau}}\, p^2 + \lambda^{(4)}_{\text{Landau}}\, p^4 + \lambda^{(6)}_{\text{Landau}}\, p^6 \ ,
\tag{2.49}
$$

which can be found by using methods of group theory, where the parameter of second order is found to be

$$
\lambda^{(2)}_{\text{Landau}} \sim T - T_0 \ .
\tag{2.50}
$$

p is one component of the polarization vector $\boldsymbol{p}$. As it was above mentioned, observable states are correlated with the minimum of the free energy, i. e. the condition

$$
\frac{\partial F_{p,1}(E,T,p)}{\partial p} = 0
\tag{2.51}
$$

holds. To get an overview about the topological structure of such a free energy, see figure 2.1. In figure 2.1 the free energy (2.49) is shown, where three different choices of the relevant parameters are considered. Both the free energy of the paraelectric state and the free energy of the ferroelectric state are given. The minima of these graphs represent possible equilibrium states. Thus, the reader can see that before a phase transition only one absolute minimum exists, which corresponds to the equilibrium state. After the transition two possible states occur, in which case a jump of the absolute minimum takes place so that this function describes a phase

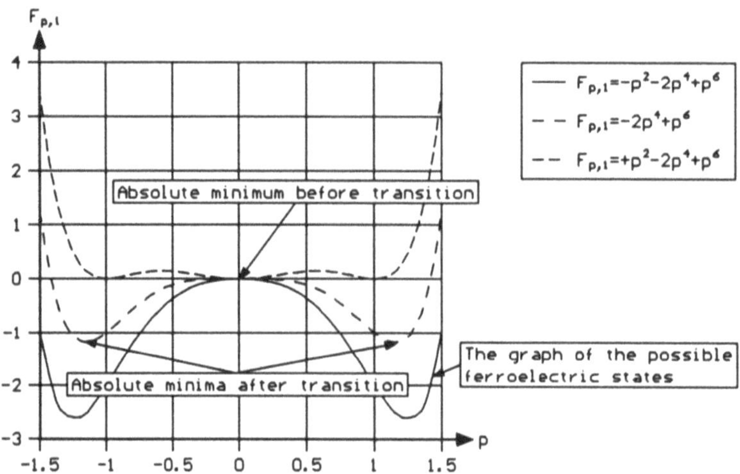

Figure 2.1 The problem of ferroelectricity. The free energy as a function of the polarization p

transition of first order. In figure 2.2 another example is given. This figure shows the problem of a ferromagnetic phase transition. As the absolute minimum changes in a continual way, figure 2.2 shows a phase transition of second order. The development of two minima of the free energy $F_{p,2}$ is called a *bifurcation*. (In this case p denotes a magnetization. Thus, it has to be noted that in this book the notation m will be used, too.) Such a bifurcation occurs by changing a control parameter (here, the temperature T). Such a function will again be used, this has to be mentioned. In this context I have to refer the reader to chapter 6, i. e. such a function is also relevant in laser theory.

This example is a special example of *Landau's* theory. Other phase transition problems can be considered as well. A parameter such as the polarization p is called an *order parameter*, because such a parameter determines the order of the system. The possibility to describe a multi-component system by using only few variables is a possibility that occurs very often. In a more general way the problem of order parameters will be discussed in the context of laser theory.

So far, fluctuating effects were neglected. But these effects exist. For example, fluctuating effects decide which minimum of the free energy in figure 2.2 will be reached. The problem of influence of fluctuations will now be discussed.

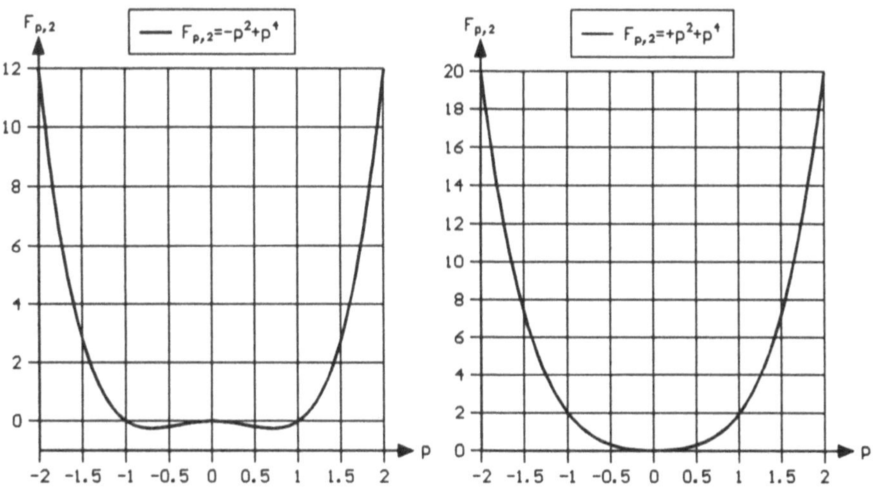

Figure 2.2   The problem of ferromagnetism. The free energy as a function of the magnetization p

2.3.2   The Correlated Statistics

The Langevin Level

In principle observable states in thermal equilibrium are correlated with the minima of the free energy, this was above mentioned. But during these considerations fluctuating effects were neglected. To introduce fluctuating effects, an evolution equation of the *Langevin* type can be used, which is a generalization of the condition (2.51), i. e. an equation of the form

$$\dot{p} = -\frac{\partial F_p(E, T, p)}{\partial p} + \text{fluctuation term} , \tag{2.52}$$

can be introduced. Then fluctuating behavior near a minimum of the free energy can be considered. (2.52) is a special statistical differential equation to gain the behavior of the specific variable p. Generalized applications to other problems are possible in the same way.

The Fokker-Planck Level

If the basic structure of the physical system is known, i. e. if the strength of the fluctuating behavior, additional information about the underlying process (for example, a *Markov* process) and basic evolution equations are known, an evolution equation can be derived, which describes the evolution of a correlated distribution function.

If one assumes that a *Fokker-Planck* statistics is the underlying statistics, a *Fokker-Planck* equation is the underlying evolution equation. Then a *Langevin* equation is necessary to put a *Fokker-Planck* equation in concrete terms. To find stationary solutions of such an equation, the stationary form of such a *Fokker-Planck* equation has to be used. The problem of *Langevin* equations, *Fokker-Planck* equations and interrelations will be discussed in chapter 5. There, only some essential facts which refer to the now discussed problem shall be pointed out.

If the basic *Langevin* equation is given by (2.52), a *Fokker-Planck* equation of the form

$$\frac{\partial \rho_{\text{Landau}}(p,t)}{\partial t} = -\frac{\partial\left[-\frac{\partial F_p(E,T,p)}{\partial p}\rho_{\text{Landau}}(p,t)\right]}{\partial p} + kT\frac{\partial^2 \rho_{\text{Landau}}(p,t)}{\partial^2 p} \qquad (2.53)$$

holds, where the factor kT describes the strength of the fluctuations. (kT is a function of the fluctuation term in (2.52).) Then the condition

$$\frac{\partial \rho_{\text{Landau}}(p,t)}{\partial t} = 0 \qquad (2.54)$$

generates a stationary form of the *Fokker-Planck* equation (2.53).

Stationary Solutions

Such a stationary equation has solutions of the form

$$\rho_{\text{Landau}}(p) = Z_{\text{Landau}}^{-1} \exp\left(-\frac{F_p(E,T,p)}{kT} \right) , \qquad (2.55)$$

where the partition function is given by

$$Z_{\text{Landau}} = \int \exp\left(-\frac{F_p(E,T,p)}{kT} \right) dp . \qquad (2.56)$$

By using two examples which are shown in figure 2.1 and figure 2.2 one obtains the distribution functions

$$\rho_{1,\text{Landau}}(p) \sim \exp\left[-\frac{1}{kT}\left(\lambda_{\text{Landau}}^{(2)}p^2 - |\lambda_{\text{Landau}}^{(4)}|p^4 + |\lambda_{\text{Landau}}^{(6)}|p^6 \right)\right] , \qquad (2.57)$$

$$\rho_{2,\text{Landau}}(p) \sim \exp\left[-\frac{1}{kT}\left(\lambda_{\text{Landau}}^{(2)}p^2 + |\lambda_{\text{Landau}}^{(4)}|p^4 \right)\right] . \qquad (2.58)$$

(2.57), (2.58) are exponential functions with non-linear exponents, with (2.57) describing the statistics of a phase transition of first order, and (2.58) being a distribution function which describes a second order transition. A phase transition occurs if the parameters of second order change their signs. Every maximum of such a distribution function corresponds to a minimum of the free energy. Illustrations of such non-linear distribution functions will be given later (see chapter 4) – in the context of a more general consideration of such exponential distribution functions.

Macroscopic Determination of Distribution Functions

In order to determine a distribution completely, i. e. to determine the form of a distribution function as well as the structure of the distribution function parameters, a microscopic access is possible. (The above described access is such a microscopic access.) Another possibility to determine exponential distribution functions (also of higher order, and also with an infinite number of variables) is the possibility to use a macroscopic determination process. Then an extreme principle, the maximum information entropy principle (MIEP), has to be used as a basis. With such a principle the form of a distribution function can be found. In order to determine the structure of the parameters, too, an additional formalism is necessary. The MIEP as well as the mentioned concept will be discussed in the next two chapters. In this context a generalization of the term *entropy* will be given as well. However, the standpoint of the following will be a more general one, i. e. in the following some aspects of system theory will be considered, and thermodynamic systems will be used as easy examples.

3 Aspects of System Theory

In order to describe multi-component systems like a thermodynamic system, a laser system, high-dimensional quantum systems or the human brain, it is useful to introduce quantities which are able to describe essential facts in a compressed way. Such a compression is useful, because such systems produce an immense amount of data. In the last chapter such expressions have been introduced. In particular, the reader has to think of the quantity *entropy*, which describes the evolution possibilities of a thermodynamic system. Other quantities are necessary to describe systems like the human brain. The measurable level then is the electromagnetic field level on the brain surface, i. e. the level of an electroencephalogram (EEG). To describe and to classify such an EEG, one needs a quantity to describe the measurement in a compressed way. Furthermore, it is desirable to have quantities which have a far-reaching meaning. In order to be able to use such quantities, it is necessary to have additional laws belonging to these quantities.

In this chapter the term *information entropy* or shorter, *information*, will be introduced. In this book this term is a direct generalization of the thermodynamic entropy. With this quantity it is possible to describe a wealth of physical systems and to classify different systems. In this chapter the basic ideas will be presented. Examples will be given in chapter 8. Using such a concept it is possible to classify different states of a laser system, to classify different states of a quantum system (like the states of a harmonic oscillator) and to classify high-dimensional quantum systems (like a *Bose-Einstein* or a *Fermi-Dirac* gas). Moreover, an extreme principle will be discussed which is a generalized form of the maximum entropy principle (MEP), namely the maximum information entropy principle (MIEP). With this principle it is possible to find the form of a wealth of relevant distribution functions of thermodynamics, laser physics and brain research by using macroscopic measurement values as constraints. Therefore, the principle to describe and to classify multi-component systems by the term *information* and by using the MIEP is a universal principle. A total measurement correlated determination of a wealth of distribution functions is possible if the distribution function parameters are given. This leads to the concept of hyper-surface equations. However, this concept will be discussed in the next chapter. In the following the term *information* shall be introduced, and some basic properties shall be discussed.

3.1 Information

In order to describe a multi-component system, it is useful to introduce a measure to characterize the system on a compressed level. Such a measure can be found by using a generalized form of the above introduced thermodynamic entropy (see subsection 2.2.2, (2.13)-(2.19) and (2.43)-(2.47)).

3.1.1 The Basic Information Expression

So far, the considered space was a phase space, i. e. a space of position and impulse coordinates. However, on closer examination such a restriction is not necessary, because the statistical considerations of subsection 2.2.2 are of a universal kind. Therefore, the considerations starting with (2.13) can be generalized so that the thermodynamic entropy (2.19) can be used in a much more general way. In particular, it is not necessary to consider ensembles of single elements, i. e. ensembles which consist of ensembles can be taken as a basis. Furthermore, any coordinates can be considered. If an expression of the form (2.19) is used in this sense, *Boltzmann's* constant k makes no sense. Thus, kN shall be replaced by a more general constant α. Therefore, we obtain the expression

$$I = -\alpha \sum_i w_i \ln w_i \ , \tag{3.1}$$

where this expression can be considered as a generalized form of the discrete entropy expression (2.19). w_i represents any probabilities. This expression shall be called *information entropy* or shorter, *information I*. A suitable choice of α generates the special case *thermodynamic entropy* if a μ-space is considered. The thermodynamic entropy of only one ensemble element can then be found by the formal choice $\alpha := k$ (this means that the expression (2.19) has to be divided by N to get the entropy of one ensemble element).

If such a derivation is taken as a basis, (3.1) represents a direct generalized form of the term *entropy*. Here it has to be remarked that such a derivation diverges from the justification of *Shannon's* information, which can be found by analyzing data processes. In this book, however, always basic physical principles shall be taken as a basis.

3.1.2 The Meaning of Information

Now the question arises in which way the expression *information* shall be applied in this book. To answer this question, one has to consider again the derivation of (2.19) (see (2.13)ff.). If general coordinates are permitted, W is the number of realization possibilities of any statistical system. In particular, W is not always identical with the number of the microscopic realization possibilities of a special macro state, i. e. W normally cannot be used in the thermodynamic sense. Therefore, W represents

a measure of the basic statistics, in which case this measure increases if the space
of the statistcal activity increases. For example, if the number of the considered
ensemble elements is equal to 4, the numbers

$$W_1 = \frac{4!}{1!1!1!1!} = 24 \ , \tag{3.2}$$

$$W_2 = \frac{4!}{2!1!1!} = 12 \ , \tag{3.3}$$

$$W_3 = \frac{4!}{2!2!} = 6 \ , \quad W_4 = \frac{4!}{3!1!} = 4 \ , \tag{3.4}$$

$$W_5 = \frac{4!}{4!} = 1 \tag{3.5}$$

(compare with (2.13)) hold, which describe the realization possibilities of special
statistical distributions. As the space of the statistical activity is given by the population of the space elements (i. e. the number of the elements of the denominators),
it can be seen that a wide statistical activity forces a large measure W. Therefore,
W is a measure of the space of the statistical activity, or, in other words, a measure
of the width of a statistical distribution. In this context it has to be noticed that
such a distribution function includes both stochastic and deterministic systems. (If
the elements of a deterministic system show different states, a statistical distribution function can be used.) As the increase of the width of a statistical distribution
is very often equivalent to the increase of disorder, the increase of W describes very
often the increase of disorder. (This fact has already been mentioned. The reader
may compare with the remarks after (2.47).) But this statement is not always valid.
If the reader compares (3.5) and (3.4), it is obvious that an increase of the width
of a distribution can be overlayed by the occurrence of patterns, which are represented by special distributions of the populations of the denominators. Therefore,
it is useful to consider W only as a measure of the space of the statistical activity.
This shall be done. Then (3.1) is such a measure, too. However, in contrary to W,
(3.1) is an additive quantity, i. e. the whole information of a physical system is the
sum of informations of the partial systems – if interactions are neglected. Due to the
fact that these properties hold, the term *information* has the meaning of the very
often used word *information*, and the notation *information* makes sense. However,
an exact definition can now be taken as a basis. This has to be emphasized.

With (3.1) in particular different states of multi-component systems can be compared, or different physical systems of one multi-component system can be characterized in a compressed way. In chapter 8 a more detailed discussion will follow. Then
examples will be given, i. e. in chapter 8 the information of a one-mode laser near
a phase transition point will be considered, the information of stationary states of
a harmonic oscillator will be considered, and the information of *Bose-Einstein* and
Fermi-Dirac ensembles will be considered. Additionally, connections with texture
analysis and the genetic code will be discussed. It has to be remarked here that also
human EEG's can be classified with such an information. Therefore, the concept

of information can also be used in the analytical physiology. Then the statistical variables are the components of the electromagnetic field, which can be measured at the surface of the skull.

Instead of probabilities, in (3.1) probability densities can be used. Then a quasi-continuous formulation of the information is possible.

3.1.3 The Measurement Integral Formalism

In order to introduce probability densities ρ_i, the relation

$$\rho_i = \frac{w_i}{\Delta\Omega} \tag{3.6}$$

has to be used, where $\Delta\Omega$ denotes an element of accuracy. Inserting (3.6) into (3.1) the equation

$$I = -\alpha \sum_i \rho_i \ln\left(\rho_i \Delta\Omega\right) \Delta\Omega \tag{3.7}$$

holds, where (3.7) can be replaced by the expression

$$I = -\alpha \sum_\Omega \rho(\Omega) \ln\left[\rho(\Omega)\Delta\Omega\right] \Delta\Omega \ . \tag{3.8}$$

$\rho(\Omega)$ represents a function, where the argument Ω is defined within a discrete codomain. (3.8) is equivalent to (3.1), and it can be seen that information depends on the size of the element of accuracy. By using the limit $\lim_{\Delta\Omega \to d\Omega}$ the information expression

$$I_m = -\alpha \lim_{\Delta\Omega \to d\Omega} \sum_\Omega \rho(\Omega) \ln\left[\rho(\Omega)\Delta\Omega\right] \Delta\Omega \tag{3.9}$$

occurs, where $d\Omega$ represents an element of accuracy which belongs to the best possible measurement. (The *best possible measurement* is a measurement which is actually possible. Thus, such an expression is directly correlated with technical development. Absolute limits are given by physical relations like *Heisenberg's* uncertainty relation.) Normally, such elements of accuracy are very small so that the expression (3.9) is a quasi-continuous information expression. (3.9) is a generalized form of the measurement integral (2.45). By using the same notation as in 2.2.2 the information (3.9) can be written in the form

$$\boxed{I_m = -\alpha \int_m \rho(\Omega) \ln\left[\rho(\Omega)d\Omega\right] d\Omega \ ,} \tag{3.10}$$

where $d\Omega$ is no operation symbol so that algebraic operations are possible. (3.10) represents a generalized form of (2.45). (3.10) is correlated with the best possible measurement. A more detailed discussion of the term *information* will follow (see chapter 8). Then examples will be discussed.

In the following the expression (3.10) will be taken as a basis, and the choice

$$\alpha := 1 \tag{3.11}$$

will be made. Such an expression has a universal meaning and can be written as a function of given measurement quantities. Such a macroscopic information expression shall now be considered (in this context, in particular see [40]).

3.1.4 Information and Measurement

To introduce an information expression which is directly connected with the macroscopic level, distribution functions of the form

$$\rho(\mathbf{\Omega}) = Z^{-1} \exp\left[-\left(\sum_{i=1}^{N} \lambda_i \Omega_i + \sum_{i,k=1}^{N} \lambda_{i,k}\, \Omega_i \Omega_k + \sum_{i,k,l=1}^{N} \lambda_{i,k,l}\, \Omega_i \Omega_k \Omega_l + \right.\right.$$
$$\left.\left. \sum_{i,k,l,m=1}^{N} \lambda_{i,k,l,m}\, \Omega_i \Omega_k \Omega_l \Omega_m + \text{tho} \right) \right] \tag{3.12}$$

(tho=terms of higher order) shall be assumed, where the partition function is given by

$$Z = \int_m \exp\left[-\left(\sum_{i=1}^{N} \lambda_i \Omega_i + \sum_{i,k=1}^{N} \lambda_{i,k}\, \Omega_i \Omega_k + \sum_{i,k,l=1}^{N} \lambda_{i,k,l}\, \Omega_i \Omega_k \Omega_l + \right.\right.$$
$$\left.\left. \sum_{i,k,l,m=1}^{N} \lambda_{i,k,l,m}\, \Omega_i \Omega_k \Omega_l \Omega_m + \text{tho} \right) \right]\, d\Omega \tag{3.13}$$

if a measurement integral is used. Such an assumption is very useful, because such exponential functions play a crucial role in this book. However, it has to be remarked that other functions are also possible. Inserting (2.12) into (2.10), under consideration of (2.11), the information

$$I_m = \int_m \rho(\mathbf{\Omega})[-\ln(d\Omega/Z)]\, d\Omega + \sum_{i=1}^{N} \lambda_i \int_m \rho(\mathbf{\Omega})\Omega_i\, d\Omega + $$
$$\sum_{i,k=1}^{N} \lambda_{i,k} \int_m \rho(\mathbf{\Omega})\Omega_i \Omega_k\, d\Omega + \sum_{i,k,l=1}^{N} \lambda_{i,k,l} \int_m \rho(\mathbf{\Omega})\Omega_i \Omega_k \Omega_l\, d\Omega + $$
$$\sum_{i,k,l,m=1}^{N} \lambda_{i,k,l,m} \int_m \rho(\mathbf{\Omega})\Omega_i \Omega_k \Omega_l \Omega_m\, d\Omega + \text{tho} \tag{3.14}$$

holds. The various measurement integrals of (2.14) represent macroscopic quantities, namely correlation functions. By using the abbreviation

$$\int_m \rho(\Omega)\big(\Omega_i \Omega_k \Omega_l \Omega_m \ldots\big)\, d\Omega = \big\langle \Omega_i \Omega_k \Omega_l \Omega_m \ldots \big\rangle \tag{3.15}$$

one obtains the expression

$$
\begin{aligned}
I_m = I_{\text{norm}} &+ \sum_{i=1}^{N} \lambda_i \langle \Omega_i \rangle + \sum_{i,k=1}^{N} \lambda_{i,k} \langle \Omega_i \Omega_k \rangle + \\
&+ \sum_{i,k,l=1}^{N} \lambda_{i,k,l} \langle \Omega_i \Omega_k \Omega_l \rangle + \\
&+ \sum_{i,k,l,m=1}^{N} \lambda_{i,k,l,m} \langle \Omega_i \Omega_k \Omega_l \Omega_m \rangle + \text{tho} ,
\end{aligned}
\tag{3.16}
$$

which is directly correlated with the macroscopic level. The first part, which is a consequence of the standardization condition, is given by

$$I_{\text{norm}} = \int_m \rho(\Omega)[-\ln(d\Omega/Z)]\, d\Omega = -\ln(d\Omega/Z) . \tag{3.17}$$

(3.16) represents a relation which connects the measurement level with the information. If it were possible to determine the parameters λ_i, $\lambda_{i,k}$, $\lambda_{i,k,l}$, $\lambda_{i,k,l,m}$, etc. as well, a totally macroscopic expression would be given. Such a determination is indeed possible. In order to find the relevant determination equations, namely the hyper-surface equations, the equation system (2.15) can be taken as a basis. The way of derivation of such hyper-surface equations will be considered in the next chapter.

On the basis of the information (3.10) a generalized form of the MEP can be introduced. This shall be considered in the following section.

3.2 The Maximum Information Entropy Principle

The maximum entropy principle (MEP) introduced in the last section allows to find relevant distribution functions of thermodynamics (see (2.20)ff.). This elementary principle can be generalized so that a wealth of relevant statistical distribution functions for systems in the thermodynamic equilibrium as well as far from thermodynamic equilibrium can be found by using an elementary principle. This generalized form will be called *maximum information entropy principle* (see [40]) and shall now 1be considered.

3.2.1 The Basic Principle

The requirement

$$\delta I_m = 0 \; , \tag{3.18}$$

under consideration of both the given constraints

$$\int_m \rho(\Omega)\,(\Omega_i\Omega_k\Omega_l\Omega_m \ldots)\,d\Omega = \langle \Omega_i\Omega_k\Omega_l\Omega_m \ldots \rangle = \text{constant} \tag{3.19}$$

and the standardization condition

$$\int_m \rho(\Omega)\,d\Omega = \langle 1 \rangle \; , \tag{3.20}$$

leads to the variational expression

$$\delta \int_m \rho(\Omega)C(\Omega)\,d\Omega = 0 \tag{3.21}$$

if the method of *Lagrangian* multipliers is used. Here $C(\Omega)$ is given by

$$C(\Omega) = \ln\left[\rho(\Omega)d\Omega\right] + \lambda_0 + \sum_{i=1}^{N} \lambda_i \Omega_i + \sum_{i,k=1}^{N} \lambda_{i,k}\Omega_i\Omega_k +$$

$$\sum_{i,k,l=1}^{N} \lambda_{i,k,l}\Omega_i\Omega_k\Omega_l + \sum_{i,k,l,m=1}^{N} \lambda_{i,k,l,m}\Omega_i\Omega_k\Omega_l\Omega_m + \text{tho} \; , \tag{3.22}$$

where

$$\lambda_0 = -\ln(d\Omega/Z) - 1 \tag{3.23}$$

holds, and wherein the parameters λ_0, λ_i, $\lambda_{i,k}$, $\lambda_{i,k,l}$, $\lambda_{i,k,l,m}$ etc. are the *Lagrangian* multipliers. The variational expression (3.21) is a generalized form of the thermodynamic relation (2.23), which represents a mathematical formulation of the MEP. Thus, (3.21) represents a more general principle, namely the maximum information entropy principle (MIEP), where only correlation functions as constraints were assumed.

3.2.2 The General Solution of the Basic Principle

(3.21) is equivalent to the relation

$$\int_m \left[C(\Omega)\delta\rho(\Omega) + \rho(\Omega)\delta C(\Omega)\right]d\Omega = \int_m \left[C(\Omega)+1\right]\delta\rho(\Omega)\,d\Omega = 0 \; . \tag{3.24}$$

As any elements $\delta\rho(\Omega)$ are allowed, the relation

$$C(\Omega) + 1 = 0 \tag{3.25}$$

holds, where the general solution of (3.25) is given by

$$\rho(\Omega) = Z^{-1} \exp\left[-\left(\sum_{i=1}^{N} \lambda_i \Omega_i + \sum_{i,k=1}^{N} \lambda_{i,k}\, \Omega_i \Omega_k + \right.\right.$$
$$\sum_{i,k,l=1}^{N} \lambda_{i,k,l}\, \Omega_i \Omega_k \Omega_l +$$
$$\left.\left. \sum_{i,k,l,m=1}^{N} \lambda_{i,k,l,m}\, \Omega_i \Omega_k \Omega_l \Omega_m + \mathrm{tho} \right) \right] . \tag{3.26}$$

As the reader can see, if correlation functions are taken as a basis, solutions of the MIEP are exponential functions with sums over easy variable products as exponents. Such exponential functions will be called *exponential functions of a special order*, where the *order* is defined by the largest occuring variable product. (For example, in (3.26) only products up to fourth order were taken down. If terms of higher order were neglected, a function of fourth order would arise.) Functions of a special order play a crucial role in this book, where (3.26) can be considered as a general expression which gives the form of the relevant distribution functions. Special cases of (3.26) are the thermodynamic functions (2.57) and (2.58) or *Boltzmann's* function. Additionally, it has to be noticed that the derived information expression (3.16) corresponds to such an exponential distribution function. Therefore, if correlation functions are taken as constraints, the first principle *MIEP* can be taken as a basis to gain the form of the relevant distribution functions as well as the relevant macroscopic information. In order to determine the distribution function parameters as well, i. e. the *Lagrange* multipliers, an additional formalism is necessary. This will lead to the concept of hyper-surface equations, which will be discussed in chapter 4. (It has to be noticed here that this problem already was introduced during the thermodynamic considerations. However, the above given considerations show that this problem is a more general one. Furthermore, it has to be noticed that the problem of determination of such *Lagrange* multipliers is identical with the determination problem which occured in 3.1.4. There, the problem of a complete determination of a relevant information was discussed.)

In order to gain other distribution functions by using the MIEP, other constraints are necessary. Then the MIEP has to be formulated in a more general form.

3.2.3 The General Principle

A more general formulation of the MIEP can be found by using arbitrary constraints, i. e. by using a more general form of (3.21). If the symbol $f_i(\Omega)$ denotes any functions of the specific vaiables Ω_k, a general formulation is given by

$$\delta \int_m \rho(\Omega) \left\{ \ln[\rho(\Omega)d\Omega] + \sum_{i=0}^{i_{max}} \kappa_i f_i(\Omega) \right\} d\Omega = 0 , \tag{3.27}$$

where the index $i = 0$ allows covering the standardization relation as well. The parameters κ_i are more general *Lagrangian* multipliers. The MIEP (3.27) includes the special case (3.21). Therefore, a quite universal principle has been introduced. It can be taken as a basis to gain a wealth of relevant statistical distribution functions, in which case the constraints have to be known. However, the following has to be noticed: If only special measurement quantities are known, the way to find the form of the distribution function by using the MIEP allows to find a function which is – in last consequence – completely determinable. Such a function might be an approximative function, however, such a function is conform with reference to the knowledge about the considered physical system. In this context it has to be remarked that the MIEP is as significant in statistical physics as the *Hamiltonian principle* (HP) (see chapter 9, section 9.3) in deterministic physics. The essential difference between these two first principles is that the MIEP needs macroscopic quantities as additional constraints and the HP needs the specific *Lagrangian* function.

3.2.4 Some Additional Remarks

As measurement integrals were used, the space of the variables is a discrete one. As it has been mentioned before, such measurement integrals are necessary in the context of information. Otherwise ordinary integrals can be used. Therefore, in the following ordinary integrals will be taken as a basis. In particular, a basic equation system of the form

$$\int \rho(\Omega)(\Omega_i \Omega_k \Omega_l \Omega_m \ldots)\, d\Omega = \langle \Omega_i \Omega_k \Omega_l \Omega_m \ldots \rangle \tag{3.28}$$

will be taken as a basis, and functions of the form (3.26) will be considered as continuous functions.

In the following chapters exponential functions of a special order will play a crucial role. By using a *Taylor* expansion non-exponential functions can also be taken into account so that the later following concepts are much more universal as it was described above. In order to understand how *Taylor* expansions can be included and to introduce a later often used functional symbol, the following considerations shall be presented.

3.3 The Taylor Approximation

3.3.1 The Expansion

A statistical problem which can be described by a statistical distribution function with

$$0 < \rho(\Omega) < \infty \tag{3.29}$$

can be written in the form

$$\rho(\Omega) = d \exp\{\ln[\rho(\Omega)/d]\} \,, \tag{3.30}$$

where d guarantees a dimensionless argument of the natural logarithm. Very often the natural logarithm can be expanded into a *Taylor* series, i. e. the formula

$$f(\Omega) = \left\{ \prod_{i=1}^{N} \sum_{M_i=0}^{\infty} \frac{1}{M_i!} \left[\partial_{\Omega_i}^{M_i} f(\Omega) \right] \Big|_{\Xi} (\Omega_i - \Xi_i)^{M_i} \right\}$$

$$= f(\Xi) + \sum_{\alpha=1}^{\infty} \left\{ \prod_{i=1}^{\alpha} \sum_{\Theta_i=1}^{N} {}^{z}\Theta_\alpha \left[\partial_{\Omega_{\Theta_i}} f(\Omega) \right] \Big|_{\Xi} (\Omega_{\Theta_i} - \Xi_{\Theta_i}) \right\} \tag{3.31}$$

with

$${}^{z}\Theta_\alpha = \frac{1}{M_{\Theta_\alpha}(\partial_{\Omega_1})! M_{\Theta_\alpha}(\partial_{\Omega_2})! \ldots M_{\Theta_\alpha}(\partial_{\Omega_N})!} \tag{3.32}$$

can be used. In this case

$$\Xi = \Xi_1, \Xi_2, \ldots, \Xi_N \tag{3.33}$$

marks the Ω-point which is the centre of the expansion, and

$$\Theta_\alpha = \Theta_1, \Theta_2, \ldots, \Theta_\alpha \tag{3.34}$$

represents a multi-component index with ordinary numbers Θ_i. $M_{\Theta_\alpha}(\partial_{\Omega_i})$ represents multiplicities, which indicate how often a special partial differential operator ∂_{Ω_i} has to be counted. The corresponding *Taylor* coefficients read

$$T_{\Theta_\alpha} = {}^{z}\Theta_\alpha \left[\left\{ \prod_{i=1}^{\alpha} \partial_{\Omega_{\Theta_i}} \right\} f(\Omega) \right] \Big|_{\Xi} . \tag{3.35}$$

In the now considered case $f(\Omega)$ has to be equal $\ln[\rho(\Omega)/d]$. If such a *Taylor* expansion is possible, the result can be written in the form

$$\rho_{\mathrm{TE}}(\Omega) = Z_{\mathrm{TE}}^{-1} \exp\left\{ -\left[\sum_{\alpha=1}^{\infty} \left\{ \prod_{i=1}^{\alpha} \sum_{\Theta_i=1}^{N} T_{\Theta_\alpha} (\Omega_{\Theta_i} - \Xi_{\Theta_i}) \right\} \right] \right\} , \tag{3.36}$$

where Z_{TE} is the standardization factor which includes also d. By using the transformation

$$\Omega_{\Theta_i} - \Xi_{\Theta_i} = \tilde{\Omega}_{\Theta_i} \tag{3.37}$$

an infinite exponential function of the kind (3.26) arises, which can be written in the form

$$\rho_{\mathrm{TE}}(\Omega) = Z_{\mathrm{TE}}^{-1} \exp\left[-\left(\sum_{\alpha=1}^{\infty} \left\{ \prod_{i=1}^{\alpha} \sum_{\Theta_i=1}^{N} \lambda_{\Theta_\alpha} \Omega_{\Theta_i} \right\} \right) \right] , \tag{3.38}$$

if instead of $\tilde{\Omega}$ the notation Ω is used again. Such a *Taylor* series very often is the exact representation of the considered function, sometimes such a series represents an approximation. As the form of such a function is equivalent to the form of the functions used before, *Taylor* expansions can be included into the considerations, too.

Within the expressions (3.31), (3.35), (3.36) and (3.38) brackets of the form $\rangle$, $\langle$ occur. In the following some remarks concerning such brackets shall be given.

3.3.2 The Product Brackets

In this book very often complicated functions of products and sums occur. Thus, it is useful to introduce structuring elements, namely product brackets. The symbol of the opening product bracket shall be the symbol $\rangle$, and the symbol of the closing product bracket shall be the symbol $\langle$. Such brackets shall delimitate the effect of a product symbol $\prod$, which has to be written after the opening product bracket. With such product brackets a useful structuring of complicated formulae is possible. Thus, the relation

$$\left. \rangle \prod_{i=1}^{\alpha} \sum_{\Theta_i=1}^{N} \lambda_{\Theta_\alpha} \Omega_{\Theta_i} \langle \right. = \sum_{\Theta_1=1}^{N} \sum_{\Theta_2=1}^{N} \cdots \sum_{\Theta_\alpha=1}^{N} \lambda_{\Theta_\alpha} \Omega_{\Theta_1} \Omega_{\Theta_2} \ldots \Omega_{\Theta_\alpha} \tag{3.39}$$

holds. Such product brackets will widely be used in this book.

In this chapter a classification principle for multi-component systems has been introduced (namely the concept of information), and a first principle has been introduced, which allows to find a wealth of relevant distribution functions of multi-component systems (namely the MIEP). Therefore, a theoretical lattice has been introduced under which the following considerations can be submitted.

The remaining problem of this chapter was the problem of determination of *Lagrangian* multipliers. Therefore, in the following the problem of determination of *Lagrangian* multipliers will be considered. The following chapter as well will show how to calculate many relevant relations of statistical systems. Therefore, in the following a special aspect of system theory will be discussed.

4   System Analysis

The statistical behavior of many classes of physical systems is governed by exponential distribution functions of a special order. Some examples are given in table 4.1. As the reader can see, thermodynamic distribution functions as well as distribution functions of the statistical laser theory are exponential distribution functions of a special order. In order to describe quantum systems, it is possible to use special path integrals, namely *Feynman* path integrals. Path integrals, too, can be incorporated into these considerations, as such integrals contain exponential functions as central parts. It has to be mentioned here that in the same way path integrals which occur in the theory of *Fokker-Planck* equations can be incorporated. As many functions can be described in a more or less approximative way by a *Taylor* expansion, many non-exponential functions can be incorporated in these considerations, too. Furthermore, it has to be mentioned here that functions usable in an approximative brain research are sometimes exponential distribution functions of a special order. All these functions can then be described by only one mathematical expression, in which case a special physical system is given by a special choice of the distribution function parameters. In particular, thermodynamic distribution function parameters are functions of the absolute temperature, specific laser distribution function parameters are functions of the inversion, of loss constants, etc., and distribution functions of brain theory may show dependences of the environmental activity, i. e. of the outer scene. One can gain the forms of all these distribution functions by using the MIEP, i. e. by using a first principle. These distribution functions are connected in a well-defined way with macroscopic quantities, namely correlation functions. In this book equations which describe such a connection are called *basic equations*. By starting from a basic equation system of a low order it is possible to calculate all statistically relevant relations like macroscopic phase transition conditions and inversion formulae, with such inversion formula describing the connection between the distribution function parameters and the macroscopic quantities. In this book such equations are called hyper-surface equations. Such hyper-surface equations determine the distribution function parameters by using macroscopic quantities, i. e. without using a microscopic theory. The calculation of such equations can be carried out without considering a special physical system, i. e. in a universal way.

Thus, in this chapter it will be presented how to analyse a complex statistical system by basic mathematical methods, and it will be shown how to get relevant relations, in particular a complete system of hyper-surface equations. This chapter is then a contribution to the macroscopic theory of statistical systems, in particular to the theory of a macroscopic determination of statistical distribution functions. The physical systems which will normally be considered are thermodynamic systems. The application to laser systems will be given in chapter 6. The extension to quantum

Table 4.1 Examples of exponential distribution functions in various fields of physics. Standardization factors are neglected

Ferroelectricity (equilibrium phase transitions in thermodynamics): $$\exp\left[-\tfrac{1}{kT}\left(\lambda^{(2)}_{\mathrm{Landau}}\,p^2 + |\lambda^{(4)}_{\mathrm{Landau}}|\,p^4\right)\right]$$ (phase transitions of second order) $$\exp\left[-\tfrac{1}{kT}\left(\lambda^{(2)}_{\mathrm{Landau}}\,p^2 - |\lambda^{(4)}_{\mathrm{Landau}}|\,p^4 + |\lambda^{(6)}_{\mathrm{Landau}}|\,p^6\right)\right]$$ (phase transitions of first order) $p \ldots$ polarization, $k \ldots$ *Boltzmann's* constant, $\lambda^{(2)}_{\mathrm{Landau}} \sim (T - T_0)$, $T_0 \ldots$ critical temperature near the *Curie* temperature
Ferromagnetism (equilibrium phase transitions in thermodynamics): $$\exp\left[-\tfrac{1}{kT}\left(\lambda^{(2)}_{\mathrm{Landau}}\,m^2 + |\lambda^{(4)}_{\mathrm{Landau}}|\,m^4\right)\right]$$ (phase transitions of second order) $m \ldots$ magnetization, $k \ldots$ *Boltzmann's* constant, $\lambda^{(2)}_{\mathrm{Landau}} \sim (T - T_0)$, $T_0 \ldots$ critical temperature near the *Curie* temperature
One-mode laser near a laser threshold: $$\exp\left[-\left(\lambda^{(2,1)}_{w}\,E^2_w + \lambda^{(4,1)}_{ww}\,E^4_w\right)\right]$$ (non-equilibrium phase transitions, phase transitions of second order)   Multi-mode laser near a laser threshold: $$\exp\left[-\left(\sum_{w_1}\lambda^{(2,N)}_{w_1}\,E^2_{w_1} + \sum_{w_1,w_2,\,w_1<w_2}\lambda^{(c,N)}_{w_1 w_2}\,E^2_{w_1}E^2_{w_2} + \lambda^{(4,N)}_{w_1 w_1}\,E^4_{w_1}\right)\right]$$ (non-equilibrium phase transitions, phase transitions of second order)   One-mode laser far from a laser threshold: $$\exp\left[-\left(\lambda^{(1,1,f)}_{w}\,E_w + \lambda^{(2,1,f)}_{w}\,E^2_w\right)\right]$$ $E_w \ldots$ amplitudes of the laser modes, $w \ldots$ wavelength
Central part of a pro-function of a Feynman path integral in the context of high-dimensional quantum systems: $$\exp\left[\frac{i}{\hbar}\left\{\frac{m_0}{2}\left[\left(\frac{q - q_{P-1}}{\Delta t}\right)^2\Delta t + \left(\frac{q_{P-1} - q_{P-2}}{\Delta t}\right)^2\Delta t + \cdots + \left(\frac{q_P - q_0}{\Delta t}\right)^2\Delta t\right]\right.\right.$$ $$\left.\left. - V(q_{P-1})\Delta t - \cdots - V(q_0)\Delta t\right\}\right]$$ $m_0 \ldots$ mass, $q \ldots$ space vector, contains all coordinates of a N-particle-system, $\Delta t \ldots$ time difference, $P \ldots$ number of time points, $V \ldots$ potential energy,   which can be written in the form of a power series

systems will be considered in chapter 7.

4.1   The Basic Equation System

In order to derive equations which connect *Lagrangian* multipliers with correlation functions, namely hyper-surface equations, a system of the kind (3.28) can be taken as a basis. Such a basic system will be called a *basic equation system*, in which case the result of the inversion of such a basic equation system are the hyper-surface equations. By using product brackets such a basic equation system can be written in the form

$$\int_{-\infty}^{+\infty} \rho(\Omega) \left\{ \prod_{i=1}^{\alpha} \Omega_{\Theta_i} \right\} d\Omega = \left\langle \left\{ \prod_{i=1}^{\alpha} \Omega_{\Theta_i} \right\} \right\rangle ,$$

(4.1)

where (4.1) is identical with (3.28) if suitable boundaries are used in (3.28). In the following the boundaries $-\infty$ and $+\infty$ shall always be used. An extension to other boundaries is possible, but shall not be discussed in this book. Then a distribution function $\rho(\Omega)$ of N $(N = 1, 2, 3, \ldots \infty)$ variables shows the form

$$\rho(\Omega) = Z^{-1} \exp \left[- \left(\sum_{\alpha=1}^{O} \left\{ \prod_{i=1}^{\alpha} \sum_{\Theta_i=1}^{N} \lambda_{\Theta_\alpha} \Omega_{\Theta_i} \right\} \right) \right] ,$$

(4.2)

where O represents the order of the function. Using the boundaries $\pm\infty$ the order O has to be even $(O = 2, 4, 6, \ldots \infty)$, otherwise divergent integrals would occur. The correlated partition function is given by

$$Z = \int_{-\infty}^{+\infty} \exp \left[- \left(\sum_{\alpha=1}^{O} \left\{ \prod_{i=1}^{\alpha} \sum_{\Theta_i=1}^{N} \lambda_{\Theta_\alpha} \Omega_{\Theta_i} \right\} \right) \right] d\Omega ,$$

(4.3)

where the *Lagrangian* multipliers shall be real, because here quantum mechanical functions shall not be considered (see chapter 7). By using such partition functions the basic equation system (4.2) obtains the form

$$-Z^{-1} \frac{\partial Z}{\partial \lambda_{\Theta_\alpha}} = \left\langle \left\{ \prod_{i=1}^{\alpha} \Omega_{\Theta_i} \right\} \right\rangle .$$

(4.4)

(4.4) represents a universal formulation to describe correlation functions, where the partition function formalism has been used. (4.4) includes all basic equation systems which will be considered in this book, i. e. in particular *Landau* systems are included. In this case a correlation function represents a polarization or a magnetization which are measurable on a macroscopic level.

In the following it shall be shown that a determination of *Lagrangian* multipliers is easily possible if measurement quantities of higher order are known. During these considerations only problems of fourth order shall be considered.

4.2 Linear Determination Equations

Using a partition function of fourth order the basic equation system has the form

$$-Z_4^{-1}\frac{\partial Z_4}{\partial \lambda_o} = \langle \Omega_o \rangle \,, \quad -Z_4^{-1}\frac{\partial Z_4}{\partial \lambda_{o,k}} = \langle \Omega_o \Omega_k \rangle \,, \quad -Z_4^{-1}\frac{\partial Z_4}{\partial \lambda_{o,k,l}} = \langle \Omega_o \Omega_k \Omega_l \rangle \,,$$

$$-Z_4^{-1}\frac{\partial Z_4}{\partial \lambda_{o,k,l,m}} = \langle \Omega_o \Omega_k \Omega_l \Omega_m \rangle \,. \tag{4.5}$$

By partial integration of this set of equations it is possible to derive a linear system of equation to determine the *Lagrangian* multipliers. An equivalent way of derivation requires the partition function as a starting point. This way shall now be considered, because then relations arise which are needed later.

By partial integration of the partition function Z_4 and by division with Z_4 afterwards, where

$$Z_4 = \int_{-\infty}^{+\infty} \exp\left[-\left(\sum_{i=1}^{N} \lambda_i \Omega_i + \sum_{i,k=1}^{N} \lambda_{i,k}\,\Omega_i \Omega_k + \sum_{i,k,l=1}^{N} \lambda_{i,k,l}\,\Omega_i \Omega_k \Omega_l + \sum_{i,k,l,m=1}^{N} \lambda_{i,k,l,m}\,\Omega_i \Omega_k \Omega_l \Omega_m \right)\right] d\Omega \tag{4.6}$$

holds, one obtains the expression

$$\lambda_o \langle \Omega_o \rangle + 2\sum_{k=1}^{N} \lambda_{o,k}\langle \Omega_o \Omega_k \rangle + 3\sum_{k,l=1}^{N} \lambda_{o,k,l}\langle \Omega_o \Omega_k \Omega_l \rangle + 4\sum_{k,l,m=1}^{N} \lambda_{o,k,l,m}\langle \Omega_o \Omega_k \Omega_l \Omega_m \rangle = 1 \,, \tag{4.7}$$

where (4.5) has been used. Summation with respect to o leads to the relation

$$\sum_{o=1}^{N} \lambda_o \langle \Omega_o \rangle + 2\sum_{o,k=1}^{N} \lambda_{o,k}\langle \Omega_o \Omega_k \rangle + 3\sum_{o,k,l=1}^{N} \lambda_{o,k,l}\langle \Omega_o \Omega_k \Omega_l \rangle + 4\sum_{o,k,l,m=1}^{N} \lambda_{o,k,l,m}\langle \Omega_o \Omega_k \Omega_l \Omega_m \rangle = N \,. \tag{4.8}$$

After partial differentiation of (4.8) with respect to all multipliers one obtains the equation

$$\sum_{\alpha=1}^{4}\left[\alpha\left\{\prod_{i=1}^{\alpha}\sum_{\Theta_i=1}^{N}\lambda_{\Theta_\alpha}\left\{CD(G,\alpha,\Theta_{\alpha+4})\right]=G\left\langle\left\{\prod_{i=5}^{G+4}\Omega_{\Theta_i}\right\}\right\rangle,\right.\right.$$

(4.9)

in which case the correlation function difference $CD(G,\alpha,\Theta_{\alpha+4})$ is given by the relation

$$CD(G,\alpha,\Theta_{\alpha+4})=$$
$$\left\langle\left\{\prod_{i=1}^{G+\alpha}\Omega_{\Theta_i}\right\}\right\rangle-\left\langle\left\{\prod_{i=1}^{\alpha}\Omega_{\Theta_i}\right\}\right\rangle\left\langle\left\{\prod_{i=5}^{G+4}\Omega_{\Theta_i}\right\}\right\rangle.$$

(4.10)

(4.9) connects all distribution function parameters (i. e. *Lagrangian* multipliers up to fourth order) with all necessary correlation functions (i. e. functions up to fourth order) and with correlation functions of higher order as necessary. G generates four different equation systems, in which case $G = 1, 2, 3, 4$ holds. For example, $G = 1$ generates

$$1\sum_{i=1}^{N}\lambda_i\left(\langle\Omega_i\Omega_n\rangle-\langle\Omega_i\rangle\langle\Omega_n\rangle\right)+2\sum_{i,k=1}^{N}\lambda_{i,k}\left(\langle\Omega_i\Omega_k\Omega_n\rangle-\langle\Omega_i\Omega_k\rangle\langle\Omega_n\rangle\right)+$$

$$3\sum_{i,k,l=1}^{N}\lambda_{i,k,l}\left(\langle\Omega_i\Omega_k\Omega_l\Omega_n\rangle-\langle\Omega_i\Omega_k\Omega_l\rangle\langle\Omega_n\rangle\right)+$$

$$4\sum_{i,k,l,m=1}^{N}\lambda_{i,k,l,m}\left(\langle\Omega_i\Omega_k\Omega_l\Omega_m\Omega_n\rangle-\langle\Omega_i\Omega_k\Omega_l\Omega_m\rangle\langle\Omega_n\rangle\right)=\langle\Omega_n\rangle.$$

(4.11)

Such a linear system of determination equations can be used to determine the *Lagrangian* multipliers. But, as the reader can see, not only a minimum set of correlation functions is necessary, functions of higher order are also needed. As a measurement of correlation functions of high order is a difficult one, the question arises whether it is possible to use only a minimum set of correlation functions. This is actually possible and shall be discussed extensively, i. e. in the following the problem of inversion of a basic equation system of the kind (4.4) shall be discussed so that *Lagrangian* multipliers can be determined by using analytical functions. Then, however, a totally non-linear problem has to be solved. This problem is a very difficult one. Therefore, it shall be solved in a successive way.

4.3 The Non-Linear Inversion Problem

In order to solve the inversion problem of a non-linear system of the type (4.4), some steps are useful. First, an easy basic equation system of fourth order and one variable shall be considered. On this occasion, principle mathematical problems will

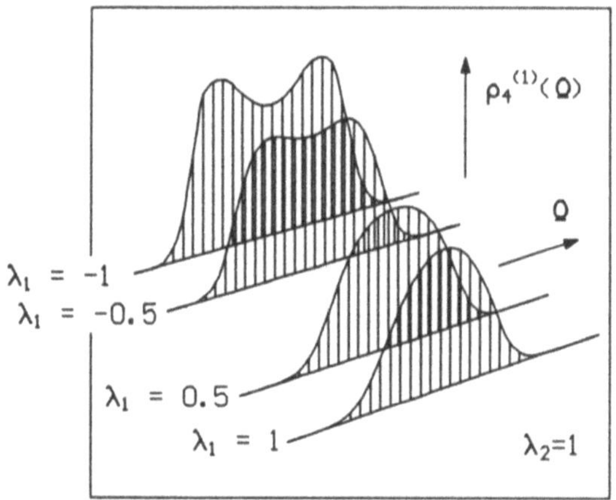

Figure 4.1
A symmetrical distribution function of fourth order and one variable Ω: $\rho_4^{(1)}(\Omega) = Z_4^{(1)\,-1} \exp[-(\lambda_1\Omega^2+\lambda_2\Omega^4)]$. $Z_4^{(1)}$ represents the partition function

be presented and approximative solutions will be derived which describe the system near a critical point (i. e. critical hyper-surface equations). Second, a systematic inversion procedure will be described, which – in principle – allows to solve the inversion problem in a successive way. Two-dimensional systems of fourth order will then be considered. However, such a systematic calculation is a very extensive one. Therefore, only the basic ideas will be presented. Such a presentation is very useful, because such an analysis allows to find the form of the exact solution of one- and two-dimensional systems (i. e. the form of hyper-surface equations of one- and two-dimensional problems). In a second step it will be shown how to determine these solutions totally. Then an extension to systems of fourth order and any number of variables will be discussed. Then it is possible to find the solutions of problems of higher order. During these considerations only discrete problems will be discussed. Therefore, it is useful to introduce continuous problems afterwards. Then, instead of ordinary integrals, path integrals arise. In this context it will be shown how to solve such continuous problems (i. e. how to gain continuous hyper-surface equations).

However, the considerations shall be started with a short overview about essential underlying distribution functions, partition functions and hyper-surface equations. This can easily be done, because such functions can be visualized by using numerical methods.

4.3.1 Some Numerical Results

Distribution Functions

Very often used distribution functions are symmetrical functions of fourth order. An example is represented the distribution function which can be used to describe the behavior of a magnetization near a critical point in the context of *Landau's* theory of phase transitions (see table 4.1). Such a function is visualized in figure

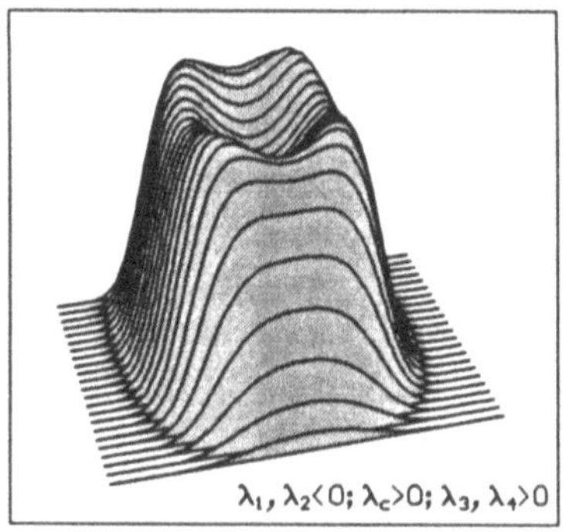

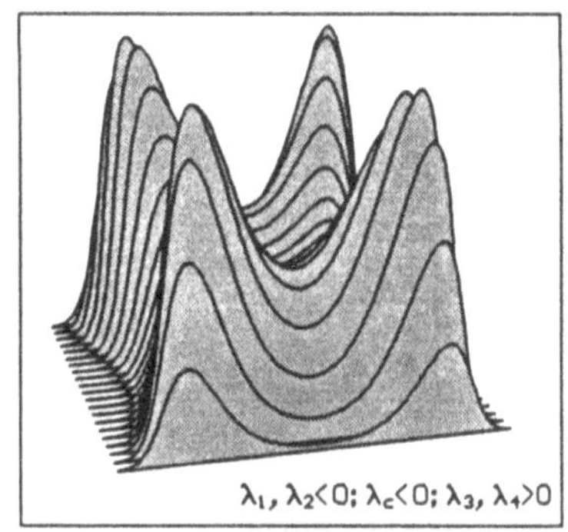

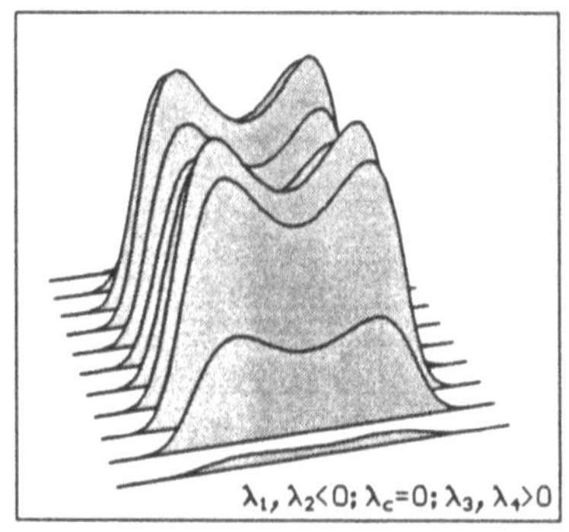

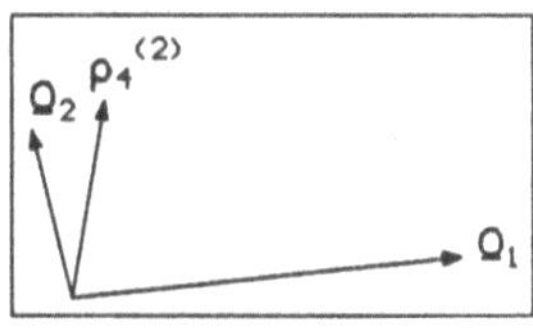

Figure 4.2 A symmetrical distribution function of fourth order and two variables Ω_1 and Ω_2: $\rho_4^{(2)}(\Omega_1, \Omega_2) = Z_4^{(2)^{-1}} \exp[-(\lambda_1 \Omega_1^2 + \lambda_2 \Omega_2^2 + \lambda_c \Omega_1^2 \Omega_2^2 + \lambda_3 \Omega_1^4 + \lambda_4 \Omega_2^4)]$. $Z_4^{(2)}$ represents the partition function

4.1. As the reader can see, the change of the sign of the parameter of second order λ_1 causes a bifurcation, i. e. two maxima arise. In *Landau's* theory these two maxima correspond to two possible states of magnetization, in which case a distribution function is necessary since fluctuating effects are very strong near a phase transition point. It has to be remarked that far away from such a critical point the system adopts one of the possible states, because then the fluctuating forces are small so that normally a *Gaussian* function is sufficient to describe the statistics of the state. This statement has a wide-ranging meaning, i. e. distribution functions of higher order occur near critical points, and far away from such points only basic distribution functions are sufficient. Before such a phase transition only one state is

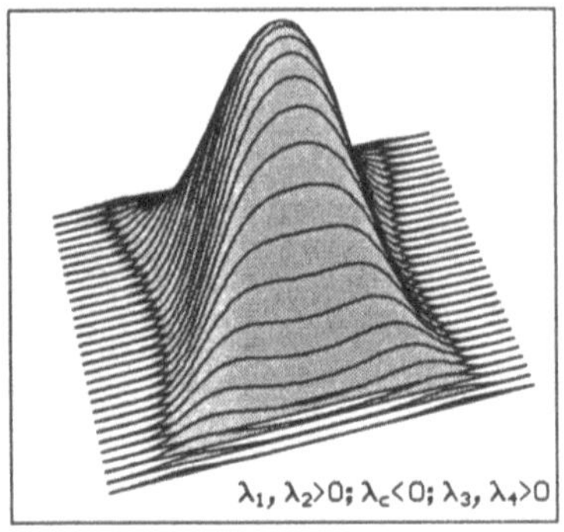

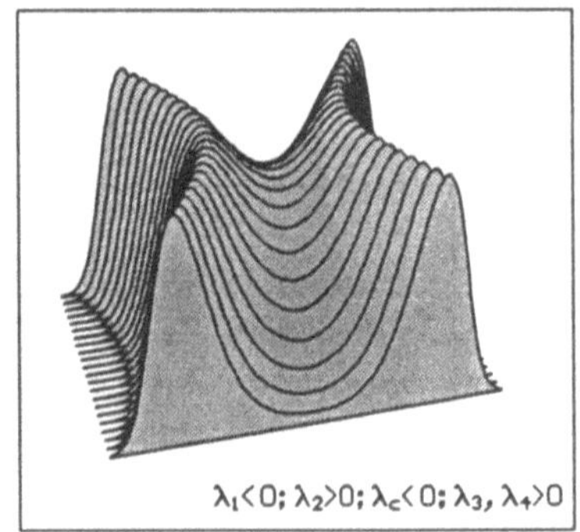

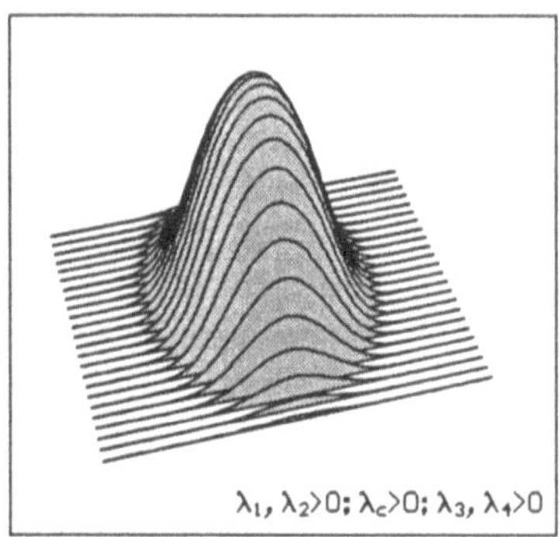

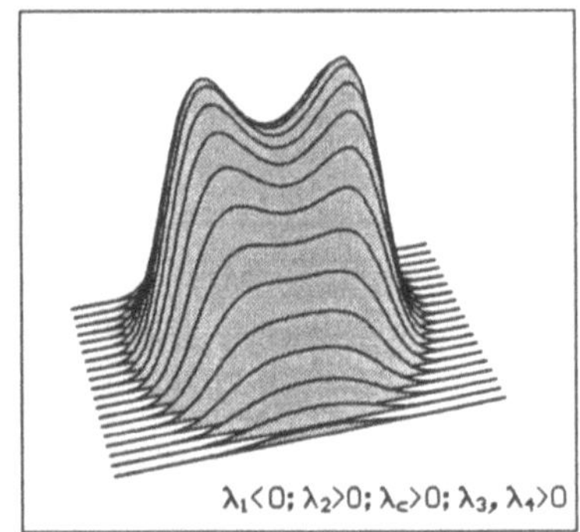

Figure 4.3   A symmetrical distribution function of fourth order and two variables Ω_1 and Ω_2: $\rho_4^{(2)}(\Omega_1, \Omega_2) = Z_4^{(2)^{-1}} \exp[-(\lambda_1 \Omega_1^2 + \lambda_2 \Omega_2^2 + \lambda_c \Omega_1^2 \Omega_2^2 + \lambda_3 \Omega_1^4 + \lambda_4 \Omega_2^4)]$. $Z_4^{(2)}$ represents the partition function. The reference frame is the same as in figure 4.2

possible which corresponds to a non-magnetic state. This can also be seen in figure 4.1. However, figure 4.2 shows a symmetrical distribution function of second order. In this context it has to be remarked that the zero point of the reference frame is identical with the centres of the bottoms of the various pictures. It can be seen that the change of the sign of the coupling term produces four emphasized maxima, i. e. four states are possible. For example, such more-dimensional problems have to be considered if not only one component of a magnetization (or a polarization) is necessary to describe the system, i. e. if vectors have to be used. Then two or three components occur. In figure 4.3 the same distribution function is shown, but additionally a phase transition is visualized, i. e. the change of the sign of one parameter of second order is considered. Furthermore, in this figure it can be seen that a bifurcation occurs if one parameter of second order changes its sign. And it can be seen that the transition of the coupling parameter λ_c causes a centralization of such a distribution function. The examples the reader should have in mind are thermodynamic examples. However, it has to be remarked that such a description is a quite universal one. For example, laser systems, too, can be described by such functions. This shall be discussed in chapter 6. However, it has to be remarked that such low-dimensional distribution functions only play the role of easy examples. The concept which will be developed will deal with an infinite number of statistical variables.

Partition Functions

A partition function of a symmetrical two-dimensional problem is shown in figure 4.4. In this picture four cases are shown. It can bee seen that the increase of a parameter of second order causes the decrease of the value of the partition function. The dependence of the change of the sign of the coupling parameter λ_c and the

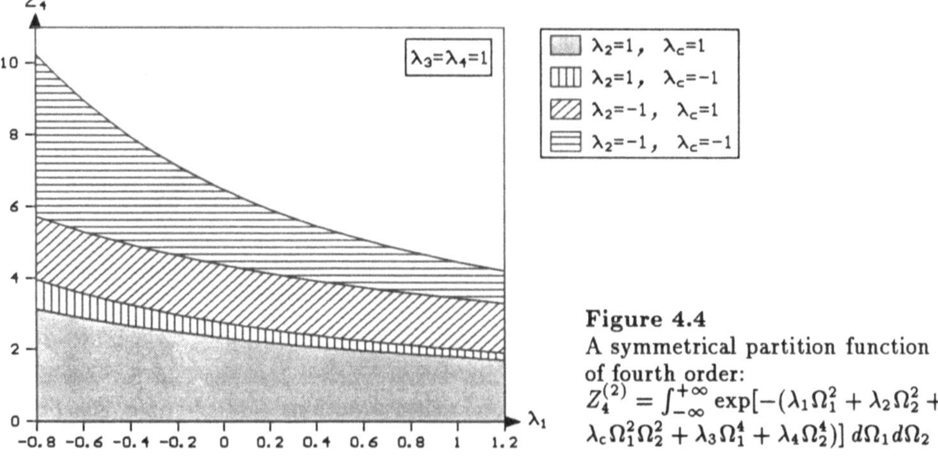

Figure 4.4
A symmetrical partition function of fourth order:
$$Z_4^{(2)} = \int_{-\infty}^{+\infty} \exp[-(\lambda_1\Omega_1^2 + \lambda_2\Omega_2^2 + \lambda_c\Omega_1^2\Omega_2^2 + \lambda_3\Omega_1^4 + \lambda_4\Omega_2^4)]\, d\Omega_1 d\Omega_2$$

influence of a phase transition can be seen, too. Such a partition function cannot be interpreted clearly, however, the reader has to know that the behavior of these graphs is directly correlated with relevant macroscopic quantities, because – this was above mentioned – such quantities can be calculated by using the partition function and suitable operators, in particular differential operators.

Hyper-Surface Equations

Functions of the form
$$\lambda_{\Theta_\alpha} = \lambda_{\Theta_\alpha}[\langle f(\Omega)\rangle] \tag{4.12}$$

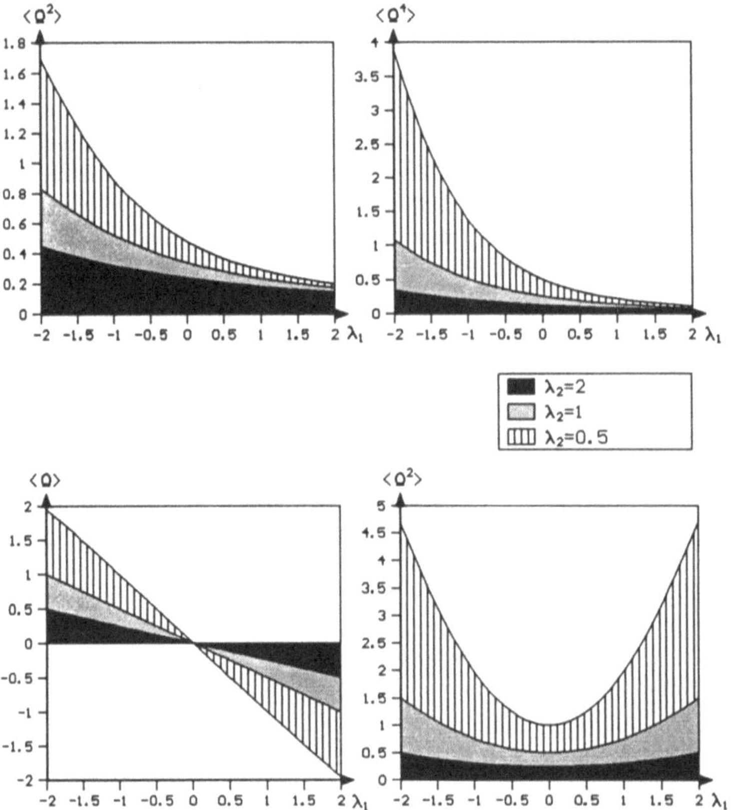

Figure 4.5 The visualization of elementary hyper-surface equations. The case of fourth order (see above), and the *Gaussian* case (see below)

are called *hyper-surface equations* in this book. Such equations can be visualized by using integral representations of corrleation functions (for example, see (4.1)), because then numerical calculations can be carried out easily. Elementary examples of hyper-surface equations are such equations which correspond to the *Gaussian* distribution function

$$\rho_2^{(1)}(\Omega) = Z_2^{(1)^{-1}} \exp[-(\lambda_1\Omega + \lambda_2\Omega^2)] \tag{4.13}$$

or to the elementary function of fourth order, the function

$$\rho_4^{(1)}(\Omega) = Z_4^{(1)^{-1}} \exp[-(\lambda_1\Omega^2 + \lambda_2\Omega^4)] . \tag{4.14}$$

If the *Gaussian* case holds, the hyper-surface equations are of the kind

$$\lambda_1 = \lambda_1(\langle\Omega\rangle, \langle\Omega^2\rangle) , \quad \lambda_2 = \lambda_2(\langle\Omega\rangle, \langle\Omega^2\rangle) , \tag{4.15}$$

and if the case of fourth order holds, the hyper-surface equations are of the kind

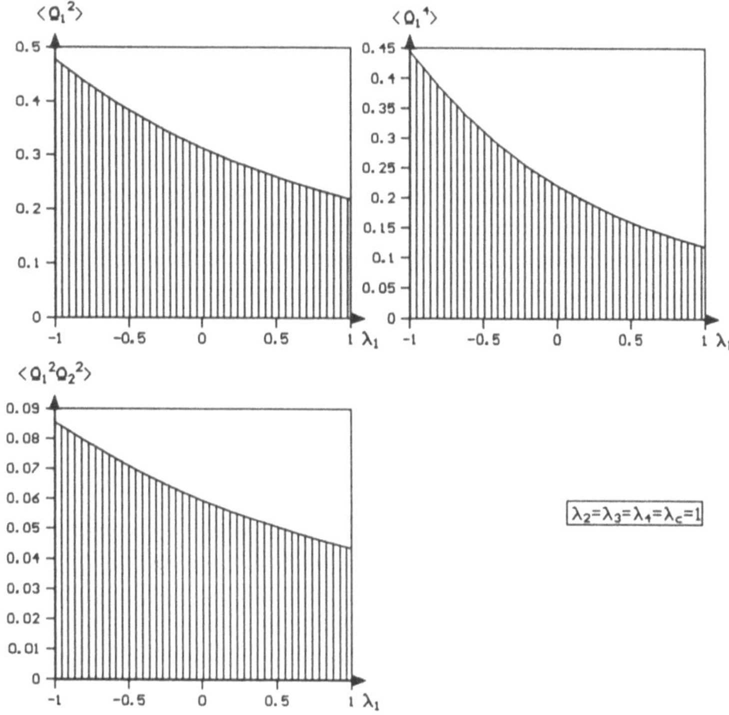

Figure 4.6 The visualization of elementary hyper-surface equations in the symmetrical case of fourth order and two variables Ω_1 and Ω_2

$$\lambda_1 = \lambda_1(\langle \Omega^2 \rangle, \langle \Omega^4 \rangle) \, , \quad \lambda_2 = \lambda_2(\langle \Omega^2 \rangle, \langle \Omega^4 \rangle) \, , \tag{4.16}$$

where the equations (4.15) and (4.16) show that in this book within a hyper-surface equation always such correlation functions shall be used which are of the same kind as the variable functions. For example, if the variable functions Ω^2 and Ω^4 occur within the exponential function, the correlation functions which shall be used to determine the parameters are of the kind $\langle \Omega^2 \rangle$ and $\langle \Omega^4 \rangle$. The functional dependence of the parameters of second order with respect to the correlation functions is shown in figure 4.5. The upper two pictures show the case of fourth order, and the pictures below show the *Gaussian* case. In the *Gaussian* case negative correlation functions are possible (because a non-symmetric term within the basic distribution function exists). Hyper-surface equations which correspond to the *Gaussian* case can be used to describe the connection between the mean magnetization and the statistical behavior far away from a critical point. Hyper-surface equations of fourth order can be used to describe such a connection near a critical point. A two-dimensional problem shows figure 4.6. Then the basic distribution function is of the kind

$$\rho_4^{(2)}(\Omega_1, \Omega_2) = Z_4^{(2)^{-1}} \exp[-(\lambda_1 \Omega_1^2 + \lambda_2 \Omega_2^2 + \lambda_c \Omega_1^2 \Omega_2^2 + \lambda_3 \Omega_1^4 + \lambda_4 \Omega_2^4)] \, , \tag{4.17}$$

in which case the hyper-surface equations are of the form

$$\lambda_{1,2,c,3,4} = \lambda_{1,2,c,3,4}(\langle \Omega_1^2 \rangle, \langle \Omega_2^2 \rangle, \langle \Omega_1^2 \Omega_2^2 \rangle, \langle \Omega_1^4 \rangle, \langle \Omega_2^4 \rangle) \; . \tag{4.18}$$

At all events such low-dimensional hyper-surface equations are of theoretical interest. Functions of practical interest are functions of more variables. For example, the reader has to think of the method of *Simon*, which is used in medcine to measure the field of the human brain surface, i. e. to measure an EEG. This method uses exactly 21 measurement points. An approximative description of the behavior of the measured signal is the exponential description introduced above. In this case hyper-surface equations which correspond to 21 variables describe the connection between the measurement level and the statistical level.

With the derivation of hyper-surface equations shall now be started.

4.3.2   The Problem of Partition Functions

In order to solve equations of the kind (4.4), integrals of the exponential type have to be calculated, i. e. partition functions have to be known in their explicit form. Symmetrical integrals of fourth order and one variable can be calculated by using integrals of the *Laplacian* type. This shall be shown first. As symmetrical integrals of fourth order and more variables can be constructed by using such an easy integral, such an integral is an elementary integral.

The Elementary Partition Function

The partition function

$$Z_4^{(1)} = \int_{-\infty}^{+\infty} \exp\left[- \left(\lambda_1 \Omega^2 + \lambda_2 \Omega^4\right)\right] d\Omega \tag{4.19}$$

can be calculated by using the well-known relation (see [23, 77])

$$\int_0^{+\infty} \exp\left[- \left(\lambda_1 \Omega + \lambda_2 \Omega^2\right)\right] \Omega^{\nu-1} \, d\Omega =$$
$$\left(2\lambda_2\right)^{-\nu/2} \Gamma(\nu) \exp\left(\lambda_1^2 / 8\lambda_2\right) D_\nu \left(\lambda_1 / \sqrt{2\lambda_2}\right)$$
$$(\nu > 0, \; \lambda_2 > 0) \; , \tag{4.20}$$

where the integral of the l. h. s. is a special integral of the *Laplacian* type, and in which case $D_\nu \left(\lambda_1 / \sqrt{2\lambda_2}\right)$ represents a special parabolic cylinder function with the argument $\lambda_1 / \sqrt{2\lambda_2}$. This parabolic cylinder function is shown in figure 4.7. Such a function can numerically be calculated by using the relation (4.20). A series representation of such a parabolic cylinder function can be found by solving the *Schrödinger* equation of the quantum mechanical harmonic oscillator, i. e. parabolic cylinder functions are solutions of a basic quantum mechanical evolution equation

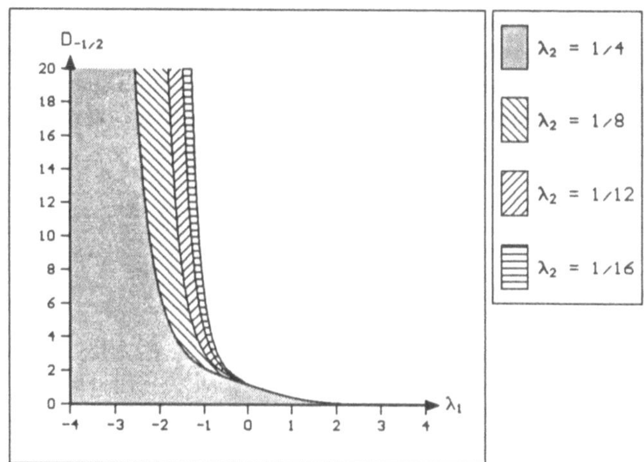

Figure 4.7
The parabolic cylinder
function $D_{-1/2}$

(see chapter 5). However, such a series representation will be considered later (see (4.47)). Γ represents the so-called *Gamma function* (see [2]). Transforming (4.20) with

$$\Omega = \tilde{\Omega}^2 \;, \quad d\Omega = 2\tilde{\Omega} d\tilde{\Omega} \tag{4.21}$$

and using

$$2\nu - 1 = \mu \;, \quad \mu := 0 \tag{4.22}$$

one obtains the relation

$$2\int_0^{+\infty} \exp\left[-\left(\lambda_1\Omega^2 + \lambda_2\Omega^4\right)\right]\Omega^{\nu-1}\, d\Omega =$$
$$\left(2\lambda_2\right)^{-1/4}\Gamma(1/2)\exp\left(\lambda_1^2/8\lambda_2\right)D_{-1/2}\left(\lambda_1/\sqrt{2\lambda_2}\right)$$
$$\left(\lambda_2 > 0\right), \tag{4.23}$$

where instead of $\tilde{\Omega}$ the symbol Ω is used again. As such an integral of fourth order is symmetrical with reference to 0, the relation

$$2\int_0^{+\infty} = \int_{-\infty}^{+\infty} \tag{4.24}$$

holds so that the partition function (4.19) yields

$$\boxed{Z_4^{(1)} = \left(2\lambda_2\right)^{-1/4}\Gamma(1/2)\exp\left(\lambda_1^2/8\lambda_2\right)D_{-1/2}\left(\lambda_1/\sqrt{2\lambda_2}\right).} \tag{4.25}$$

Apart from the parabolic cylinder function $D_{-1/2}$, (4.25) represents the explicit form of the elementary partition function of fourth order. In this relation the inequality $\lambda_2 > 0$ has to be taken as a basis. It has to be remarked here that this inequality is no pure mathematical restriction but a physical restriction, i. e. this restriction guarantees convergent (i. e. physical) partition functions. Such a representation can be used to construct more complicated solutions, i. e. solutions of more-variable problems. This shall be shown by considering a two-variable problem.

The Coupling of Partition Functions

A two-dimensional partition function of the kind

$$Z_4^{(2)} = \int_{-\infty}^{+\infty} \exp\left[-\left(\lambda_1 \Omega_1^2 + \lambda_2 \Omega_2^2 + \lambda_c \Omega_1^2 \Omega_2^2 + \lambda_3 \Omega_1^4 + \lambda_4 \Omega_2^4\right)\right] d\Omega_1 d\Omega_2 \quad (4.26)$$

can be calculated. In order to do this, the series representation of the exponential coupling part has to be used, i. e. the relation

$$\exp\left[-\left(\lambda_c \Omega_1^2 \Omega_2^2\right)\right] = \sum_{\nu=0}^{\infty} \left[(-1)^\nu (1/\nu!) \left(\Omega_1^2 \Omega_2^2\right)^\nu\right] \lambda_c^\nu \quad (4.27)$$

has to be used. Inserting (4.27) into (4.26), the relation (4.26) yields

$$Z_4^{(2)} = \sum_{\nu=0}^{\infty} \left\{ (-1)^\nu (1/\nu!) \int_{-\infty}^{+\infty} \exp\left[-\left(\lambda_1 \Omega_1^2 + \lambda_3 \Omega_1^4\right)\right] \Omega_1^{2\nu} \, d\Omega_1 \right.$$

$$\left. \int_{-\infty}^{+\infty} \exp\left[-\left(\lambda_2 \Omega_2^2 + \lambda_4 \Omega_2^4\right)\right] \Omega_2^{2\nu} \, d\Omega_2 \right\} \lambda_c^\nu \ . \quad (4.28)$$

Using derivatives of the elementary partition functions

$$\boxed{\begin{aligned} Z_m &= \int_{-\infty}^{+\infty} \exp\left[-\left(\lambda_m \Omega_m^2 + \lambda_{m+2} \Omega_m^4\right)\right] d\Omega_m \\ &= \left(2\lambda_{m+2}\right)^{-1/4} \Gamma(1/2) \exp\left(\lambda_m^2 / 8\lambda_{m+2}\right) \\ & \quad D_{-1/2}\left(\lambda_m / \sqrt{2\lambda_{m+2}}\right) \quad (m = 1, 2) \end{aligned}} \quad (4.29)$$

the representation

$$Z_4^{(2)} = \sum_{\nu=0}^{\infty} \left[(-1)^\nu (1/\nu!) \left(\frac{\partial^\nu Z_1}{\partial \lambda_1^\nu}\right) \left(\frac{\partial^\nu Z_2}{\partial \lambda_2^\nu}\right)\right] \lambda_c^\nu \quad (4.30)$$

holds. By introducing a coupling factor of the form

$$K_c(\lambda_c) = \sum_{\nu=0}^{\infty} \left[(-1)^\nu (1/\nu!) \left(Z_1^{-1} \frac{\partial^\nu Z_1}{\partial \lambda_1^\nu}\right) \left(Z_2^{-1} \frac{\partial^\nu Z_2}{\partial \lambda_2^\nu}\right)\right] \lambda_c^\nu$$

$$= \sum_{\nu=0}^{\infty} \left[(-1)^\nu (1/\nu!) \langle \Omega_1^{2\nu}\rangle_{dc} \langle \Omega_2^{2\nu}\rangle_{dc}\right] \lambda_c^\nu \quad (4.31)$$

the partition function (4.30) takes on the form

$$Z_4^{(2)} = K_c(\lambda_c) Z_1 Z_2 \ ,$$
(4.32)

where the functions $\left\langle \Omega_m^{2\nu} \right\rangle_{dc}$ are correlation functions of the totally decoupled problem. Such a coupling coefficient shall be written in the form

$$K_c(\lambda_c) = \sum_{\nu=0}^{\infty} \left(k_\nu^{(1)} k_\nu^{(2)} \right) \lambda_c^\nu \ ,$$
(4.33)

where the coefficients

$$\boxed{\begin{aligned} k_\nu^{(m)} &= k_\nu^{(m)}(\lambda_m, \lambda_{m+2}) = \mathrm{i}^\nu \sqrt{(1/\nu!)} \left(Z_m^{-1} \frac{\partial^\nu Z_m}{\partial \lambda_m^{\ \nu}} \right) \\ (\mathrm{i} &= \sqrt{-1} \ , \ m = 1, 2) \end{aligned}}$$
(4.34)

have to be considered, which shall be called *complex elementary coefficients*. Such complex elementary coefficients contain all *Lagrangian* multipliers of one single system. With such complex elementary coefficients the partition function (4.32) can be written in the form

$$\boxed{Z_4^{(2)} = \sum_{\nu=0}^{\infty} \left(k_\nu^{(1)} k_\nu^{(2)} Z_1 Z_2 \right) \lambda_c^\nu \ .}$$
(4.35)

Apart from the parabolic cylinder function $D_{-1/2}$, (4.35) represents the explicit form of the symmetrical partition function of fourth order and two variables. As the reader can see, such a function consists of the elementary partition functions and complex elementary coefficients. The coupling parameter λ_c guarantees the coupling of the single systems. In the case of decoupling the coupling factor $K_c(\lambda_c)$ is equal to one so that a product of the elementary partition functions arises, i. e. the case of decoupling is given by the relation

$$Z_{4,dc}^{(2)} = K_{c,dc}(\lambda_c) Z_1 Z_2 \ , \quad K_{c,dc}(\lambda_c) = 1 \ .$$
(4.36)

This method of solution can be extended to multi-dimensional symmetrical problems. This shall now be considered.

Multi-Dimensional Symmetrical Partition Functions

A partition function of a large number of variables Ω_i can be solved by using a generalized form of the above introduced concept. In order to have a more graphic access to such a problem, the reader may interpret the statistical variables Ω_i as measured field components of the human skull at points i, or the reader may think of a multi-component laser so that such variables are field amplitudes of various laser modes. The partition function which have to be considered are of the form

$$Z_4^{(N)} = \int_{-\infty}^{+\infty} \exp\left[-\left(\sum_{i=1}^{N} \lambda_i^{(2)}\Omega_i^2 + \sum_{i,k=1,i<k}^{N} \lambda_{i,k}^{(c)}\Omega_i^2\Omega_k^2 + \sum_{i=1}^{N} \lambda_i^{(4)}\Omega_i^4\right)\right] d\Omega \;,$$

$$(4.37)$$

where the inequality guarantees that only independent terms arise, and in which case N prescribes the number of variables. (Such a concept can be extended so that operators instead of variables Ω_i can be considered. Then it has to be borne in mind that not all operators are interchangeable. Then an inequality is not useful.) In order to solve such a multi-dimensional integral problem, instead of (4.27) the series representation

$$\exp\left[-\left(\lambda_{i,k}^{(c)}\Omega_i^2\Omega_k^2\right)\right] = \sum_{\nu_{i,k}=0}^{\infty} \left[(-1)^{\nu_{i,k}}(1/\nu_{i,k}!)\left(\Omega_i^2\Omega_k^2\right)^{\nu_{i,k}}\right] \lambda_{i,k}^{(c)\,\nu_{i,k}}$$

$$(4.38)$$

has to be used. Inserting (4.38) into (4.37) the relation (4.37) yields

$$Z_4^{(N)} = \left\{\prod_{i,k=1,i<k}^{N} \sum_{\nu_{i,k}=0}^{\infty}\right\} \int_{-\infty}^{+\infty} \left\{\exp\left[-\left(\sum_{i=1}^{N}\lambda_i^{(2)}\Omega_i^2 + \sum_{i=1}^{N}\lambda_i^{(4)}\Omega_i^4\right)\right]\right.$$
$$\left.\left\{\prod_{i,k=1,i<k}^{N}\left[(-1)^{\nu_{i,k}}(1/\nu_{i,k}!)\left(\Omega_i^2\Omega_k^2\right)^{\nu_{i,k}}\right]\lambda_{i,k}^{(c)\,\nu_{i,k}}\right\}\right\} d\Omega \;,$$

$$(4.39)$$

where product brackets are used to achieve a clear structuring. Such an expression is equivalent with the formula

$$Z_4^{(N)} = \left\{\prod_{i,k=1,i<k}^{N} \sum_{\nu_{i,k}=0}^{\infty}\right\}\left\{\prod_{i,k=1,i<k}^{N} (-1)^{\nu_{i,k}}(1/\nu_{i,k}!)\right\}$$
$$\left\{\prod_{i=1}^{N}\int_{-\infty}^{+\infty}\exp\left[-\left(\lambda_i^{(2)}\Omega_i^2 + \lambda_i^{(4)}\Omega_i^4\right)\right]\left(\Omega_i^2\right)^{\mu_i} d\Omega_i\right\}$$
$$\left\{\prod_{i,k=1,i<k}^{N} \lambda_{i,k}^{(c)\,\nu_{i,k}}\right\} \;.$$

$$(4.40)$$

(Such a conversion is possible, because $\nu_{i,k} = \nu_{k,i}$ holds.) (4.40) is a generalized form of (4.28), in which case

$$\mu_i = \sum_{k=1,i\neq k}^{N} \nu_{i,k}$$

$$(4.41)$$

holds. By using the generalized form of (4.29), namely

$$\boxed{\begin{aligned}
Z_i &= \int_{-\infty}^{+\infty} \exp\left[-\left(\lambda_i^{(2)}\Omega_i^2 + \lambda_i^{(4)}\Omega_i^4\right)\right] d\Omega_i \\
&= \left(2\lambda_i^{(4)}\right)^{-1/4} \Gamma(1/2)\exp\left(\lambda_i^{(2)^2}/8\lambda_i^{(4)}\right) \\
&\quad D_{-1/2}\left(\lambda_i^{(2)}/\sqrt{2\lambda_i^{(4)}}\right) \quad (i = 1\ldots N)\,,
\end{aligned}}$$

$$(4.42)$$

the relation (4.40) can be written in the form

$$Z_4^{(N)} = \left\{\prod_{i,k=1,i<k}^{N}\sum_{\nu_{i,k}=0}^{\infty}\right\}\left\{\prod_{i,k=1,i<k}^{N}(-1)^{\nu_{i,k}}(1/\nu_{i,k}!)\right\}$$
$$\left\{\prod_{i=1}^{N}(-1)^{\mu_i}\frac{\partial^{\mu_i}Z_i}{\partial\lambda_i^{(2)^{\mu_i}}}\right\}\left\{\prod_{i,k=1,i<k}^{N}\lambda_{i,k}^{(c)^{\nu_{i,k}}}\right\}. \tag{4.43}$$

Introducing real elementary coefficients

$$\boxed{k_{\mu_i}^{(i)} = z_{\mu_i}^{(i)} Z_i^{-1}\frac{\partial^{\mu_i}Z_i}{\partial\lambda_i^{(2)^{\mu_i}}}} \tag{4.44}$$

with the numbers

$$z_{\mu_i}^{(i)} = (-1)^{\mu_i}\left\{\prod_{k=1,i<k}^{N}(-1)^{\nu_{i,k}}(1/\nu_{i,k}!)\right\}\,,\quad z_{\mu_i}^{(i)} = z_{\mu_i+n}^{(i)}\,,$$
$$\mu_i = \sum_{k=1,i\neq k}^{N}\nu_{i,k}\,,\quad n = 0,1,2 \tag{4.45}$$

the partition function (4.43) takes on the form

$$\boxed{Z_4^{(N)} = \left\{\prod_{i,k=1,i<k}^{N}\sum_{\nu_{i,k}=0}^{\infty}\left\{\prod_{i=1}^{N}k_{\mu_i}^{(i)}\right\}\lambda_{i,k}^{(c)^{\nu_{i,k}}}\right\}\left\{\prod_{i=1}^{N}Z_i\right\}.} \tag{4.46}$$

(4.46) is the partition function of the multi-dimensional symmetrical problem. Apart from the parabolic cylinder function $D_{-1/2}$, an explicit form is now available. As the reader may not be very familiar with the product formalism, the explicit form of (4.46) shall be considered.

In an explicit form the last product term reads

$$\left\{\prod_{i=1}^{N}Z_i\right\} = Z_1 Z_2\ldots Z_N\,. \tag{4.47}$$

As the inner product term is of the form

$$\left\{\prod_{i=1}^{N} k_{\mu_i}^{(i)}\right\} = k_{\mu_1}^{(1)} k_{\mu_2}^{(2)} \ldots k_{\mu_N}^{(N)} \,, \tag{4.48}$$

the first product term reads

$$\left\{\prod_{i,k=1,i<k}^{N} \sum_{\nu_{i,k}=0}^{\infty} \left\{\prod_{i=1}^{N} k_{\mu_i}^{(i)}\right\} \lambda_{i,k}^{(c)\,\nu_{i,k}}\right\} =$$

$$\sum_{\nu_{1,2}=0}^{\infty} \sum_{\nu_{1,3}=0}^{\infty} \cdots \sum_{\nu_{2,3}=0}^{\infty} \sum_{\nu_{2,4}=0}^{\infty} \cdots \sum_{\nu_{N-1,N}=0}^{\infty} \left(k_{\mu_1}^{(1)} k_{\mu_2}^{(2)} \ldots k_{\mu_N}^{(N)} \right.$$

$$\left. \lambda_{1,2}^{(c)\,\nu_{1,2}} \lambda_{1,3}^{(c)\,\nu_{1,3}} \cdots \lambda_{2,3}^{(c)\,\nu_{2,3}} \lambda_{2,4}^{(c)\,\nu_{2,4}} \cdots \lambda_{N-1,N}^{(c)\quad \nu_{N-1,N}} \right) . \tag{4.49}$$

In order to have the explicit form of the parabolic cylinder function, this function shall now be considered.

The Parabolic Cylinder Function $D_{-1/2}$

A series representation of the parabolic cylinder function is given by the relation (see [2])

$$\boxed{D_{-1/2}\left(\lambda_i^{(2)}/\sqrt{2\lambda_i^{(4)}}\right) = \sum_{k=0}^{\infty} D^{(k)} \left(\lambda_i^{(2)}/\sqrt{2\lambda_i^{(4)}}\right)^k} \,, \tag{4.50}$$

where the coefficients are definable by

$$D^{(0)} = \cos\left(\frac{\pi}{4}\right) \frac{1}{2^{1/4}\sqrt{\pi}} \Gamma(1/4) = 1.21628021 \,,$$

$$D^{(1)} = -\sin\left(\frac{\pi}{4}\right) \frac{1}{2^{-1/4}\sqrt{\pi}} \Gamma(3/4) = -0.581368317 \,,$$

$$D^{(k=2,3+4d)} = 0 \,, \quad D^{(k=4+4d)} = \frac{D^{(0)}}{k!} a_k \,, \quad D^{(k=5+4d)} = \frac{D^{(1)}}{k!} a_k \,,$$

$$a_4 = 1/2 \,, \quad a_5 = 3/2 \,, \quad a_k = \frac{4}{(k+2)(k+1)} a_{k+4} \,,$$

$$d = 0,1,2,3,\ldots\infty \,. \tag{4.51}$$

Using this series representation the above derived partition functions can be formulated in an explicit way. (It has to be remarked that $D_{-1/2}$ is only one special parabolic cylinder function. The whole class of parabolic cylinder functions one can find by solving a quantum mechanical evolution equation, namely the *Schrödinger* equation of the quantum mechanical harmonic oscillator. This problem will be considered in chapter 5.)

Non-Symmetrical Partition Functions of Higher Order

The concept which has been introduced bases on the well-known solution of an elementary *Lagrangian* integral. Now the question arises in which way it is possible to calculate integrals with non-symmetrical terms in the exponent of an exponential function. Furthermore, the question arises whether it is possible to calculate integrals with exponential functions of higher order. Actually, this is possible. This can be done by using series representations of such exponential parts which do not guarantee convergent integrals. Then infinit sums of easy integrals arise, in which case these integrals can be transformed into *Gaussian* form. All these operations can be formulated in a universal way. This will now be shown. Then the reader will see that the above described solutions are special cases of this general procedure. Here product brackets are useful to have an obvious formulation.

The partition function which has to be considered is of the form

$$Z = \int_{-\infty}^{+\infty} \exp\left[-\left(\sum_{\alpha=1}^{O} \left\{ \prod_{i=1}^{\alpha} \sum_{\Theta_i=1}^{N} \lambda_{\Theta_\alpha} \Omega_{\Theta_i} \right\} \right) \right] d\Omega \tag{4.52}$$

(order $O = 2, 4, 6, \ldots \infty$, number of variables $N = 1, 2, 3, \ldots \infty$), where a separation of all exponential parts which guarantee a convergent integral leads to the formulation

$$Z = \int_{-\infty}^{+\infty} \left\{ \exp\left[-\left(\sum_{\alpha=1}^{O} \left\{ \prod_{\substack{i=1 \\ \Theta_O \neq \Theta_{conv}}}^{\alpha} \sum_{\Theta_i=1}^{N} \lambda_{\Theta_\alpha} \Omega_{\Theta_i} \right\} \right) \right] \right.$$
$$\left. \exp\left[-\left(\sum_{conv=1}^{N} \lambda_{\Theta_{conv}} \Omega_{conv}^{O} \right) \right] \right\} d\Omega , \tag{4.53}$$

where the multipliers $\lambda_{\Theta_{conv}}$ are given by

$$\lambda_{\Theta_{conv}} = \underbrace{\lambda_{conv,conv,\ldots conv}}_{O \text{ times}} \quad (conv = 1, 2, \ldots N) , \tag{4.54}$$

i. e. the relation

$$\Theta_{conv} = \underbrace{conv, conv, \ldots conv}_{O \text{ times}} \tag{4.55}$$

holds. (4.53) represents a general partition function with two essential exponential parts, namely a part which guarantees the convergence of the partition function (the second part), and a part which can be used in form of a series.

In order to get the mentioned series representation of the first exponential part of (4.53), the relation

$$\exp\left[-\left(\sum_{\alpha=1}^{O}\left\{\prod_{i=1}^{\alpha}\lambda_{\Theta_\alpha}\Omega_{\Theta_i}\right\}\right)\right] =$$
$$\sum_{\Lambda_{\Theta_\alpha}=0}^{\infty}\left[(-1)^{\Lambda_{\Theta_\alpha}}\left(1/\Lambda_{\Theta_\alpha}!\right)\left\{\prod_{i=1}^{\alpha}\Omega_{\Theta_i}\right\}^{\Lambda_{\Theta_\alpha}}\lambda_{\Theta_\alpha}^{\Lambda_{\Theta_\alpha}}\right] \tag{4.56}$$

has to be used. (4.56) is a generalized form of (4.38).

Inserting (4.56) into (4.53) one obtains the infinite integral series

$$Z = \left\{\prod_{\substack{\alpha=1 \\ \Theta_O\neq\Theta_{conv}}}^{O}\prod_{i=1}^{\alpha}\prod_{\Theta_i=1}^{N}\sum_{\Lambda_{\Theta_\alpha}=0}^{\infty} z_{\Lambda}^{(1)}\left\{\prod_{conv=1}^{N} I_{conv,\Lambda}\right\}\lambda_{\Theta_\alpha}^{\Lambda_{\Theta_\alpha}}\right\}, \tag{4.57}$$

where Λ denotes all summation indices Λ_{Θ_α}. The numbers $z_{\Lambda}^{(1)}$ are defined by

$$z_{\Lambda}^{(1)} = \left\{\prod_{\substack{\alpha=1 \\ \Theta_O\neq\Theta_{conv}}}^{O}\prod_{i=1}^{\alpha}\prod_{\Theta_i=1}^{N}(-1)^{\Lambda_{\Theta_\alpha}}\left(1/\Lambda_{\Theta_\alpha}!\right)\right\}. \tag{4.58}$$

(4.57) contains integrals $I_{conv,\Lambda}$ of higher order, in which case

$$I_{conv,\Lambda} = \int_{-\infty}^{+\infty}\exp\left[-\left(\lambda_{\Theta_{conv}}\Omega_{conv}^{O}\right)\right]\Omega_{conv}^{\xi(\Lambda)}\,d\Omega_{conv} \tag{4.59}$$

holds. In order to define $\xi(\Lambda)$, a direct formula can be used, however, in this book the relation

$$\left\{\prod_{\substack{\alpha=1 \\ \Theta_O\neq\Theta_{conv}}}^{O}\prod_{i=1}^{\alpha}\prod_{\Theta_i=1}^{N}\left\{\prod_{i=1}^{\alpha}\Omega_{\Theta_i}\right\}^{\Lambda_{\Theta_\alpha}}\right\} = \left\{\prod_{conv=1}^{N}\Omega_{conv}^{\xi(\Lambda)}\right\} \tag{4.60}$$

shall be used to define $\xi(\Lambda)$.

As $\xi(\Lambda)$ can take on values between 0 and ∞, (4.59) can be separated into two parts so that

$$I_{conv,\Lambda} = \begin{cases} 0 \text{ if } \xi(\Lambda) = 1+2n = \xi_{odd}(\Lambda)\,, \quad n = 0,1,2,\ldots\infty \\ 2\int_{0}^{+\infty}\exp\left[-\left(\lambda_{\Theta_{conv}}\Omega_{conv}^{O}\right)\right]\Omega_{conv}^{\xi(\Lambda)}\,d\Omega_{conv} \\ \text{if } \xi(\Lambda) = 2+2n = \xi_{even}(\Lambda)\,, \quad n = 0,1,2,\ldots\infty \end{cases} \tag{4.61}$$

holds. The transformation

$$\Omega_{conv} = \tilde{\Omega}_{conv}^{2/O} \ , \quad d\Omega_{conv} = \frac{2}{O}\tilde{\Omega}_{conv}^{\frac{2-O}{O}}d\tilde{\Omega}_{conv} \tag{4.62}$$

leads to a *Gaussian* integral, i. e. one obtains the relation

$$I_{conv,\boldsymbol{\Lambda}} = \begin{cases} 0 \ \text{if} \ \xi(\boldsymbol{\Lambda}) = 1 + 2n = \xi_{\text{odd}}(\boldsymbol{\Lambda}) \ , \quad n = 0,1,2,\ldots\infty \\ \dfrac{4}{O}\displaystyle\int_0^{+\infty} \exp\left[-\left(\lambda_{\boldsymbol{\Theta}_{conv}}\Omega_{conv}^2\right)\right]\Omega_{conv}^{\Xi}\,d\Omega_{conv} \\ \qquad \text{if} \ \xi(\boldsymbol{\Lambda}) = 2 + 2n = \xi_{\text{even}}(\boldsymbol{\Lambda}) \ , \quad n = 0,1,2,\ldots\infty \end{cases} \tag{4.63}$$

if the original symbol Ω_{conv} is used again. In (4.63) the relation

$$\Xi = \frac{2\xi(\boldsymbol{\Lambda}) + 2 - O}{O} \tag{4.64}$$

has to be considered. Using the well-known solutions of *Gaussian* integrals, i. e.

$$\int_0^{+\infty} \exp\left[-\left(\lambda_{\boldsymbol{\Theta}_{conv}}\Omega_{conv}^2\right)\right]\Omega_{conv}^{\Xi}\,d\Omega_{conv} = \frac{1}{2}\Gamma\left(\frac{\Xi+1}{2}\right)\lambda_{\boldsymbol{\Theta}_{conv}}^{-\left(\frac{\Xi+1}{2}\right)} \ , \tag{4.65}$$

one obtains the relation

$$I_{conv,\boldsymbol{\Lambda}} = \begin{cases} 0 \ \text{if} \ \xi(\boldsymbol{\Lambda}) = 1 + 2n = \xi_{\text{odd}}(\boldsymbol{\Lambda}) \ , \quad n = 0,1,2,\ldots\infty \\ \dfrac{2}{O}\Gamma\left[\dfrac{\xi(\boldsymbol{\Lambda})+1}{O}\right]\lambda_{\boldsymbol{\Theta}_{conv}}^{-\left[\frac{\xi(\boldsymbol{\Lambda})+1}{O}\right]} \\ \qquad \text{if} \ \xi(\boldsymbol{\Lambda}) = 2 + 2n = \xi_{\text{even}}(\boldsymbol{\Lambda}) \ , \quad n = 0,1,2,\ldots\infty \end{cases} \tag{4.66}$$

which is replaceable by

$$I_{conv,\boldsymbol{\Lambda}} = \delta_{\xi_{\text{even}}(\boldsymbol{\Lambda}),\xi(\boldsymbol{\Lambda})}\frac{2}{O}\Gamma\left[\frac{\xi(\boldsymbol{\Lambda})+1}{O}\right]\lambda_{\boldsymbol{\Theta}_{conv}}^{-\left[\frac{\xi(\boldsymbol{\Lambda})+1}{O}\right]} \ . \tag{4.67}$$

$\delta_{\xi_{\text{even}}(\boldsymbol{\Lambda}),\xi(\boldsymbol{\Lambda})}$ represents a *Kronecker* delta, i. e. the relation

$$\delta_{\xi_{\text{even}}(\boldsymbol{\Lambda}),\xi(\boldsymbol{\Lambda})} = \begin{cases} 0 \ \text{if} \ \xi(\boldsymbol{\Lambda}) = \xi_{\text{odd}}(\boldsymbol{\Lambda}) \\ 1 \ \text{if} \ \xi(\boldsymbol{\Lambda}) = \xi_{\text{even}}(\boldsymbol{\Lambda}) \end{cases} \tag{4.68}$$

holds.

Inserting (4.67) into (4.57) and introducing the real numbers

$$z_{\boldsymbol{\Lambda}}^{(2)} = \left(\frac{2}{O}\right)^N \left\{\prod_{\alpha=1}^{O}\prod_{i=1}^{\alpha}\prod_{\substack{\Theta_i=1 \\ \Theta_O \neq \Theta_{conv}}}^{N}(-1)^{\Lambda_{\boldsymbol{\Theta}_\alpha}}\left(1/\Lambda_{\boldsymbol{\Theta}_\alpha}!\right)\right\}$$
$$\left\{\prod_{conv=1}^{N}\delta_{\xi_{\text{even}}(\boldsymbol{\Lambda}),\xi(\boldsymbol{\Lambda})}\Gamma\left[\frac{\xi(\boldsymbol{\Lambda})+1}{O}\right]\right\} \tag{4.69}$$

one obtains the expression

$$Z = \left\{ \prod_{conv=1}^{N} \lambda_{\Theta_{conv}}^{-1/O} \right\}$$
$$\left\{ \prod_{\substack{\alpha=1 \\ \Theta_O \neq \Theta_{conv}}}^{O} \prod_{i=1}^{\alpha} \prod_{\Theta_i=1}^{N} \prod_{conv=1}^{N} \sum_{\Lambda_{\Theta_\alpha}=0}^{\infty} z_{\Lambda}^{(2)} \lambda_{\Theta_\alpha}^{\Lambda_{\Theta_\alpha}} \lambda_{\Theta_{conv}}^{-\left[\frac{\varepsilon(\Lambda)}{O}\right]} \right\} . \tag{4.70}$$

(4.70) represents the explicit form of a general partition function. This expression includes the elementary case (4.25). This can be seen without any calculation, because the first product term of (4.70) is equal to the factor of (4.25) if the case $N = 1$, $O = 4$ is taken into account, i. e. the relation

$$\left\{ \prod_{conv=1}^{N} \lambda_{\Theta_{conv}}^{-1/O} \right\}\Bigg|_{N=1,\,O=4} = \lambda_{\Theta_1}^{-1/4} = \lambda_2^{-1/4} \tag{4.71}$$

holds. Furthermore, (4.70) includes both the multi-dimensional symmetrical partition function (4.46) and the two-dimensional case (4.35). Thus, it has to be mentioned that solutions of partition functions with exponential kernels can be calculated by the above described method, i. e. by using a series which contains an infinite number of *Gauß* integrals. This means that it is possible to construct the solution of a complicated mathematical problem by using elementary functions which are correlated to elementary mathematical problems. I would like to draw the reader's attention to this fact, because such a method of construction is a universal principle in nature – not only on the mathematical level, in the directly observable world, too. (As mathematics describes the observable world, this is not surprising.) In this context the reader may think of the human brain, which consists of a large number of cells, namely neurons, which interact in a self-organizing way.

Thus, the partition function problem is solved. Now the problem of inversion of a non-linear system of the kind (4.4) can be considered.

4.3.3   Critical Hyper-Surface Equations

A special case of the non-linear basic equation system (4.4) is given by

$$-Z_4^{(1)-1} \frac{\partial Z_4^{(1)}}{\partial \lambda_1} = \langle \Omega^2 \rangle \,, \quad -Z_4^{(1)-1} \frac{\partial Z_4^{(1)}}{\partial \lambda_2} = \langle \Omega^4 \rangle \,. \tag{4.72}$$

Such a symmetrical basic equation system of fourth order and one variable can be solved if the explicit formulation of the partition function is used, i. e. if (4.25) is used, in which case *solution* means that analytical functions of the kind

$$\lambda_1 = \lambda_1(\langle \Omega^2 \rangle, \langle \Omega^4 \rangle) \,, \quad \lambda_2 = \lambda_2(\langle \Omega^2 \rangle, \langle \Omega^4 \rangle) \tag{4.73}$$

have to be found. The correlated distribution function is of the kind (4.14). Now it shall be shown how to find a linear solution of this basic problem, i. e. to find solutions which hold near a critical point. Such solutions shall be called *critical hyper-surface equations*. In this context the principle mathematical problems shall be considered, and ways to solve these problems shall be introduced.

The Non-Linear Basic Equation System

Inserting (4.25) into (4.72) one obtains the basic equation system

$$-\frac{\partial_{\lambda_1}\left[\exp\left(\lambda_1^2/8\lambda_2\right)D_{-1/2}\left(\lambda_1/\sqrt{2\lambda_2}\right)\right]}{\exp\left(\lambda_1^2/8\lambda_2\right)D_{-1/2}\left(\lambda_1/\sqrt{2\lambda_2}\right)} = \left\langle\Omega^2\right\rangle,$$

$$-\frac{\partial_{\lambda_2}\left[(2\lambda_2)^{-1/4}\exp\left(\lambda_1^2/8\lambda_2\right)D_{-1/2}\left(\lambda_1/\sqrt{2\lambda_2}\right)\right]}{(2\lambda_2)^{-1/4}\exp\left(\lambda_1^2/8\lambda_2\right)D_{-1/2}\left(\lambda_1/\sqrt{2\lambda_2}\right)} = \left\langle\Omega^4\right\rangle. \tag{4.74}$$

The explicit formulation of the occuring parabolic cylinderfunction is given by (4.50) if the now relevant abbreviation is used, i. e. the relation

$$D_{-1/2}\left(\lambda_1/\sqrt{2\lambda_2}\right) = \sum_{k=0}^{\infty} D^{(k)}\left(\lambda_1/\sqrt{2\lambda_2}\right)^k \tag{4.75}$$

holds. Inserting (4.75) into (4.74), multiplying this basic equation system by $D_{-1/2}$ and executing the derivations one obtains a basic equation system of the kind

$$\sum_{i=0}^{\infty} g_\kappa^{(i)}\lambda_1^i = 0 \ (\kappa = 1, 2), \tag{4.76}$$

where the coefficients are definable by

$$g_1^{(i)} = g_{1,1}^{(i)}\left(\frac{1}{\sqrt{2\lambda_2}}\right)^i + g_{1,2}^{(i)}\left(\frac{1}{\sqrt{2\lambda_2}}\right)^{i+1},$$

$$g_2^{(i)} = g_{2,1}^{(i)}\left(\frac{1}{\sqrt{2\lambda_2}}\right)^i + g_{2,2}^{(i)}\left(\frac{1}{\sqrt{2\lambda_2}}\right)^{i+2}. \tag{4.77}$$

The coefficients $g_{1,1}^{(i)}, g_{1,2}^{(i)}, g_{2,1}^{(i)}, g_{2,2}^{(i)}$ are defined by

$$g_{1,1}^{(i)} = \begin{cases} 0, & i = m_4 + 4n, \ m_4 = 2, 3 \\ D^{(i)}\langle\Omega^2\rangle, & \text{all } i \end{cases},$$

$$g_{1,2}^{(i)} = \begin{cases} (i+1)D^{(i+1)}, & i = m_1 + 4n, \ m_1 = 0, 3 \\ (1/2)D^{(i-1)}, & i = m_2 + 4n, \ m_2 = 1, 2 \end{cases},$$

$$g_{2,1}^{(i)} = \begin{cases} 0, & i = m_4 + 4n, \ m_4 = 2, 3 \\ D^{(i)}\langle\Omega^4\rangle, & \text{all } i \end{cases},$$

$$g_{2,2}^{(i)} = \begin{cases} -[(2i+1)/2]D^{(i)} \,, & i = m_3 + 4n \,, \quad m_3 = 0,1 \\ -(1/2)D^{(i-2)} \,, & i = m_4 + 4n \,, \quad m_4 = 2,3 \end{cases} \,,$$
$$n = 0,1,2,3,\ldots\infty \,. \tag{4.78}$$

The non-linear system (4.76) represents the basic equation system in an explicit form.

An Inversion Formula

In order to invert (4.76), the formula

$$\lambda_1 = \sum_{k=1}^{\infty} h_\kappa^{(k)} \lambda_{1,\kappa}^{(\mathrm{lin})^k} \quad (\kappa = 1,2) \tag{4.79}$$

is possible, where $\lambda_{1,\kappa}^{(\mathrm{lin})}$ represents the linear solutions of (4.76), i. e. $\lambda_{1,\kappa}^{(\mathrm{lin})}$ is defined by

$$g_\kappa^{(0)} + g_\kappa^{(1)} \lambda_{1,\kappa}^{(\mathrm{lin})} = 0 \,. \tag{4.80}$$

This can be shown by inserting (4.79) into (4.76). Then a recursive scheme occurs, which defines the coefficients of (4.79). This scheme has the form

$$h_\kappa^{(1)} = 1 \,,$$
$$h_\kappa^{(k>1)} = -\sum_{l=2}^{k} \frac{g_\kappa^{(l)}}{g_\kappa^{(1)}} \left\{ \prod_{n=1}^{l-1} \sum_{\nu_n=n}^{\nu_{n+1}-1} \right\} \left\{ \prod_{n=0}^{l-1} h_\kappa^{(-\nu_n+\nu_{n+1})} \right\}$$
$$(\nu_0 = 0, \; \nu_l = k) \,. \tag{4.81}$$

As (4.79) fulfills (4.76), (4.79) can be considered as a special formulation of the basic equation system.

Elimination of λ_1

In order to eliminate the *Lagrangian* multiplier λ_1, the relation (4.79), which represents two equations, can be used. Equating these two equations one obtains the relation

$$\sum_{k=1}^{\infty} \left(h_1^{(k)} \lambda_{1,1}^{(\mathrm{lin})^k} - h_2^{(k)} \lambda_{1,2}^{(\mathrm{lin})^k} \right) = 0 \,. \tag{4.82}$$

(4.82) represents a non-linear equation to determine the *Lagrangian* multiplier λ_2.

The Region Near a Threshold

In the following only the linear form of (4.82) shall be considered, i. e.

$$h_1^{(1)}\lambda_{1,1}^{(\mathrm{lin})} - h_2^{(1)}\lambda_{1,2}^{(\mathrm{lin})} = 0 \tag{4.83}$$

shall be used. This means considering only the narrow environment of a critical point, because a critical behavior is correlated with the change of the sign of λ_1 so that the values of λ_1 are small (see figure 4.1) and only linear terms are necessary to describe the behavior of the system near the threshold. Inserting the linear solutions $\lambda_{1,1}^{(\mathrm{lin})}$ and $\lambda_{1,2}^{(\mathrm{lin})}$ into (4.83), i. e. inserting

$$\lambda_{1,\kappa}^{(\mathrm{lin})} = -\frac{g_\kappa^{(0)}}{g_\kappa^{(1)}} \,, \tag{4.84}$$

and inserting the coefficients $h_1^{(1)}$ and $h_2^{(1)}$, i. e. inserting

$$h_\kappa^{(1)} = 1 \,, \tag{4.85}$$

one obtains the equation

$$g_1^{(0)}g_2^{(1)} - g_2^{(0)}g_1^{(1)} = 0 \,. \tag{4.86}$$

Using the definitions (4.77) one obtains the polynomial

$$p_0 + p_1 \left(\frac{1}{\sqrt{2\lambda_2}}\right) + p_2 \left(\frac{1}{\sqrt{2\lambda_2}}\right)^2 = 0 \,, \tag{4.87}$$

where the definition

$$p_0 = D^{(1)2} \left[1 - \frac{1}{4}\frac{\Gamma^2(1/4)}{\Gamma^2(3/4)}\right] \langle\Omega^4\rangle \,, \quad p_1 = D^{(1)2} \left[\frac{1}{\sqrt{2}}\frac{\Gamma(1/4)}{\Gamma(3/4)}\right] \langle\Omega^2\rangle \,,$$

$$p_2 = D^{(1)2} \left[-\frac{3}{2} + \frac{1}{8}\frac{\Gamma^2(1/4)}{\Gamma^2(3/4)}\right] \tag{4.88}$$

(use (4.51)) has to be considered.

Solutions

Using the critical equation (4.87) two solutions are possible. This ambiguity is a consequence of the approximation. (A full treatment would lead to non-ambiguous solutions. This will be shown later.) The possible two solutions are of the form

$$\left(\frac{1}{\sqrt{2\lambda_2}}\right)_\pm = C_1\langle\Omega^2\rangle \pm \sqrt{C_1^2\langle\Omega^2\rangle^2 - C_2\langle\Omega^4\rangle} \,, \tag{4.89}$$

where the coefficients are given by

$$C_1 = -\frac{1}{2}\left[\frac{1}{\sqrt{2}}\frac{\Gamma(1/4)}{\Gamma(3/4)}\right]\left[-\frac{3}{2} + \frac{1}{8}\frac{\Gamma^2(1/4)}{\Gamma^2(3/4)}\right]^{-1} = 2.57787256 \,,$$

$$C_2 = \left[1 - \frac{1}{4}\frac{\Gamma^2(1/4)}{\Gamma^2(3/4)}\right]\left[-\frac{3}{2} + \frac{1}{8}\frac{\Gamma^2(1/4)}{\Gamma^2(3/4)}\right]^{-1} = 2.92877681 \,. \tag{4.90}$$

In order to decide which of these solutions is the useful one, a comparision with solutions which hold in the critical point has to be done.

The Critical Point

The critical point is characterized by the relation

$$\lambda_1 = 0 \, , \tag{4.91}$$

i. e. a dramatic change of the distribution function occurs (see figure 4.1). The correlated critical mean values are given by (4.74) if the constraint (4.91) is considered, i. e. in the critical point the relation

$$\left[\frac{1}{\sqrt{2}}\frac{\Gamma(1/4)}{\Gamma(3/4)}\right]^{-1}\left(\frac{1}{\sqrt{2\lambda_2}}\right) = \left\langle\Omega^2\right\rangle_{\mathrm{crit}} \, , \quad \frac{1}{2}\left(\frac{1}{\sqrt{2\lambda_2}}\right)^2 = \left\langle\Omega^4\right\rangle_{\mathrm{crit}} \tag{4.92}$$

holds. Therefore, the relation

$$\boxed{\left\langle\Omega^4\right\rangle_{\mathrm{crit}} = C_{\mathrm{crit}}\left\langle\Omega^2\right\rangle_{\mathrm{crit}}^2} \tag{4.93}$$

has to be considered, where the critical coefficient C_{crit} is given by

$$C_{\mathrm{crit}} = \left[\frac{1}{4}\frac{\Gamma^2(1/4)}{\Gamma^2(3/4)}\right] = 2.18843962 \, . \tag{4.94}$$

(4.93) represents a phase transition condition, which separates the critical from the non-critical region. As the reader can see, in the critical point the correlation function of fourth order is proportional to the square of the correlation function of second order, i. e. if thermodynamic problems are considered, the mean value of the fourth moment of the magnetization (or polarization) is proportional to the mean value of the square of the second moment of the magnetization (or polarization).

The Second Critical Hyper-Surface Equation

In order to decide which of the solutions (4.89) is the useful one, the critical case of (4.89) has to be considered, i. e.

$$\left(\frac{1}{\sqrt{2\lambda_2}}\right)_\pm = C_1\left\langle\Omega^2\right\rangle_{\mathrm{crit}} \pm \sqrt{C_1^2\left\langle\Omega^2\right\rangle_{\mathrm{crit}}^2 - C_2\left\langle\Omega^4\right\rangle_{\mathrm{crit}}} \tag{4.95}$$

has to be considered. Using the phase transition condition (4.94) one obtains the equation

$$\left(\frac{1}{\sqrt{2\lambda_2}}\right)_\pm = C_{\mathrm{crit}\pm}\left\langle\Omega^2\right\rangle_{\mathrm{crit}} \tag{4.96}$$

with the coefficients

$$C_{\mathrm{crit}-} = \left[\frac{1}{4}\frac{\Gamma^2(1/4)}{\Gamma^2(3/4)}\right] \, ,$$

$$C_{\mathrm{crit}+} = \left[1 - \frac{1}{4}\frac{\Gamma^2(1/4)}{\Gamma^2(3/4)}\right]\left[3 - \frac{1}{4}\frac{\Gamma^2(1/4)}{\Gamma^2(3/4)}\right] C_{\mathrm{crit}-} \, . \tag{4.97}$$

As the reader can easily calculate, only the first coefficient is identical with the correct coefficient (4.94) so that the negative sign is the useful one. Using the negative sign in (4.89) one gets the equation

$$\lambda_2 = \frac{1}{\langle\Omega^2\rangle^2}\left(\sqrt{2}C_1 - \sqrt{2}\sqrt{C_1^2 - C_2\frac{\langle\Omega^4\rangle}{\langle\Omega^2\rangle^2}}\right)^{-2} , \tag{4.98}$$

which is nothing but the second critical hyper-surface equation. This hyper-surface equation connects a macroscopic level of description with a microscopic level, i. e. (4.98) describes the correlation between mean values and one relevant statistical parameter. The validity is restricted to a narrow environment of a critical point.

The First Critical Hyper-Surface Equation

In order to get the first critical hyper-surface equation, it is not necessary to consider again the basic equation system (4.76), because the general relation (4.8) holds in this special case, too. Using (4.8) in the now considered case ($N = 1$, $\lambda_o = 0$, $\lambda_{o,k} = \delta_{o,1}\delta_{k,1}\lambda_1$, $\lambda_{o,k,l} = 0$, $\lambda_{o,k,l,m} = \delta_{o,1}\delta_{k,1}\delta_{l,1}\delta_{m,1}\lambda_2$), i. e. using the relation

$$2\lambda_1\langle\Omega^2\rangle + 4\lambda_2\langle\Omega^4\rangle = 1 , \tag{4.99}$$

a direct correlation between the two *Lagrangian* multipliers is given so that the first critical hyper-surface equation can be written in the form

$$\lambda_1 = \frac{1}{\langle\Omega^2\rangle}\left[\frac{1}{2} - \frac{\langle\Omega^4\rangle}{\langle\Omega^2\rangle^2}\left(C_1 - \sqrt{C_1^2 - C_2\frac{\langle\Omega^4\rangle}{\langle\Omega^2\rangle^2}}\right)^{-2}\right] . \tag{4.100}$$

Together with the second hyper-surface equation the first hyper-surface equation describes the correlation between the macroscopic and the microscopic system level near a threshold. If thermodynamic systems are considered, this means that moments of magnetization (or polarization) are directly correlated with distribution function parameters. Such distribution function parameters depend on specific parameters such as temperature (see table 4.1). Therefore, such hyper-surface equations can be used to determine the structure of a distribution function in detail if such dependences have been measured on the macroscopic level.

Further Comments

As the reader can see, the critical hyper-surface equations consist of dimensionless parts which contain the dimensionless corrleation function quotient $\langle\Omega^4\rangle/\langle\Omega^2\rangle^2$ and parts which guarantee the correct dimension of the two *Lagrangian* multipliers, namely the parts $1/\langle\Omega^2\rangle$, $1/\langle\Omega^2\rangle^2$. As the following will show, this structure is the basic structure of any exact solutions, i. e. typical of the structure of hyper-surface equations are two parts, namely correlation function parts which guarantee the dimension of the parameters and parts which base on dimensionless correlation function quotients.

Due to the fact that the basic equations are power series, the exact solutions of the inversion problem have to be power series as well. Therefore, to solve the problem of hyper-surface equations exactly, one needs special mathematical tools, namely addition, multiplication and division of power series. As it was above shown, a useful starting point of an exact inversion procedure is represented by an explicit form of a basic equation system.

Additionally, it shall be remarked that the name *hyper-surface equation* is used, because parameter functions of the kind $\lambda = \lambda(\langle f(\Omega_i) \rangle)$ describe surfaces which are hyper-surfaces with respect to the trajectories of all possible dynamical processes.

In the following it will be shown how to solve one- and two-dimensional symmetrical inversion problems in an exact way.

4.3.4 The Problem of One- and Two-Dimensional Hyper-Surface Equations of Fourth Order

In the following the basic equation system

$$-Z_4^{(2)^{-1}}\frac{\partial Z_4^{(2)}}{\partial \lambda_1} = \langle \Omega_1^2 \rangle \,, \quad -Z_4^{(2)^{-1}}\frac{\partial Z_4^{(2)}}{\partial \lambda_2} = \langle \Omega_2^2 \rangle \,, \quad +Z_4^{(2)^{-1}}\frac{\partial^2 Z_4^{(2)}}{\partial \lambda_1^{\,2}} = \langle \Omega_1^4 \rangle \,,$$

$$+Z_4^{(2)^{-1}}\frac{\partial^2 Z_4^{(2)}}{\partial \lambda_2^{\,2}} = \langle \Omega_2^4 \rangle \,, \quad +Z_4^{(2)^{-1}}\frac{\partial Z_4^{(2)}}{\partial \lambda_c} = \langle \Omega_1^2 \Omega_2^2 \rangle \tag{4.101}$$

will be taken as a basis. (4.101) represents a symmetrical basic equation system of fourth order and two variables. The partition function $Z_4^{(2)}$ is defined by (4.35), i. e. the correlated distribution function is of the kind (4.17). In this subsection the way of derivation of exact hyper-surface equations of one- and two-dimensional problems shall be considered. In particular, it shall be shown that the structure of hyper-surface equations is the structure of a power series which bases on special dimensionless correlation function quotients.

The Series Representation of the Basic Equation System

Inserting the partition function (4.35) into (4.101) one obtains the basic equation system

$$\sum_{i=0}^{\infty} \left(-\mathrm{i}\sqrt{i+1}k_{i+1}^{(1)}k_i^{(2)} + \langle \Omega_1^2 \rangle k_i^{(1)} k_i^{(2)} \right) \lambda_c^i = 0 \,,$$

$$\sum_{i=0}^{\infty} \left(-\mathrm{i}\sqrt{i+1}k_{i+1}^{(2)}k_i^{(1)} + \langle \Omega_2^2 \rangle k_i^{(1)} k_i^{(2)} \right) \lambda_c^i = 0 \,,$$

$$\sum_{i=0}^{\infty} \left[-\mathrm{i}^2 \sqrt{(i+1)(i+2)} k_{i+2}^{(1)} k_i^{(2)} + \langle \Omega_1^4 \rangle k_i^{(1)} k_i^{(2)} \right] \lambda_c^i = 0 \ ,$$

$$\sum_{i=0}^{\infty} \left[-\mathrm{i}^2 \sqrt{(i+1)(i+2)} k_{i+2}^{(2)} k_i^{(1)} + \langle \Omega_2^4 \rangle k_i^{(1)} k_i^{(2)} \right] \lambda_c^i = 0 \ ,$$

$$\sum_{i=0}^{\infty} \left[(i+1) k_{i+1}^{(1)} k_{i+1}^{(2)} + \langle \Omega_1^2 \Omega_2^2 \rangle k_i^{(1)} k_i^{(2)} \right] \lambda_c^i = 0 \ , \tag{4.102}$$

which is totally symmetrical with respect to the exchange

$$\lambda_1, \ \lambda_3, \ \Omega_1 \leftrightarrow \lambda_2, \ \lambda_4, \ \Omega_2 \ , \tag{4.103}$$

i. e. the exchange (4.103) generates again the basic equation system (4.102). Introducing coupling terms of the kind

$$\Gamma_c^{(1)} = \sum_{i=1}^{\infty} \left(-\mathrm{i}\sqrt{i+1} k_{i+1}^{(1)} k_i^{(2)} + \langle \Omega_1^2 \rangle k_i^{(1)} k_i^{(2)} \right) \lambda_c^i \ ,$$

$$\Gamma_c^{(2)} = \sum_{i=1}^{\infty} \left(-\mathrm{i}\sqrt{i+1} k_{i+1}^{(2)} k_i^{(1)} + \langle \Omega_2^2 \rangle k_i^{(1)} k_i^{(2)} \right) \lambda_c^i \ ,$$

$$\Gamma_c^{(3)} = \sum_{i=1}^{\infty} \left[-\mathrm{i}^2 \sqrt{(i+1)(i+2)} k_{i+2}^{(1)} k_i^{(2)} + \langle \Omega_1^4 \rangle k_i^{(1)} k_i^{(2)} \right] \lambda_c^i \ ,$$

$$\Gamma_c^{(4)} = \sum_{i=1}^{\infty} \left[-\mathrm{i}^2 \sqrt{(i+1)(i+2)} k_{i+2}^{(2)} k_i^{(1)} + \langle \Omega_2^4 \rangle k_i^{(1)} k_i^{(2)} \right] \lambda_c^i \ ,$$

$$\Gamma_c^{(c)} = \sum_{i=1}^{\infty} \left[(i+1) k_{i+1}^{(1)} k_{i+1}^{(2)} + \langle \Omega_1^2 \Omega_2^2 \rangle k_i^{(1)} k_i^{(2)} \right] \lambda_c^i \tag{4.104}$$

one obtains the system

$$\mathrm{i} k_1^{(m)} - \Gamma_c^{(m)} = \langle \Omega_m^2 \rangle \ , \quad \mathrm{i}^2 \sqrt{2} k_2^{(m)} - \Gamma_c^{(m+2)} = \langle \Omega_m^4 \rangle \ ,$$

$$-k_1^{(1)} k_2^{(2)} - \Gamma_c^{(c)} = \langle \Omega_1^2 \Omega_2^2 \rangle \ , \tag{4.105}$$

where m can be equal to 1 or 2. (4.102) or (4.105) are the series representations of the symmetrical basic equation system of fourth order and two variables.

The Case of Decoupling

In the case of decoupling, i. e. the relation

$$\lim_{\lambda_c \to 0} \Gamma_c^{(\epsilon)} = 0 \ (\epsilon = 1, 2, 3, 4, c) \tag{4.106}$$

holds, two independent systems of the kind

$$\mathrm{i} k_1^{(m)} = \langle \Omega_m^2 \rangle \ , \quad \mathrm{i}^2 \sqrt{2} k_2^{(m)} = \langle \Omega_m^4 \rangle \tag{4.107}$$

have to be considered. The coefficients $k_i^{(m)}$ are the complex elementary coefficients (see (4.34)). They contain the *Lagrangian* parameters λ_1, λ_2, λ_3, λ_4.

The Structure of the Complex Elementary Coefficients

In order to get the explicit structure of such a basic equation system, the structure of the complex elementary coefficients has to be be derived. This shall now be done.

The complex elementary coefficients $k_\nu^{(m)}$ are defined by (4.34), in which case the partition function is of the form (4.29). Inserting (4.29) into (4.34) and using the abbreviations

$$f_1^{(m)} = D_{-1/2}\left(\lambda_m/\sqrt{2\lambda_{m+2}}\right), \quad f_2^{(m)} = \exp\left(\lambda_m^2/8\lambda_{m+2}\right) \tag{4.108}$$

one obtains the relation

$$\begin{aligned}
k_\nu^{(m)} &= i^\nu \sqrt{(1/\nu!)} \left(f_1^{(m)} f_2^{(m)}\right)^{-1} \left(\partial_{\lambda_m}^\nu f_1^{(m)} f_2^{(m)}\right) \\
&= i^\nu \sqrt{(1/\nu!)} \left(f_1^{(m)} f_2^{(m)}\right)^{-1} \sum_{\mu=0}^{\nu} g_{\nu,\mu}^{(1)} \left(\partial_{\lambda_m}^\mu f_1^{(m)}\right) \left(\partial_{\lambda_m}^{\nu-\mu} f_2^{(m)}\right) .
\end{aligned} \tag{4.109}$$

The definition of the binomial coefficients $g_{\nu,\mu}^{(1)}$ shall be considered later. Using the series representation of the parabolic cylinder function, i. e.

$$D_{-1/2}\left(\lambda_m/\sqrt{2\lambda_{m+2}}\right) = \sum_{k=0}^{\infty} D^{(k)}\left(\lambda_m/\sqrt{2\lambda_{m+2}}\right)^k , \tag{4.110}$$

the derivatives can easily be calculated, in which case

$$\partial_{\lambda_m}^\mu f_1^{(m)} = \sum_{\epsilon=0}^{\infty} g_{\mu,\epsilon}^{(2)} \left(\frac{1}{\sqrt{\lambda_{m+2}}}\right)^\epsilon \lambda_m^{\epsilon-\mu} ,$$

$$\partial_{\lambda_m}^{\nu-\mu} f_2^{(m)} = f_2^{(m)} \sum_{\kappa=0}^{\kappa_{max}} g_{\nu,\mu,\kappa}^{(3)} \left(\frac{1}{\sqrt{\lambda_{m+2}}}\right)^{\nu-\mu-\kappa} \lambda_m^{\nu-\mu-2\kappa} \tag{4.111}$$

holds. (The validity of these relations can be shown by using the method of complete induction. The coefficients $g_{\mu,\epsilon}^{(2)}$, $g_{\nu,\mu,\kappa}^{(3)}$ shall be defined later. These coefficients guarantee that no terms of the kind $1/\lambda_m$ occur.) The upper boundary is given by

$$\kappa_{max} = \frac{\nu - \mu - \delta_{odd,(\nu-\mu)}}{2} , \tag{4.112}$$

where *Kronecker's* delta is defined by

$$\delta_{odd,(\nu-\mu)} = \begin{cases} 0 \text{ if } \nu - \mu = \text{even} \\ 1 \text{ if } \nu - \mu = \text{odd} \end{cases} . \tag{4.113}$$

Inserting (4.111) into (4.109) a function arises which contains products and quotients of a power series. Under consideration of the general formulae

$$\frac{\sum_{i=0}^{\infty} s_{1,i} x^i}{\sum_{i=0}^{\infty} s_{2,i} x^i} = \sum_{i=0}^{\infty} s_{3,i} x^i \ , \quad s_{3,0} = \frac{s_{1,0}}{s_{2,0}}, \quad s_{3,i>0} = \frac{s_{1,i}}{s_{2,0}} - \sum_{k=0}^{i-1} \frac{s_{3,k} s_{2,i-k}}{s_{2,0}} \ ,$$

$$\sum_{i=0}^{\infty} s_{1,i} x^i \cdot \sum_{i=0}^{\infty} s_{2,i} x^i = \sum_{i=0}^{\infty} s_{4,i} x^i \ , \quad s_{4,i} = \sum_{k=0}^{i} s_{1,k} s_{2,i-k} \tag{4.114}$$

(see [6]) the complex elementary coefficients take on the form of a power series, i. e. the relation

$$\boxed{ k_\nu^{(m)} = i^\nu \sqrt{(1/\nu!)} \left(\frac{1}{\sqrt{\lambda_{m+2}}} \right)^\nu \sum_{\xi=0}^{\infty} g_{\nu,\xi}^{(6)} \left(\frac{1}{\sqrt{\lambda_{m+2}}} \right)^\xi \lambda_m^\xi } \tag{4.115}$$

holds. (4.115) describes an explicit representation of the complex elementary coefficients. As the reader can see, the structure of these coefficients is the structure of power series, in which case parameter quotients $\lambda_m / \sqrt{\lambda_{m+2}}$ have to be considered.

The numbers $g_{\nu,\xi}^{(6)}$ are functions of the coefficients of the parabolic cylinder function. In order to define the numbers $g_{\nu,\xi}^{(6)}$, first the binomial coefficients

$$g_{\nu,\mu}^{(1)} = \frac{\nu!}{\mu!(\nu-\mu)!} \tag{4.116}$$

have to be calculated. Then the numbers

$$g_{\mu,\epsilon}^{(2)} = \frac{D^{(\epsilon)}}{\epsilon+1} \left(\frac{1}{\sqrt{2}} \right)^\epsilon \prod_{\xi=0}^{\mu} (\epsilon + 1 - \xi) \tag{4.117}$$

as well as the numbers

$$g_{\nu,\mu,\kappa}^{(3)} = \left(\frac{1}{4} \right)^{\nu-\mu-\kappa} g_{\nu,\mu,\kappa}^{(4)}, \quad g_{\nu^*,\mu^*,\epsilon}^{(4)} = g_{\nu,\mu,\epsilon}^{(4)} + (\nu^* - \mu^* - 2\epsilon + 1) g_{\nu,\mu,\epsilon-1}^{(4)} \ ,$$

$$g_{\nu,\mu=\nu,\epsilon=0}^{(4)} = 1 \ , \quad g_{\nu,\mu,\epsilon>\kappa_{\max}}^{(4)} = 0 \ , \quad g_{\nu,\mu,\epsilon-1<0}^{(4)} = 0 \ ,$$

$$\nu^* - \mu^* = \nu - \mu + 1 \tag{4.118}$$

have to be considered. Using the recursive definition

$$g_{\epsilon=0,\mu}^{(5)} = \frac{g_{\mu,\mu}^{(2)}}{g_{\mu=0,\epsilon=0}^{(2)}} \ , \quad g_{\epsilon>0,\mu}^{(5)} = \frac{g_{\mu,\epsilon+\mu}^{(2)}}{g_{\mu=0,\epsilon=0}^{(2)}} - \sum_{\xi=0}^{\epsilon-1} g_{\xi,\mu}^{(5)} \frac{g_{\mu=0,\epsilon-\xi}^{(2)}}{g_{\mu=0,\epsilon=0}^{(2)}} \tag{4.119}$$

the numbers $g_{\nu,\xi}^{(6)}$ can be defined by the relation

$$g_{\nu,\xi}^{(6)} = \sum_{\substack{\epsilon=0 \\ \epsilon+\nu-\mu-2\kappa=\xi}}^{\infty} \sum_{\mu=0}^{\nu} \sum_{\kappa=0}^{\kappa_{\max}} g_{\nu,\mu}^{(1)} g_{\nu,\mu,\kappa}^{(3)} g_{\epsilon,\mu}^{(5)} \ . \tag{4.120}$$

This scheme is the scheme which one gets if the above mentioned operations (series division, series multiplication) are realized.

Now it is possible to calculate the explicit structure of the basic equation system. First, the decoupled systems shall be considered.

The Explicit Form of the One-Dimensional Basic Equation System

The basic equation system is given by (4.107) if the case of decoupling holds. Inserting the elementary coefficients (4.115) one obtains the relations

$$\boxed{\begin{aligned}
\left(\frac{1}{\sqrt{\lambda_{m+2}}} \right) \sum_{\xi=0}^{\infty} \left(-g_{1,\xi}^{(6)} \right) \left(\frac{\lambda_m}{\sqrt{\lambda_{m+2}}} \right)^{\xi} &= \langle \Omega_m^2 \rangle \ , \\[2ex]
\left(\frac{1}{\sqrt{\lambda_{m+2}}} \right)^2 \sum_{\xi=0}^{\infty} \left(g_{2,\xi}^{(6)} \right) \left(\frac{\lambda_m}{\sqrt{\lambda_{m+2}}} \right)^{\xi} &= \langle \Omega_m^4 \rangle \ ,
\end{aligned}} \tag{4.121}$$

which are nothing but infinite power series of special dimensionless parameter quotients.

Now the question arises which structure the solutions of the basic equation system (4.121) have. Then it would be possible to solve the basic equation system.

The Structure of the One-Dimensional Hyper-Surface Equations

In order to solve an infinite power series of the kind (4.121), one needs power series, too, with correlation functions being relevant. Furthermore, one needs dimensionless additive terms, because otherwise different physical quantities would be added up. Therefore, the assumption is obvious that series terms of correlation function quotients are necessary to formulate the solutions of (4.121). One possibility which makes sense is given by the choice $\langle \Omega_m^4 \rangle / \langle \Omega_m^2 \rangle^2$, because such a correlation function quotient is dimensionless, and such a quotient guarantees a structure which is symmetrical to the structure of the basic equation system, i. e. instead of dimensionless parameter quotients, dimensionless correlation function quotients are used. Therefore, the formulation

$$\begin{aligned}
\lambda_m &= \frac{1}{\langle \Omega_m^2 \rangle} \sum_{i=0}^{\infty} \Lambda_1^{(i)} \left(\frac{\langle \Omega_m^4 \rangle}{\langle \Omega_m^2 \rangle^2} \right)^i , \\
\lambda_{m+2} &= \frac{1}{\langle \Omega_m^2 \rangle^2} \sum_{i=0}^{\infty} \Lambda_2^{(i)} \left(\frac{\langle \Omega_m^4 \rangle}{\langle \Omega_m^2 \rangle^2} \right)^i
\end{aligned}$$

(4.122)

is useful, in which case the correlation function factors $1/\langle \Omega_m^2 \rangle$, $1/\langle \Omega_m^2 \rangle^2$ guarantee the correct dimension of the power series. Such a formulation of one-dimensional hyper-surface equations is totally symmetrical with respect to the structure of the basic equation system. Additionally, this formulation guarantees the correct limit case which holds in the critical point, i. e. using the phase transition condition $\langle \Omega_m^4 \rangle_{\text{crit}} \sim \langle \Omega_m^2 \rangle_{\text{crit}}^2$ (see (4.93)) one obtains the correct critical relation $\lambda_{m+2} \sim \langle \Omega_m^2 \rangle_{\text{crit}}^2$. In order to show additionally that in this case $\lambda_m = 0$ holds, the unknown coefficients have to be known. In order to define these coefficients, and to show that this formulation is valid, (4.122) has to be inserted into the basic equation system (4.121). In the following this shall be done.

The Definition Scheme for the One-Dimensional Hyper-Surface Equations

In order to find the definition scheme of the coefficients of (4.122) (and thus to demonstrate the validity of (4.122)), one needs the square root of the second hyper-surface formulation, i. e. one needs

$$\begin{aligned}
\lambda_m &= \frac{1}{\langle \Omega_m^2 \rangle} \sum_{i=0}^{\infty} \Lambda_1^{(i)} \left(\frac{\langle \Omega_m^4 \rangle}{\langle \Omega_m^2 \rangle^2} \right)^i , \\
\sqrt{\lambda_{m+2}} &= \frac{1}{\langle \Omega_m^2 \rangle} \sum_{i=0}^{\infty} \Lambda_{2,r}^{(i)} \left(\frac{\langle \Omega_m^4 \rangle}{\langle \Omega_m^2 \rangle^2} \right)^i
\end{aligned}$$

(4.123)

instead of (4.122). The coefficients $\Lambda_{2,r}^{(i)}$ and the coefficients $\Lambda_2^{(i)}$ are connected by the relation (4.114), i. e.

$$\Lambda_2^{(i)} = \sum_{k=0}^{i} \Lambda_{2,r}^{(k)} \Lambda_{2,r}^{(i-k)}$$

(4.124)

holds.

Inserting (4.123) into (4.121) one obtains the power series

$$\sum_{i=0}^{\infty} DEF_\nu^{(i)} \left(\frac{\langle \Omega_m^4 \rangle}{\langle \Omega_m^2 \rangle^2} \right)^i = 0 \ (\nu = 1, 2) ,$$

(4.125)

where the choice

$$DEF_\nu^{(i)} = 0 \tag{4.126}$$

is possible so that the formulation (4.123) is valid. (4.126) represents the basic definition scheme of the unknown coefficients. In the following the explicit structure of this definition scheme shall be considered. The steps to get such a definition scheme shall not be presented, because such a calculation is not of interest, i. e. only power series operations have to be done. Then it has to be noticed that such a basic definition scheme consists of two essential parts, namely the system of the starting coefficients and the system of the other coefficients.

The equation of the starting coefficients is given by

$$\delta_{1,\nu}\left(\Lambda_{2,r}^{(0)}\right)^\nu + \sum_{\xi=0}^\infty g_{\nu,\xi}^{(6)}\left(\frac{\Lambda_1^{(0)}}{\Lambda_{2,r}^{(0)}}\right)^\xi = 0 \ (\nu = 1,2)\,, \tag{4.127}$$

i. e. the two coefficients $\Lambda_1^{(0)}$, $\Lambda_{2,r}^{(0)}$ are defined by (4.127). (4.127) consists of two coupled power series, in which case $\delta_{1,\nu}$ is *Kronecker's* delta. By using the method introduced before (see (4.79)) the second power series of (4.127) ($\nu = 2$) can be inverted, i. e. the formulation

$$\frac{\Lambda_1^{(0)}}{\Lambda_{2,r}^{(0)}} = \sum_{\xi=1}^\infty (-1)^\xi \alpha_\xi \left(\frac{g_{2,0}^{(6)}}{g_{2,1}^{(6)}}\right)^\xi := \Xi_1 \tag{4.128}$$

is possible, with $g_{2,0}^{(6)}/g_{2,1}^{(6)}$ being the linear solution of (4.127) ($\nu = 2$). The coefficients of (4.128) are given by the recursive scheme

$$\alpha_1 = 1,$$

$$\alpha_{\xi>1} = -\sum_{i=2}^\xi \frac{g_{2,i}^{(6)}}{g_{2,1}^{(6)}} \left\{\prod_{k=1}^{i-1}\sum_{\kappa_k=k}^{\kappa_{k+1}-1}\right\}\left\{\prod_{k=0}^{i-1}\alpha_{-\kappa_k+\kappa_{k+1}}\right\}$$

$$(\kappa_0 = 0\,,\ \kappa_i = \xi) \tag{4.129}$$

(compare with (4.81)). By using (4.128) and the first power series of (4.127) ($\nu = 1$) the starting coefficient $\Lambda_{2,r}^{(0)}$ can be calculated, i. e.

$$\Lambda_{2,r}^{(0)} = -\sum_{\xi=0}^\infty g_{1,\xi}^{(6)}\,\Xi_1^\xi := \Xi_2 \tag{4.130}$$

holds. Therefore, the other starting coefficient is given by

$$\Lambda_1^{(0)} = \Xi_1\Xi_2\,. \tag{4.131}$$

By using these starting coefficients the other coefficients of (4.123) can be calculated.

The other coefficients of (4.123) can be calculated by using the inhomogenous system of equations

$$C_\nu^{(1)}\Lambda_1^{(i)} + C_\nu^{(2)}\Lambda_{2,r}^{(i)} + IH_\nu^{(i)} = 0 \quad (\nu = 1, 2, \ i > 0) \,, \tag{4.132}$$

where the starting coefficients are needed, i. e.

$$C_\nu^{(1)} = \frac{1}{\Lambda_1^{(0)}\Lambda_{2,r}^{(0)\nu}} \sum_{\xi=0}^{\infty} \xi \, g_{\nu,\xi}^{(6)} \left(\frac{\Lambda_1^{(0)}}{\Lambda_{2,r}^{(0)}}\right)^{\xi} \,,$$

$$C_\nu^{(2)} = -\frac{1}{\Lambda_{2,r}^{(0)\nu+1}} \sum_{\xi=0}^{\infty} (\xi + \nu) \, g_{\nu,\xi}^{(6)} \left(\frac{\Lambda_1^{(0)}}{\Lambda_{2,r}^{(0)}}\right)^{\xi} \tag{4.133}$$

holds. (4.132) is the second essential part of (4.126). The inhomogenous term is given by

$$IH_\nu^{(i)} = -\delta_{1,i}\delta_{2,\nu} + \sum_{\xi=0}^{\infty} g_{\nu,\xi}^{(6)} I_{1,\nu,\xi}^{(i)} \,, \tag{4.134}$$

where the recursive relation

$$I_{1,\nu,\xi}^{(0)} = \sum_{\xi=0}^{\infty} g_{\nu,\xi}^{(6)} \frac{1}{\Lambda_{2,r}^{(0)\nu}} \left(\frac{\Lambda_1^{(0)}}{\Lambda_{2,r}^{(0)}}\right)^{\xi} \,,$$

$$I_{1,\nu,\xi}^{(1)} = 0 \,, \quad I_{1,\nu,\xi}^{(i>1)} = I_{2,\nu,\xi}^{(i)} - \sum_{k=1}^{i-1} I_{1,\nu,\xi}^{(k>0)} I_{3,\nu,\xi}^{(k)} \tag{4.135}$$

holds. In order to define the numbers $I_{2,\nu,\xi}^{(i)}$, $I_{3,\nu,\xi}^{(k)}$, one needs additional relations. These relations are given by

$$I_{2,\nu,\xi}^{(i)} = I_{4,\nu,\xi}^{(i)} - \sum_{k=1}^{i-1} \left[I_{3,\nu,\xi}^{(k)} \frac{1}{\Lambda_{2,r}^{(0)\nu}} \left(\frac{\Lambda_1^{(0)}}{\Lambda_{2,r}^{(0)}}\right)^{\xi} \left(\xi\frac{\Lambda_1^{(k)}}{\Lambda_1^{(0)}} - (\xi + \nu)\frac{\Lambda_{2,r}^{(k)}}{\Lambda_{2,r}^{(0)}}\right) \right] \,,$$

$$I_{3,\nu,\xi}^{(k)} = \frac{I_{5,\nu,\xi}^{(i-k>0)}}{\Lambda_{2,r}^{(0)\xi+\nu}} + (\xi + \nu)\frac{\Lambda_{2,r}^{(i-k)}}{\Lambda_{2,r}^{(0)}} \,,$$

$$I_{4,\nu,\xi}^{(i)} = \frac{I_{6,\xi}^{(i>0)}}{\Lambda_{2,r}^{(0)\xi+\nu}} - \frac{\Lambda_1^{(0)\xi}}{\Lambda_{2,r}^{(0)2\xi+2\nu}} I_{5,\nu,\xi}^{(i>0)} \tag{4.136}$$

and

$$I_{6,\xi}^{(i>0)} = \left\{ \prod_{k=1}^{\xi-1} \sum_{\kappa_k=0}^{\kappa_{k+1}} \right\} \left\{ \prod_{k=0}^{\xi-1} \Lambda_1^{(-\kappa_k+\kappa_{k+1})} \right\}\Bigg|_{-\kappa_k+\kappa_{k+1}\neq i}$$
$$\left(1 - \delta_{0,\xi}\right)\left(1 - \delta_{1,\xi}\right) \ (\kappa_0 = 0 \,, \ \kappa_\xi = i) \,,$$

$$I_{5,\nu,\xi}^{(i>0)} = \left\{ \prod_{k=1}^{\xi+\nu-1} \sum_{\kappa_k=0}^{\kappa_{k+1}} \right\} \left\{ \prod_{k=0}^{\xi+\nu-1} \Lambda_{2,r}^{(-\kappa_k+\kappa_{k+1})} \right\} \Bigg|_{-\kappa_k+\kappa_{k+1}\neq i}$$
$$\left(1 - \delta_{0,(\xi+\nu)}\right)\left(1 - \delta_{1,(\xi+\nu)}\right) \; \left(\kappa_0 = 0 \,, \;\; \kappa_{\xi+\nu} = i\right) \,. \tag{4.137}$$

The formulation (4.123) fulfills the basic equation system (4.121) if the definition scheme is used, in which case the inhomogenous equation system (4.132) defines the coefficients of the formulations (4.123) in a recursive way. Using the relation (4.124) the one-dimensional hyper-surface equations (4.122) are defined as well. The hyper-surface equations (4.122) are the exact solutions of the one-dimensional symmetrical problem of fourth order, i. e. they replace the critical hyper-surface equations (4.98) and (4.100) outside the critical region. (In order to use the equations (4.122), the identification λ_m, λ_{m+2}, $\Omega_m \leftrightarrow \lambda_1$, λ_2, Ω has to be made.) Such power series can be used to describe the connection between the macroscopic and the microscopic level outside the critical region. For example, the fluctuating behavior of magnetism outside the critical region can be described by using such hyper-surface equations.

The question which structure the hyper-surface equations of the two-dimensional problem have can easily be answered.

The Structure of the Two-Dimensional Hyper-Surface Equations. Exchange Symmetry

The relevant basic equation system of the considered two-dimensional problem is the basic equation system (4.102). This equation system shows a special symmetry, namely exchange symmetry (compare with (4.103)). The exact solutions of such a basic equation system need to have the same symmetry. Moreover, they need to contain the limit case (4.122), i. e. in the case of decoupling all parts which consist of correlation functions of the respective other system have to vanish so that the limit case (4.122) occurs. (The parts which contain the correlation function $\langle \Omega_1^2 \Omega_2^2 \rangle$ have to vanish as well, in which case it has to be taken into account that in the case of decoupling the correlation function $\langle \Omega_1^2 \Omega_2^2 \rangle$ decomposes into two parts, i. e.

$$\langle \Omega_1^2 \Omega_2^2 \rangle \overset{\text{decoupling}}{=} \langle \Omega_1^2 \rangle \langle \Omega_2^2 \rangle \tag{4.138}$$

holds.) This requires the relations

$$\lambda_m = \frac{1}{\langle \Omega_m^2 \rangle} \sum_{i=0}^{\infty} \sum_{k=0}^{\infty} \Lambda_{1,m}^{(i,k)} \left(\frac{\langle \Omega_1^4 \rangle}{\langle \Omega_1^2 \rangle^2} \right)^i \left(\frac{\langle \Omega_2^4 \rangle}{\langle \Omega_2^2 \rangle^2} \right)^k \,,$$

$$\lambda_{m+2} = \frac{1}{\langle \Omega_m^2 \rangle^2} \sum_{i=0}^{\infty} \sum_{k=0}^{\infty} \Lambda_{2,m+2}^{(i,k)} \left(\frac{\langle \Omega_1^4 \rangle}{\langle \Omega_1^2 \rangle^2} \right)^i \left(\frac{\langle \Omega_2^4 \rangle}{\langle \Omega_2^2 \rangle^2} \right)^k \,, \tag{4.139}$$

with $\Lambda_{1,m}^{(i,k)}$, $\Lambda_{2,m+2}^{(i,k)}$ $(m = 1,2)$ containing the correlation function $\langle \Omega_1^2 \Omega_2^2 \rangle$ which describes the coupling of the two systems. Additionally, the exchange symmetry requires

$$\Lambda_{1,1}^{(i,k)} = \Lambda_{1,2}^{(k,i)} \ , \ \ \Lambda_{2,3}^{(i,k)} = \Lambda_{2,4}^{(k,i)} \ . \tag{4.140}$$

Therefore, the structure of four hyper-surface equations is known. Then the question arises which structure the coupling parameter λ_c has to show.

Inserting the complex elementary coefficients (4.115) and the hyper-surface equations (4.139) into the last equation of the basic equation system (4.102) and using an inversion formula of the kind (4.79) one can easily show that the structure of the coupling parameter is of the same kind, i. e.

$$\lambda_c = \frac{1}{\langle \Omega_1^2 \rangle \langle \Omega_2^2 \rangle} \sum_{i=0}^{\infty} \sum_{k=0}^{\infty} \Lambda_c^{(i,k)} \left(\frac{\langle \Omega_1^4 \rangle}{\langle \Omega_1^2 \rangle^2} \right)^i \left(\frac{\langle \Omega_2^4 \rangle}{\langle \Omega_2^2 \rangle^2} \right)^k \tag{4.141}$$

holds. Then the question arises which structure the various coupling coefficients $\Lambda_\epsilon^{(i,k)}$ ($\epsilon = (1,m), (2,m+2), c$) have.

By using power series the influence of both systems can mathematically be described, this show the considerations above. Thus, the assumption is very obvious that the coupling coefficients need to be power series as well, i. e. the assumption

$$\Lambda_\epsilon^{(i,k)} = \sum_{l=0}^{\infty} \Lambda_\epsilon^{(i,k,l)} \left(\frac{\langle \Omega_1^2 \Omega_2^2 \rangle}{\langle \Omega_1^2 \rangle \langle \Omega_2^2 \rangle} \right)^l \tag{4.142}$$

is obvious. Actually, this structure of the hyper-surface equation system is the correct one. This can be shown in a systematical way.

By starting from the basic equation system (4.102) the various *Lagrangian* parameters can be eliminated in a successive way. Then basic equation systems of higher order occur which contain less *Lagrangian* parameters. Then the structure of the two-dimensional hyper-surface equation system can totally be justified. However, to do this is an immense work. Therefore, these remarks shall be enough.

Further Comments

In order to demonstrate the validity of the formulation (4.139)-(4.142), and to define the unknown coefficients, the formulation has to be inserted into the basic equation system (4.102). However, this shall not be done in this work. Instead of such a demonstration the problem of any number of variables shall be solved. The two-dimensional problem is then a special case of such a high-dimensional problem. In order to solve such a high-dimensional problem, the considerations of this subsection are helpful. In particular, the way of calculation of hyper-surface equations is now obvious, i. e. by using a suitable basic equation system and a suitable formulation of hyper-surface equations such an inversion problem can be solved. Furthermore, the considerations of this subsection show that the solutions need to be power series, in which case the coefficients are power series, too. In a nutshell, hyper-surface equations are power series functions. Such a high-dimensional problem shall be considered later. A classification scheme to classify power series functions shall now be considered.

4.3.5 Self-Similarity in Mathematical Terms

In order to describe the structure of the relevant hyper-surface equations, the name *mathematical self-similarity* or shorter, *self-similarity*, is useful. The choice of the term *self-similarity* is useful, because now the same phenomenon occurs as in the theory of non-linear pattern evolution, however, on a mathematical level. An example of self-similarity in the context of pattern evolution is represented by the *Mandelbrot* set. If such a set is visualized, the same patterns occur on different lenght scales. This behavior can also be observed in the animated nature. For example, ferns are well-known self-similar objects. However, now a mathematical form of such a self-similarity occurs, i. e. the coefficients of power series are again power series. Another example of self-similarity is represented by a cascade of dependent systems (see figure 1.1), i. e. a physical system can be decomposed into a set of elementary systems, where every elementary system can be decomposed, too, into a set of other elementary systems. This is also a kind of self-similarity.

In order to get an exact definition of self-similar power series functions, a definition scheme of the kind

$$\Psi = \sum_{\nu_1 = b_1}^{\infty} \psi_1^{(\nu_1)} \Phi_1^{\nu_1} \, ,$$

$$\psi_1^{(\nu_1)} = \sum_{\nu_2 = b_2}^{\infty} \psi_2^{(\nu_2)} \Phi_2^{\nu_2} \, ,$$

$$\cdot \qquad \cdot \qquad \cdot$$
$$\cdot \qquad \cdot \qquad \cdot$$
$$\cdot \qquad \cdot \qquad \cdot$$

$$\psi_{O-1}^{(\nu_{O-1})} = \sum_{\nu_O = b_O}^{\infty} \psi_O^{(\nu_O)} \Phi_O^{\nu_O} \, ,$$

$$\psi_O^{(\nu_O)} = \sum_{\nu_{O+1} = b_{O+1}}^{\infty} \psi_{O+1}^{(\nu_{O+1})} \Phi_{O+1}^{\nu_{O+1}} \tag{4.143}$$

is useful. (The quantities Φ_i are specific functions. The inner coefficients $\psi_{O+1}^{(\nu_{O+1})}$ are numbers. b_i describes boundaries. The index O characterizes the order of the self-similar power series function, i. e. an ordinary power series is of the order 0. Ψ can be equal 0.) Both the basic equation systems and the hyper-surface equations can be classified by using this scheme. For example, the hyper-surface equation system

(4.139)-(4.142) represents self-similar power series functions of second order. Due to the fact that the property *self-similarity* cannot change during the transformation process, the basic equation system (4.101) as well represents self-similar power series functions. The order of self-similarity increases if problems of more variables are considered. Such high-dimensional inversion problems shall be considered in the following subsection.

4.3.6 The Problem of High-Dimensional Hyper-Surface Equations of Fourth Order

In the following the basic equation system

$$
-Z_4^{(N)^{-1}} \frac{\partial Z_4^{(N)}}{\partial \lambda_i^{(2)}} = \langle \Omega_i^2 \rangle \ , \quad +Z_4^{(N)^{-1}} \frac{\partial^2 Z_4^{(N)}}{\partial \lambda_i^{(2)^2}} = \langle \Omega_i^4 \rangle \ ,
$$

$$
-Z_4^{(N)^{-1}} \frac{\partial Z_4^{(N)}}{\partial \lambda_{i,k}^{(c)}} = \langle \Omega_i^2 \Omega_k^2 \rangle \ (i,k = 1 \ldots N \ , \ i < k) \tag{4.144}
$$

will be taken as a basis. (4.144) represents a symmetrical basic equation system of fourth order and N variables. The partition function $Z_4^{(N)}$ is defined by (4.37) or (4.46), respectively, i. e.

$$
Z_4^{(N)} = \int_{-\infty}^{+\infty} \exp\left[-\left(\sum_{i=1}^{N} \lambda_i^{(2)} \Omega_i^2 + \sum_{i,k=1,i<k}^{N} \lambda_{i,k}^{(c)} \Omega_i^2 \Omega_k^2 + \sum_{i=1}^{N} \lambda_i^{(4)} \Omega_i^4 \right) \right] d\Omega
$$

$$
= \left\{ \prod_{i,k=1,i<k}^{N} \sum_{\nu_{i,k}=0}^{\infty} \left\{ \prod_{i=1}^{N} k_{\mu_i}^{(i)} \right\} \lambda_{i,k}^{(c)^{\nu_{i,k}}} \right\} \left\{ \prod_{i=1}^{N} Z_i \right\} \tag{4.145}
$$

holds, with μ_i being a function of $\nu_{i,k}$ (see (4.45)). The correlated distribution function is defined by

$$
\rho_4^{(N)}(\Omega) = Z_4^{(N)^{-1}} \exp\left[-\left(\sum_{i=1}^{N} \lambda_i^{(2)} \Omega_i^2 + \sum_{i,k=1,i<k}^{N} \lambda_{i,k}^{(c)} \Omega_i^2 \Omega_k^2 + \sum_{i=1}^{N} \lambda_i^{(4)} \Omega_i^4 \right) \right] \ , \tag{4.146}
$$

i.e. a symmetrical exponential function of fourth order and N variables has to be considered. The calculation of corresponding hyper-surface equations shall now be described. In this context a set of indices will be formulated by using set brackets $\{ \}$. For example the set of elements $\nu_{i,k}$ will be used in the form $\{\nu_{i,k}\}$.

The Series Representation of the Basic Equation System

Inserting the partition function (4.145) into the basic equation system (4.144) one obtains a relatively explicit formulation of the basic equation system, namely the power series functions

$$\left\{\prod_{i,k=1,i<k}^{N}\sum_{\nu_{i,k}=0}^{\infty}\left\{\prod_{i=1,i\neq\alpha}^{N}k_{\mu_i}^{(i)}\right\}\left(k_{\mu_\alpha+1}^{(\alpha)}+k_{\mu_\alpha}^{(\alpha)}\langle\Omega_\alpha^2\rangle\right)\lambda_{i,k}^{(c)^{\nu_{i,k}}}\right\}=0\ ,$$

$$\left\{\prod_{i,k=1,i<k}^{N}\sum_{\nu_{i,k}=0}^{\infty}\left\{\prod_{i=1,i\neq\alpha}^{N}k_{\mu_i}^{(i)}\right\}\left(k_{\mu_\alpha+2}^{(\alpha)}-k_{\mu_\alpha}^{(\alpha)}\langle\Omega_\alpha^4\rangle\right)\lambda_{i,k}^{(c)^{\nu_{i,k}}}\right\}=0\ ,$$

$$\left\{\prod_{i,k=1,i<k}^{N}\sum_{\nu_{i,k}=0}^{\infty}\left\{\prod_{i=1,i\neq\alpha,\beta}^{N}k_{\mu_i}^{(i)}\right\}\left(k_{\mu_\alpha+1}^{(\alpha)}k_{\mu_\beta+1}^{(\beta)}+k_{\mu_\alpha}^{(\alpha)}k_{\mu_\beta}^{(\beta)}\langle\Omega_\alpha^2\Omega_\beta^2\rangle\right)\right.$$

$$\left.\lambda_{i,k}^{(c)^{\nu_{i,k}}}\right\}=0\ ,\quad \alpha,\beta=1\ldots N\ ,\quad \alpha\neq\beta\ . \tag{4.147}$$

(4.147) is the generalized form of the basic equation system (4.102), i. e. such a high-dimensional system shows exchange symmetry, too. The real elementary coefficients $k_{\mu_i}^{(i)}$ are defined by (4.44). Apart from the numbers $z_{\mu_i}^{(i)}$ and the notation such real elementray coefficients are identical with the complex elementary coefficients (4.34). Therefore, the result (4.115) can be used again (this result represents the explicit form of the complex elementary coefficients, for this calculation see (4.108)ff.), i. e. the real elementary coefficients $k_{\mu_i+n}^{(i)}$ $(n=0,1,2)$ need to have the form

$$\boxed{k_{\mu_i+n}^{(i)}=z_{\mu_i}^{(i)}\left(\frac{1}{\sqrt{\lambda_i^{(4)}}}\right)^{\mu_i+n}\sum_{\xi_i=0}^{\infty}g_{\mu_i+n,\xi_i}^{(6)}\left(\frac{1}{\sqrt{\lambda_i^{(4)}}}\right)^{\xi_i}\lambda_i^{(2)^{\xi_i}}\ .} \tag{4.148}$$

(4.148) is the adapted relation (4.115). Inserting (4.148) into (4.147) one obtains a basic equation system of the form

$$\boxed{\left\{\prod_{i,k=1,i<k}^{N}\sum_{\nu_{i,k}=0}^{\infty}\sum_{\xi_i=0}^{\infty}KERNEL_{4,\alpha,\gamma}^{(\{\nu_{i,k}\},\{\xi_i\})}\left(\frac{1}{\sqrt{\lambda_i^{(4)}}}\right)^{\xi_i+\mu_i}\lambda_i^{(2)^{\xi_i}}\lambda_{i,k}^{(c)^{\nu_{i,k}}}\right\}=0\ ,}$$

$$\tag{4.149}$$

where the kernel $KERNEL_{4,\alpha,\gamma}^{(\{\nu_{i,k}\},\{\xi_i\})}$ is defined by

$$\boxed{\begin{aligned}KERNEL_{4,\alpha,\gamma}^{(\{\nu_{i,k}\},\{\xi_i\})}&=C_{4,1,\alpha,\gamma}^{(\{\nu_{i,k}\},\{\xi_i\})}\left(\frac{1}{\langle\Omega_\alpha^2\rangle\sqrt{\lambda_\alpha^{(4)}}}\right)\left(\frac{1}{\langle\Omega_\gamma^2\rangle\sqrt{\lambda_\gamma^{(4)}}}\right)+\\[2mm]&\quad C_{4,2,\alpha,\gamma}^{(\{\nu_{i,k}\},\{\xi_i\})}\left(\frac{\langle\Omega_\alpha^2\Omega_\gamma^2\rangle}{\langle\Omega_\alpha^2\rangle\langle\Omega_\gamma^2\rangle}\right)\\[2mm]&(\alpha=1,2,\ldots N\ ,\ \gamma=0,\beta=0,1,2,\ldots N)\ ,\end{aligned}}$$

$$\tag{4.150}$$

with the numbers being defined by

$$C_{4,1,\alpha,\gamma=0}^{(\{\nu_{i,k}\},\{\xi_i\})} = z_{\mu_\alpha}^{(\alpha)} \left\{ \prod_{i=1,i\neq\alpha}^{N} z_{\mu_i}^{(i)} g_{\mu_i,\xi_i}^{(6)} \right\} g_{\mu_\alpha+1,\xi_\alpha}^{(6)} ,$$

$$C_{4,1,\alpha,\alpha}^{(\{\nu_{i,k}\},\{\xi_i\})} = z_{\mu_\alpha}^{(\alpha)} \left\{ \prod_{i=1,i\neq\alpha}^{N} z_{\mu_i}^{(i)} g_{\mu_i,\xi_i}^{(6)} \right\} g_{\mu_\alpha+2,\xi_\alpha}^{(6)} ,$$

$$C_{4,1,\alpha,\beta}^{(\{\nu_{i,k}\},\{\xi_i\})} = z_{\mu_\alpha}^{(\alpha)} z_{\mu_\beta}^{(\beta)} \left\{ \prod_{i=1,i\neq\alpha,\beta}^{N} z_{\mu_i}^{(i)} g_{\mu_i,\xi_i}^{(6)} \right\} g_{\mu_\alpha+1,\xi_\alpha}^{(6)} g_{\mu_\beta+1,\xi_\beta}^{(6)} ,$$

$$C_{4,2,\alpha,\gamma=0}^{(\{\nu_{i,k}\},\{\xi_i\})} = z_{\mu_\alpha}^{(\alpha)} \left\{ \prod_{i=1,i\neq\alpha}^{N} z_{\mu_i}^{(i)} g_{\mu_i,\xi_i}^{(6)} \right\} g_{\mu_\alpha,\xi_\alpha}^{(6)} ,$$

$$C_{4,2,\alpha,\alpha}^{(\{\nu_{i,k}\},\{\xi_i\})} = -z_{\mu_\alpha}^{(\alpha)} \left\{ \prod_{i=1,i\neq\alpha}^{N} z_{\mu_i}^{(i)} g_{\mu_i,\xi_i}^{(6)} \right\} g_{\mu_\alpha,\xi_\alpha}^{(6)} ,$$

$$C_{4,2,\alpha,\beta}^{(\{\nu_{i,k}\},\{\xi_i\})} = z_{\mu_\alpha}^{(\alpha)} z_{\mu_\beta}^{(\beta)} \left\{ \prod_{i=1,i\neq\alpha,\beta}^{N} z_{\mu_i}^{(i)} g_{\mu_i,\xi_i}^{(6)} \right\} g_{\mu_\alpha,\xi_\alpha}^{(6)} g_{\mu_\beta,\xi_\beta}^{(6)} . \tag{4.151}$$

In this context the relation

$$\lambda_{\kappa=0}^{(4)} := 1 \, , \quad \Omega_{\kappa=0} := 1 \tag{4.152}$$

has been used which guarantees that one kernel represents all basic equations, i. e. if one uses a function $\langle \Omega_\alpha^2 \Omega_\gamma^2 \rangle$ to describe basic equations which contain correlation functions of the kind $\langle \Omega_\alpha^2 \rangle$, one needs the relation (4.152). (It has to be noted here that this method is only one method to gain a compressed formulation of such a kernel. In the next subsection another method will be introduced. Then non-symmetrical problems of higher order will be considered.) (4.149) is the power series representation of the symmetrical basic equation system of fourth order and any variables. Such a basic equation system is a generalized form of the one-dimensional basic equation system (4.121). (4.149) represents a universal basic equation system, because every symmetrical problem of fourth order can be described by such a basic equation system. The statistical parameters as well as the correlation functions can be any parameters and correlation functions, respectively. For example, such a basic equation system is the basic equation system of many thermodynamic problems in the context of *Landau's* phase transition theory. If the analysis of a human EEG is carried out by using symmetrical exponential functions in an approximative way, such a basic equation system is the relevant one. Then the correlation functions are mean values which describe the measured field of the brain surface. But what is the sense of such a basic equation system? The answer has already been given, i. e. such a basic equation system can be used to find the correlated hyper-surface equations. In order to solve this problem, a suitable formulation of the hyper-surface equations has to be known. However, such a formulation can easily be found if one looks at the kernel (4.150).

Structure Kernels

In order to find the basic equation system as well as the kernel (4.150) and the definition (4.151), one has to insert the real elementary coefficients into the basic equation system (4.147), this was mentioned above. If the reader looks at the kernel (4.150), and if the reader compares with the above calculated results, it is obvious that a kernel of the form (4.150) shows the strucure of the needed hyper-surface equations, i. e. the correlation function quotient $\langle \Omega_\alpha^2 \Omega_\gamma^2 \rangle / \langle \Omega_\alpha^2 \rangle \langle \Omega_\gamma^2 \rangle$ shows the structure of the needed hyper-surface formulation. In other words, the kernel (4.150) represents a *structure kernel*. However, a straightforward calculation does not lead to a kernel of the form (4.150), it leads to a kernel of the form $KERNEL = C_1 \left(1/\sqrt{\lambda_\alpha^{(4)}} \right) \left(1/\sqrt{\lambda_\gamma^{(4)}} \right) + C_2 \langle \Omega_\alpha^2 \Omega_\gamma^2 \rangle$. Therefore, the question arises which equivalence operation leads to a structure kernel. Knowing the results of the calculations before, one can postulate three requirements which lead to the equivalence operation which makes sense. Firstly, it has to be required that a structure kernel has to be a dimensionless expression. (The principle of dimension neutrality.) Secondly, it has to be required that one of the addends needs to be a function of correlation functions only. (The principle of structural homogeneity.) These two requirements give rise to a division with a function which consists of products of correlation functions. Therefore, the inverse of this function represents a special kind of multiplier. In order to decide which multiplier has to be used, a third principle has to be considered, namely the principle that the set of indices of the multiplier has to be equal to the set of the indices of the given correlation function. (The principle of index symmetry.) In the considered case this means that the multiplier has to be of the form $1/\langle \Omega_\alpha^2 \rangle \langle \Omega_\gamma^2 \rangle$. Therefore, the structure kernel has been calculated. However, to prove the validity of the name *structure kernel*, i. e. to show that such a kernel shows the structure of the hyper-surface formulation, the formulation has to be inserted into the basic equation system. In the following it will be shown that the given structure is useful, i. e. the kernel (4.150) is actually a structure kernel.

High-Dimensional Hyper-Surface Equations of Fourth Order

Using the notation

$$\lambda_{i,0} := \lambda_i^{(2)} , \; \lambda_{i,i} := \lambda_i^{(4)} , \; \lambda_{i,k} := \lambda_{i,k}^{(c)} \; (i < k) , \tag{4.153}$$

and assuming that the structure kernel (4.150) represents the necessary correlation function quotients as well as the necessary correlation function factors, the formulation

$$\lambda_{i,\gamma} = \frac{1}{\langle\Omega_i^2\rangle\langle\Omega_\gamma^2\rangle}\left\{\prod_{l=1,l\leq m}^{N}\prod_{m=1}^{N}\sum_{\epsilon_{l,m}=0}^{\infty}\Lambda_{i,\gamma}^{(\{\epsilon_{l,m}\})}\left(\frac{\langle\Omega_l^2\Omega_m^2\rangle}{\langle\Omega_l^2\rangle\langle\Omega_m^2\rangle}\right)^{\epsilon_{l,m}}\right\}$$
$$(i = 1, 2, \ldots N , \ \gamma = 0, 1, 2, \ldots N)$$

(4.154)

is useful. (4.154) represents all hyper-surface equations in a compressed form. Separating this compressed expression into three basic parts, and separating into correlation function terms of the kind $(\langle\Omega_m^2\Omega_n^2\rangle/\langle\Omega_m^2\rangle\langle\Omega_n^2\rangle$ and $\langle\Omega_l^4\rangle/\langle\Omega_l^2\rangle^2$, one obtains three classes of hyper-surface equations, i. e. one obtains

$$\lambda_i^{(2)} = \frac{1}{\langle\Omega_i^2\rangle}\left\{\prod_{l=1}^{N}\prod_{m=0,m<n}^{N}\prod_{n=1}^{N}\sum_{\epsilon_l=0}^{\infty}\sum_{\epsilon_{m,n}=0}^{\infty}\Lambda_{i,0}^{(\{\epsilon_{l,l}\},\{\epsilon_{m,n}\})}\right.$$
$$\left(\frac{\langle\Omega_m^2\Omega_n^2\rangle}{\langle\Omega_m^2\rangle\langle\Omega_n^2\rangle}\right)^{\epsilon_{m,n}}\left(\frac{\langle\Omega_l^4\rangle}{\langle\Omega_l^2\rangle^2}\right)^{\epsilon_{l,l}}\Bigg\} ,$$

$$\lambda_i^{(4)} = \frac{1}{\langle\Omega_i^2\rangle^2}\left\{\prod_{l=1}^{N}\prod_{m=0,m<n}^{N}\prod_{n=1}^{N}\sum_{\epsilon_l=0}^{\infty}\sum_{\epsilon_{m,n}=0}^{\infty}\Lambda_{i,i}^{(\{\epsilon_{l,l}\},\{\epsilon_{m,n}\})}\right.$$
$$\left(\frac{\langle\Omega_m^2\Omega_n^2\rangle}{\langle\Omega_m^2\rangle\langle\Omega_n^2\rangle}\right)^{\epsilon_{m,n}}\left(\frac{\langle\Omega_l^4\rangle}{\langle\Omega_l^2\rangle^2}\right)^{\epsilon_{l,l}}\Bigg\} ,$$

$$\lambda_{i,k}^{(c)} = \frac{1}{\langle\Omega_i^2\rangle\langle\Omega_k^2\rangle}\left\{\prod_{l=1}^{N}\prod_{m=0,m<n}^{N}\prod_{n=1}^{N}\sum_{\epsilon_l=0}^{\infty}\sum_{\epsilon_{m,n}=0}^{\infty}\Lambda_{i,k}^{(\{\epsilon_{l,l}\},\{\epsilon_{m,n}\})}\right.$$
$$\left(\frac{\langle\Omega_m^2\Omega_n^2\rangle}{\langle\Omega_m^2\rangle\langle\Omega_n^2\rangle}\right)^{\epsilon_{m,n}}\left(\frac{\langle\Omega_l^4\rangle}{\langle\Omega_l^2\rangle^2}\right)^{\epsilon_{l,l}}\Bigg\} .$$

(4.155)

(4.154) and (4.155), respectively, represent the formulation of the hyper-surface equations. An infinite number of variables is possible. In order to find a definition scheme of the unknown coefficients, such a formulation has to be inserted into the basic equation system (4.149). This shall be done.

The Definition Scheme of the Coefficients

Inserting the formulation (4.155) into the basic equation system (4.149) one obtains the power series function

$$\left\{\prod_{l=0,l<m}^{N}\prod_{m=1}^{N}\sum_{\epsilon_{l,m}=0}^{\infty}DEF_{\alpha,\gamma}^{(\{\epsilon_{l,m}\})}\left(\frac{\langle\Omega_l^2\Omega_m^2\rangle}{\langle\Omega_l^2\rangle\langle\Omega_m^2\rangle}\right)^{\epsilon_{l,m}}\right\} = 0 ,$$

(4.156)

where the requirement

$$DEF_{\alpha,\gamma}^{(\{\epsilon_{l,m}\})} := 0 \tag{4.157}$$

$(i, \alpha = 1, 2, \ldots N)$ can be used as a definition scheme of the coefficients $\Lambda_{i,\gamma}^{(\{\epsilon_{l,m}\})}$. In the following the structure of the components $DEF_{\alpha,\gamma}^{(\{\epsilon_{l,m}\})}$ shall be considered.

In the following the relation

$$\{\epsilon_{l,m}\} = \{\epsilon_{l,m}, \epsilon_{\gamma,\alpha}\}; l, m \neq \gamma, \alpha \tag{4.158}$$

has to be taken into account, which decribes the connection between two used sets of exponents. Then the definition

$$DEF_{\alpha,\gamma}^{(\{\epsilon_{l,m},\epsilon_{\gamma,\alpha}\};l,m\neq\gamma,\alpha)} = \begin{cases} DEF_{1,\alpha,\gamma}^{(\{\epsilon_{l,m},\epsilon_{\gamma,\alpha}\};l,m\neq\gamma,\alpha)} & \text{if } \epsilon_{\gamma,\alpha} = 0 \\ DEF_{1,\alpha,\gamma}^{(\{\epsilon_{l,m},\epsilon_{\gamma,\alpha}\};l,m\neq\gamma,\alpha)} + \\ DEF_{2,\alpha,\gamma}^{(\{\epsilon_{l,m},\epsilon_{\gamma,\alpha}-1\};l,m\neq\gamma,\alpha)} & \text{if } \epsilon_{\gamma,\alpha} > 0 \end{cases} \tag{4.159}$$

has to be considered, where the relations

$$DEF_{1,\alpha,\gamma}^{(\{\epsilon_{l,m}\})} = \left\{ \prod_{i,k=1,i<k}^{N} \sum_{\nu_{i,k}=0}^{\infty} \sum_{\xi_i=0}^{\infty} \right\} C_{4,1,\alpha,\gamma}^{(\{\nu_{i,k}\},\{\xi_i\})} D_{1,\{\epsilon_{l,m}\}}^{(\{\nu_{i,k}\},\{\xi_i\})} \tag{4.160}$$

and

$$DEF_{2,\alpha,\gamma}^{(\{\epsilon_{l,m}\})} = \left\{ \prod_{i,k=1,i<k}^{N} \sum_{\nu_{i,k}=0}^{\infty} \sum_{\xi_i=0}^{\infty} \right\} D_{2,\{\epsilon_{l,m}\},\alpha,\gamma}^{(\{\nu_{i,k}\},\{\xi_i\})} \tag{4.161}$$

with

$$D_{2,\{\epsilon_{l,m}\},\alpha,\gamma}^{(\{\nu_{i,k}\},\{\xi_i\})} = \left\{ \prod_{l=0,l\leq m}^{N} \prod_{m=1}^{N} \sum_{\epsilon_{1,l,m}=0}^{\infty} \sum_{\epsilon_{2,l,m}=0}^{\infty} \right\}\Bigg|_{\epsilon_{1,l,m}+\epsilon_{2,l,m}=\epsilon_{l,m}}$$
$$C_{4,2,\alpha,\gamma}^{(\{\nu_{i,k}\},\{\xi_i\})} D_{3,\{\epsilon_{1,l,m}\}}^{(\alpha,\gamma)} D_{1,\{\epsilon_{2,l,m}\}}^{(\{\nu_{i,k}\},\{\xi_i\})} \tag{4.162}$$

have to be used. The numbers $D_{1,\{\epsilon_{l,m}\}}^{(\{\nu_{i,k}\},\{\xi_i\})}$ and $D_{1,\{\epsilon_{2,l,m}\}}^{(\{\nu_{i,k}\},\{\xi_i\})}$, respectively, are defined by the recursive relation

$$D_{4,\{\epsilon_{l,m}\}}^{(\{M_1\})} = \left\{ \prod_{l=0,l\leq m}^{N} \prod_{m=1}^{N} \sum_{\epsilon_{1,l,m}=0}^{\infty} \sum_{\epsilon_{2,l,m}=0}^{\infty} \right\}\Bigg|_{\epsilon_{1,l,m}+\epsilon_{2,l,m}=\epsilon_{l,m}}$$
$$D_{3,\{\epsilon_{1,l,m}\}}^{(\{M_2\})} D_{1,\{\epsilon_{2,l,m}\}}^{(\{\nu_{i,k}\},\{\xi_i\})}, \tag{4.163}$$

where the relation

$$D_{4,\{\epsilon_{m,m}\},\{\epsilon_{n,o}\}}^{(\{M_1=empty\})} = \left\{ \prod_{m=1}^{N} \prod_{n=0,n<o}^{N} \prod_{o=1}^{N} \delta_{0,\epsilon_{m,m}} \delta_{0,\epsilon_{n,o}} \right\},$$

$$D_{4,\{\epsilon_{m,m}\},\{\epsilon_{n,o}\}}^{(\{M_1\})} = \left\{ \prod_{\{i,k,l=M_1\}} \prod_{m=1}^{N} \prod_{n=0,n<o}^{N} \prod_{o=1}^{N} \sum_{\epsilon_{l,m,m},\epsilon_{l,n,o},\epsilon_{i,k,m,m},\epsilon_{i,k,n,o}=0}^{\infty} \right]$$

$$\left[\prod_{\{i,k,l=M_1\}} \Lambda_{l,0}^{(\{\epsilon_{l,m,m}\},\{\epsilon_{l,n,o}\})} \Lambda_{i,k}^{(\{\epsilon_{i,k,m,m}\},\{\epsilon_{i,k,n,o}\})} \right\}$$

$$\left(\text{condition} : \sum_{\{i,k,l=M_1\}} \epsilon_{l,m,m},\epsilon_{l,n,o} + \sum_{\{i,k,l=M_1\}} \epsilon_{i,k,m,m},\epsilon_{i,k,n,o} = \epsilon_{m,m},\epsilon_{n,o}\right)$$

$$(4.164)$$

holds. The numbers $D_{3,\{\epsilon_{1,l,m}\}}^{(\{M_2\})}$ and $D_{3,\{\epsilon_{1,l,m}\}}^{(\alpha,\gamma)}$ are defined by the high-dimensional expression

$$D_{3,\{\epsilon_{l,l}\},\{\epsilon_{m,n}\}}^{(\{M=\text{empty}\})} = \left\{ \prod_{l=1}^{N} \prod_{m=0,m<n}^{N} \prod_{n=1}^{N} \delta_{0,\epsilon_{l,l}} \delta_{0,\epsilon_{m,n}} \right\},$$

$$D_{3,\{\epsilon_{l,l}\},\{\epsilon_{m,n}\}}^{(\{M\})} = \left\{ \prod_{\{i=M\}} \prod_{l=1}^{N} \prod_{m=0,m<n}^{N} \prod_{n=1}^{N} \sum_{\epsilon_{i,l,l},\epsilon_{i,m,n}=0}^{\infty} \right.$$

$$\left. \left\{ \prod_{\{i=M\}} \Lambda_{i,i,r}^{(\{\epsilon_{i,l,l}\},\{\epsilon_{i,m,n}\})} \right\} \right\},$$

$$\left(\text{condition} : \sum_{\{i=M\}} \epsilon_{i,l,l},\epsilon_{i,m,n} = \epsilon_{l,l},\epsilon_{m,n}\right) \qquad (4.165)$$

where the set $\{M\}$ can be equal to the set $\{M_2\}$ or to α,γ. The set $\{M_2\}$ is defined by the relation

$$\{M_2\} = \{i\} , \quad i = \left\{ \prod^{\mu_1+\xi_1} 1, \right\} \left\{ \prod^{\mu_2+\xi_2} 2, \right\} \cdots \left\{ \prod^{\mu_N+\xi_N} N, \right\}, \qquad (4.166)$$

where μ_i is defined by (4.45). The set $\{M_1\}$ is defined by the relation

$$\{M_1\} = \{(i,k),l\} , \quad (i,k) = \left\{ \prod^{\nu_{1,2}} (1,2), \right\} \left\{ \prod^{\nu_{1,3}} (1,3), \right\} \cdots \left\{ \prod^{\nu_{1,N}} (1,N), \right\}$$

$$\left\{ \prod^{\nu_{2,3}} (2,3), \right\} \left\{ \prod^{\nu_{2,4}} (2,4), \right\} \cdots \left\{ \prod^{\nu_{2,N}} (2,N), \right\}$$

$$\cdots \left\{ \prod^{\nu_{N-1,N}} (N-1,N), \right\},$$

$$l = \left\{ \prod^{\xi_1} 1, \right\} \left\{ \prod^{\xi_2} 2, \right\} \cdots \left\{ \prod^{\xi_N} N, \right\}. \qquad (4.167)$$

The product signs are defined by

$$
\left[\begin{array}{c} \text{upper boundary} \\ \succ \!\!\! \prod \quad \text{number}, \prec \end{array} \right] = \underbrace{\text{number}, \text{number}, \ldots \text{number}}_{\text{upper boundary}} . \tag{4.168}
$$

The numbers $\Lambda_{i,i,r}^{(\{\epsilon_i,l,l\},\{\epsilon_i,m,n\})}$ are the numbers of the square root of the second expression of (4.155), i. e.

$$
\Lambda_{i,i}^{(\{\epsilon_l,l\},\{\epsilon_m,n\})} = \left\{ \prod_{l=1}^{N} \prod_{m=0,m<n}^{N} \prod_{n=1}^{N} \sum_{\epsilon_{1,l,l},\epsilon_{1,m,n},\epsilon_{2,l,l},\epsilon_{2,m,n}=0}^{\infty} \right\}
$$
$$
\Lambda_{i,i,r}^{(\{\epsilon_{1,l,l}\},\{\epsilon_{1,m,n}\})} \Lambda_{i,i,r}^{(\{\epsilon_{2,l,l}\},\{\epsilon_{2,m,n}\})}
$$
$$
(\text{condition}: \ \epsilon_{1,l,l}, \epsilon_{1,m,n} + \epsilon_{2,l,l}, \epsilon_{2,m,n} = \epsilon_{l,l}, \epsilon_{m,n}) \tag{4.169}
$$

holds. Therefore, the coefficients of the hyper-surface equations (4.155) are defined. It has to be remarked that an explicit scheme of the kind (4.127)-(4.137) can be calculated. However, this shall not be done. In this book the definition (4.157) shall be a sufficient definition.

Very often problems of fourth order occur in statistical physics. However, sometimes equations of higher order are necessary to describe the system. In table 4.1 an example is shown. As the reader can see, problems of sixth order occur in the context of *Landau's* theory. Additionally, it has to be remarked that problems with non-symmetrical terms are necessary very often. Therefore, the question arises whether it is possible to deal also with such problems. This is actually possible and will be shown in the following subsection.

4.3.7 Universal Hyper-Surface Equations

In the following the basic equation system

$$
-Z^{-1}\frac{\partial Z}{\partial \lambda_{\Theta_\alpha}} = \left\langle \left[\begin{array}{c} \alpha \\ \succ\!\!\prod_{i=1}\Omega_{\Theta_i}\prec \end{array} \right] \right\rangle \tag{4.170}
$$

will be taken as a basis. (4.144) represents a universal basic equation system of any order O and N variables. The partition function Z is defined by (4.52) or (4.70), respectively, i. e. the relation

$$
Z = \int_{-\infty}^{+\infty} \exp\left[-\left(\sum_{\alpha=1}^{O} \left[\succ\!\!\prod_{i=1}^{\alpha} \sum_{\Theta_i=1}^{N} \lambda_{\Theta_\alpha}\Omega_{\Theta_i} \prec \right] \right) \right] d\Omega
$$
$$
= \left\{ \prod_{conv=1}^{N} \lambda_{\Theta_{conv}}^{-1/O} \right\} \left[\succ\!\! \prod_{\substack{\alpha=1 \\ \Theta_O \neq \Theta_{conv}}}^{O} \prod_{i=1}^{\alpha} \prod_{\Theta_i=1}^{N} \prod_{conv=1}^{N} \sum_{\Lambda_{\Theta_\alpha}=0}^{\infty} z_\Lambda^{(2)} \lambda_{\Theta_\alpha}^{\Lambda_{\Theta_\alpha}} \lambda_{\Theta_{conv}}^{-\left[\frac{\xi(\Lambda)}{O}\right]} \prec \right] \tag{4.171}
$$

holds. ($\xi(\mathbf{\Lambda})$ is defined by the relation (4.60). $\mathbf{\Lambda}$ represents all components Λ_{Θ_α}. The numbers $z_{\mathbf{\Lambda}}^{(2)}$ are defined by (4.69).) The correlated distribution function is defined by

$$\rho(\mathbf{\Omega}) = Z^{-1} \exp\left[-\left(\sum_{\alpha=1}^{O} \left\{ \prod_{i=1}^{\alpha} \sum_{\Theta_i=1}^{N} \lambda_{\Theta_\alpha} \Omega_{\Theta_i} \right\} \right) \right], \tag{4.172}$$

i.e. a non-symmetrical exponential function of any order O and N variables has to be considered. These formulations are very compressed formulations, in which case product brackets are used. Additionally, it has to be remarked that in this subsection instead of set brackets { } bold symbols will be used to describe sets of indices. For example, instead of $\{\Omega_{\Theta_i}\}$ the bold symbol $\mathbf{\Omega}$ will be used. In order to understand this representation much better, the reader may have a look at formula (3.26). This formula is an explicit representation of the now used distribution function (4.172). Such a compressed formulation is necessary in order to obtain a general view of the mathematical problem. The following equations will be formulated in the same way. It has to be remarked that now no inequalitiy constraints are used (compare with (4.146)) so that *Lagrangian* multipliers occur which correspond to terms of the same physical meaning. (For example, $\lambda_{\Theta_\alpha} := \lambda_{i,k}$, $\lambda_{i,k} = \lambda_{k,i}$.) Therefore, universal hyper-surface equations can be calculated. This calculation shall now be described.

The Series Representation of the Basic Equation System

Inserting the partition function (4.171) into the basic equation system (4.170) an explicit form of the basic equation system can be calculated. This equation system is of the form

$$\left[\left\{ \prod_{\substack{\alpha=1 \\ \Theta_O \neq \Theta_{conv}}}^{O} \prod_{i=1}^{\alpha} \prod_{\Theta_i=1}^{N} \prod_{conv=1}^{N} \sum_{\Lambda_{\Theta_\alpha}=0}^{\infty} KERNEL_{O,\Upsilon_\sigma}^{(\mathbf{\Lambda})}\ \lambda_{\Theta_\alpha}^{\Lambda_{\Theta_\alpha}} \lambda_{\Theta_{conv}}^{-\left[\frac{\xi(\mathbf{\Lambda})}{O}\right]} \right\} \right] = 0, \tag{4.173}$$

where the kernel is defined by

$$KERNEL_{O,\Upsilon_\sigma}^{(\Lambda)} = C_{O,1,\Upsilon_\sigma}^{(\Lambda)} \left(\frac{1}{\left[\prod_{i=1}^{\sigma} \lambda_{\Theta_{\Upsilon_i}}^{1/O} \sqrt{\langle \Omega_{\Upsilon_i}^2 \rangle} \right]} \right) +$$
$$C_{O,2,\Upsilon_\sigma}^{(\Lambda)} \left(\frac{\left\langle \left[\prod_{i=1}^{\sigma} \Omega_{\Upsilon_i} \right] \right\rangle}{\left[\prod_{i=1}^{\sigma} \sqrt{\langle \Omega_{\Upsilon_i}^2 \rangle} \right]} \right)$$

(4.174)

with

$$\Upsilon_\sigma = \Upsilon_1 , \Upsilon_2, \ldots \Upsilon_\sigma, \quad \sigma = 1, 2, \ldots O . \tag{4.175}$$

In order to calculate this kernel, the three introduced requirements (see subsubsection *Structure Kernels* in subsection 4.3.6) were used. These requirements guaranteed a structure kernel. Therefore, the assumption is obvious that the kernel (4.174) is a structure kernel as well, i. e. that the structure of the formulation of the correlated hyper-surface equations is identical with the structure of the structure kernel (4.174). Actually, this assumption is correct. Therefore, the basic equation system (4.173) is the generalized form of the basic equation system (4.149), and the structure kernel (4.174) replaces the structure kernel (4.150). However, in this form product signs are used so that no additional constraint of the form (4.152) is necessary. The relation (4.175) occurs instead of the indices α and γ, i. e. an index Υ_σ represents one special index Θ_α ($\alpha = 1, 2, \ldots O$). The numbers in (4.174) are defined by

$$C_{O,1,\Upsilon_\sigma}^{(\Lambda)} = \begin{cases} -\frac{\xi(\Lambda)+1}{O} z_\Lambda^{(2)} & \text{if } \Upsilon_\sigma = \Theta_{conv} \\ (\Lambda_{\Upsilon_\sigma} + 1) z_{\Lambda \neq \Lambda_{\Upsilon_\sigma}, \Lambda_{\Upsilon_\sigma}+1}^{(2)} & \text{if } \Upsilon_\sigma \neq \Theta_{conv} \end{cases} \tag{4.176}$$

and

$$C_{O,2,\Upsilon_\sigma}^{(\Lambda)} = C_{O,2}^{(\Lambda)} = z_\Lambda^{(2)}. \tag{4.177}$$

In the following the correlated hyper-surface equations will be considered. In this context it will be elucidated that the kernel (4.174) is a structure kernel.

The Correlated Hyper-Surface Equations

If the kernel (4.174) actually is a structure kernel, the formulation

$$\lambda_{\Theta_\alpha} = \left[\prod_{i=1}^{\alpha} \sqrt{\langle \Omega_{\Theta_i}^2 \rangle} \right]^{-1} \left[\prod_{\beta=1}^{O} \prod_{i=1}^{\beta} \prod_{\Xi_i=1}^{N} \sum_{\Gamma_{\Xi_\beta}=0}^{\infty} \Lambda_{\Theta_\alpha}^{(\Gamma)} \, CORR_\beta^{\Gamma_{\Xi_\beta}} \right] \tag{4.178}$$

needs to be the correct formulation, where the part $CORR_\beta$ represents *standard quotients* which are defined by

$$CORR_\beta = \frac{\left\langle \left\{ \left[\prod_{i=1}^{\beta} \Omega_{\Xi_i} \right] \right\} \right\rangle}{\left\{ \left[\prod_{i=1}^{\beta} \sqrt{\langle \Omega_{\Xi_i}^2 \rangle} \right] \right\}} \,, \tag{4.179}$$

and the relation

$$\boldsymbol{\Gamma} = \{\Gamma_{\Xi_\beta}\}, \quad \Xi_\beta = \Xi_1, \Xi_2, \ldots \Xi_\beta \tag{4.180}$$

has to be considered. Then any variables $(N = 1, 2, 3, \ldots \infty)$ are possible, and any even order $(O = 2, 4, 6, \ldots \infty)$ is possible. In order to show that the formulation (4.178) actually represents the hyper-surface equations, and to show that the kernel (4.174) represents a structure kernel, the formulation has to be inserted into the basic equation system (4.173). As it was already mentioned, such calculations are not very interesting and additionally it is an immense work. Therefore, in the following the basic results shall be presented, i. e. a definition scheme of the numbers $\Lambda_{\Theta_\alpha}^{(\boldsymbol{\Gamma})}$ shall be presented.

The Definition Scheme of the Numbers

Inserting the formulation (4.178) into the basic equation system (4.173) one obtains the power series function

$$\left\{ \prod_{\beta=1}^{O} \prod_{i=1}^{\beta} \prod_{\Xi_i=1}^{N} \sum_{\Gamma_{\Xi_\beta}=0}^{\infty} DEF_{\Upsilon_\sigma}^{(\boldsymbol{\Gamma})} \; CORR_\beta^{\Gamma_{\Xi_\beta}} \right\} = 0 \tag{4.181}$$

where the requirement

$$DEF_{\Upsilon_\sigma}^{(\boldsymbol{\Gamma})} := 0 \tag{4.182}$$

can be used as a basic definition scheme of the numbers $\Lambda_{\Theta_\alpha}^{(\boldsymbol{\Gamma})}$.

In the following the relation

$$\boldsymbol{\Gamma} = \boldsymbol{\Gamma}_r, \Gamma_{\Upsilon_\sigma} \tag{4.183}$$

has to be taken into account, which decribes the connection between two used sets of exponents. $\boldsymbol{\Gamma}_r$ represents all elements Γ_{Ξ_β} without the element Γ_{Υ_σ}. Then the definition

$$
DEF_{\Upsilon_\sigma}^{(\Gamma_r,\Gamma_{\Upsilon_\sigma})} =
\begin{cases}
DEF_{1,\Upsilon_\sigma}^{(\Gamma_r,\Gamma_{\Upsilon_\sigma})} & \text{if } \Gamma_{\Upsilon_\sigma} = 0 \\[2ex]
DEF_{1,\Upsilon_\sigma}^{(\Gamma_r,\Gamma_{\Upsilon_\sigma})} + \\[2ex]
DEF_{2,\Upsilon_\sigma}^{(\Gamma_r,\Gamma_{\Upsilon_\sigma}-1)} & \text{if } \Gamma_{\Upsilon_\sigma} > 0
\end{cases}
\tag{4.184}
$$

has to be considered, in which case the relations

$$
DEF_{1,\Upsilon_\sigma}^{(\Gamma)} = \left\{ \prod_{\substack{\alpha=1 \\ \Theta_O \neq \Theta_{conv}}}^{O} \prod_{i=1}^{\alpha} \prod_{\Theta_i=1}^{N} \sum_{\Lambda_{\Theta_\alpha}=0}^{\infty} \right\} C_{O,1,\Upsilon_\sigma}^{(\Lambda)} D_{1,\Gamma}^{(\Lambda)}
\tag{4.185}
$$

and

$$
DEF_{2,\Upsilon_\sigma}^{(\Gamma)} = \left\{ \prod_{\substack{\alpha=1 \\ \Theta_O \neq \Theta_{conv}}}^{O} \prod_{i=1}^{\alpha} \prod_{\Theta_i=1}^{N} \sum_{\Lambda_{\Theta_\alpha}=0}^{\infty} \right\} D_{2,\Gamma}^{(\Lambda,\Upsilon_\sigma)}
\tag{4.186}
$$

hold. Furthermore, the relation

$$
D_{2,\Gamma}^{(\Lambda,\Upsilon_\sigma)} = \left\{ \prod_{\beta=1}^{O} \prod_{i=1}^{\beta} \prod_{\Xi_i=1}^{N} \sum_{\Gamma_{1,\Xi_\beta}=0}^{\infty} \sum_{\Gamma_{2,\Xi_\beta}=0}^{\infty} \right\}\Bigg|_{\Gamma_{1,\Xi_\beta}+\Gamma_{2,\Xi_\beta}=\Gamma_{\Xi_\beta}}
$$

$$
C_{O,2,\Upsilon_\sigma}^{(\Lambda)} D_{3,\Gamma_1}^{(\Upsilon_\sigma)} D_{1,\Gamma_2}^{(\Lambda)}
\tag{4.187}
$$

has to be considered. The numbers $D_{1,\Gamma}^{(\Lambda)}$ and $D_{1,\Gamma_2}^{(\Lambda)}$ are defined by the recursive relation

$$
D_{4,\Gamma}^{(\mathbf{M}_3)} = \left\{ \prod_{\beta=1}^{O} \prod_{i=1}^{\beta} \prod_{\Xi_i=1}^{N} \sum_{\Gamma_{1,\Xi_\beta}=0}^{\infty} \sum_{\Gamma_{2,\Xi_\beta}=0}^{\infty} \right\}\Bigg|_{\Gamma_{1,\Xi_\beta}+\Gamma_{2,\Xi_\beta}=\Gamma_{\Xi_\beta}} D_{3,\Gamma_1}^{(\mathbf{M}_4)} D_{1,\Gamma_2}^{(\Lambda)} \, ,
\tag{4.188}
$$

where

$$
D_{4,\Gamma}^{(\mathbf{M}_3=\text{empty})} = \left\{ \prod_{\beta=1}^{O} \prod_{i=1}^{\beta} \prod_{\Xi_i=1}^{N} \delta_{0,\Gamma_{\Xi_\beta}} \right\} ,
$$

$$
D_{4,\Gamma}^{(\mathbf{M}_3)} = \left\{ \prod_{\mathbf{k}=\mathbf{M}_3} \prod_{\beta=1}^{O} \prod_{i=1}^{\beta} \prod_{\Xi_i=1}^{N} \sum_{\Gamma_{\mathbf{k},\Xi_\beta}=0}^{\infty} \right\}\Bigg|_{\sum_{\mathbf{k}=\mathbf{M}_3} \Gamma_{\mathbf{k},\Xi_\beta}=\Gamma_{\Xi_\beta}} \left\{ \prod_{\mathbf{k}=\mathbf{M}_3} \Lambda_{\mathbf{k}}^{(\Gamma_{\mathbf{k}})} \right\}
\tag{4.189}
$$

holds. The numbers $D_{3,\Gamma_1}^{(\mathbf{M}_4)}$ and $D_{3,\Gamma_1}^{(\Upsilon_\sigma)}$ are defined by the relation

$$D_{3,\Gamma}^{(\mathbf{M}=\text{empty})} = \left\{ \prod_{\beta=1}^{O} \prod_{i=1}^{\beta} \prod_{\Xi_i=1}^{N} \delta_{0,\Gamma_{\Xi_\beta}} \right\} ,$$

$$D_{3,\Gamma}^{(\mathbf{M})} = \left\{ \prod_{conv=\mathbf{M}} \prod_{\beta=1}^{O} \prod_{i=1}^{\beta} \prod_{\Xi_i=1}^{N} \sum_{\Gamma_{conv,\Xi_\beta}=0}^{\infty} \right\} \Bigg|_{\sum_{conv=\mathbf{M}} \Gamma_{k,\Xi_\beta}=\Gamma_{\Xi_\beta}}$$
$$\left\{ \prod_{conv=\mathbf{M}} \Lambda_{\Theta_{conv},r}^{(\Gamma_{conv})} \right\} , \tag{4.190}$$

where the set $\mathbf{M}$ can be equal the set $\mathbf{M}_4$ or Υ_σ. The set $\mathbf{M}_4$ is defined by

$$\mathbf{M}_4 = \left\{ \prod_{l=1}^{N} \prod^{\xi(\Lambda)} l, \right\} , \tag{4.191}$$

and the set $\mathbf{M}_3$ is defined by

$$\mathbf{M}_3 = \left\{ \prod_{\alpha=1}^{O} \prod_{i=1}^{\alpha} \prod_{\Theta_i=1}^{N} \prod_{\substack{\Theta_O \neq \Theta_{conv}}}^{\Lambda\Theta_\alpha} \Theta_\alpha, \right\} , \tag{4.192}$$

where the product signs with only upper boundaries are defined by (4.168). Within the sets $\mathbf{M}_3$ and $\mathbf{M}_4$ special values are possible several times. In such a case the indices $\mathbf{k}$ and $conv$ have to show two components, i. e. in such a case the relation

$$\mathbf{k}, conv = \mathbf{k}_v^*, conv_v^* \tag{4.193}$$

holds. The additional index v marks the multiplicity of the considered values $\mathbf{k}^*, conv^*$ Otherwise the same notation would occur several times. The numbers $\Lambda_{\Theta_{conv},r}^{(\Gamma_{conv})}$ are numbers of roots of power series functions. (The reader has to compare with the kernel (4.174)). Within this kernel roots of power series functions of the kind $\lambda_{\Theta_{conv}}^{1/O}$ occur.) The connection of such numbers with the original numbers is given by

$$\Lambda_{\Theta_{conv}}^{(\Gamma)} = \left\{ \prod_{k=1}^{O} \prod_{\beta=1}^{O} \prod_{i=1}^{\beta} \prod_{\Xi_i=1}^{N} \sum_{\Gamma_{k,\Xi_\beta}=0}^{\infty} \right\} \Bigg|_{\sum_{k=1}^{O} \Gamma_{k,\Xi_\beta}=\Gamma_{\Xi_\beta}} \left\{ \prod_{k=1}^{O} \Lambda_{\Theta_{conv},r}^{(\Gamma_k)} \right\} . \tag{4.194}$$

Therefore, the numbers of the formulation (4.178) are defined, and it has been shown that the kernel (4.178) represents a structure kernel. This definition scheme is symmetical to the definition scheme of the numbers of the symmetrical hyper-surface equations, the scheme (4.156)-(4.169), i. e. the now calculated scheme represents a generalized form of the scheme (4.156)-(4.169). In the following some further comments shall be given.

Further Comments

In this subsection it was shown that the problem of inversion of a basic equation system is solvable in a totally analytical way. By using product brackets a clear formulation of the relevant equations is possible, and by using product signs which generate products of sums and parameters a short formulation is guaranteed. The basic equation system (4.170) as well as the hyper-surface equations (4.178) are self-similar power series functions. The borderline cases *critical hyper-surface equations* and *symmetrical hyper-surface equations of fourth order* are special cases of the now derived hyper-surface equations. Strictly speaking, such a hyper-surface equation describes the connection between a macroscopic and a microscopic level of a multi-component system, in which case the macroscopic level is represented by the correlation functions and the microscopic level is represented by the *Lagrangian* multipliers. Using such an analytical function one can determine an exponential distribution function in a totally analytical way. As a starting point the maximum information entropy principle can be used, i. e. a totally macroscopic determination scheme was introduced. Essential quantities of system theory such as information can be transformed into a form which contains only macroscopic quantities, i. e. quantities of system theory can be calculated by starting from a macroscopic level. Such hyper-surface equations are universal equations, because such hyper-surface equations are valid for a large class of distribution functions and many different physical systems. For example, the temperature dependence of distribution functions of thermodynamic systems can be determined if this dependence has been measured on a macroscopic level, and the dependence of the pump parameter of distribution functions which occur in laser theory can be determined as well. Not only low-dimensional systems, in particular high-dimensional physical systems can be described by using the concept of hyper-surface equations. On every level of consideration a compressed formulation of the basic facts is possible. By using the above introduced scheme of calculation many relevant relations can be calculated. For example, macroscopic phase transition conditions of the kind (4.93) can systematically be calculated, and specific partition functions of high-dimensional systems can be calculated as well. In a nutshell, the above introduced concept is a universal concept of *system analysis*.

Sometimes continuous functions are necessary to describe the considered physical system. For example, to describe a continuous multi-mode laser system, exponential functions which contain integrals instead of sums have to be used. Then functional integrals instead of ordinary partition functions occur. Thus, it makes sense to consider the extension of the concept of system analysis to functional problems. In the following this will be done.

4.3.8 Functional Hyper-Surface Equations

Exponential functions of the kind (4.172) are functions with sums in the exponent. Such an exponential function is identical with the function

$$\rho(\Omega) = Z^{-1} \exp\left[-\left(\sum_{\alpha=1}^{O} \left\{ \prod_{i=1}^{\alpha} \sum_{\Theta_i=1}^{N} \frac{\Delta\lambda(\Theta_\alpha)}{\Delta\Theta_i} \Omega(\Theta_i)\Delta\Theta_i \right\} \right) \right], \qquad (4.195)$$

where $\Omega(\Theta_i)$ represents a variable function which have to be used at discrete points Θ_i, and $\Delta\lambda(\Theta_\alpha)$ represents a multiplier functions which also have to be used at discrete points. This means that the set of multipliers λ_{Θ_α} is replaced by the multiplier function $\lambda(\Theta_\alpha)$, and the set of variables Ω_{Θ_i} is replaced by the variable function $\Omega(\Theta_i)$, without changing the meaning of the distribution function. Such multiplier functions can be extensive quantities, i. e. an increase of the number of points Θ_i can be correlated with a decrease of the values of the multiplier functions. Therefore, the symbol $\Delta\lambda(\Theta_\alpha)$ instead of $\lambda(\Theta_\alpha)$ shall be used. The distance between the considered points shall be $\Delta\Theta_i = \Delta\Theta$. In the case of laser theory $\Theta_i := \Theta_1$ represents frequencies, and $\Omega_{\Theta_i} := \Omega_{\Theta_1}$ represents mode amplitudes. In this case $\Delta\Theta$ represents a frequency difference ω/N, with ω being the total frequency band. Other examples can be found in the theory of statistical particle dynamics. Then $\Theta_i := \Theta_1$ represents time points, and $\Omega_{\Theta_i} := \Omega_{\Theta_1}$ represents the correlated positions. In this case $\Delta\Theta$ represents the time difference t/N, with t being the total time. Then the partition function Z has to be written in the form

$$Z = \int_{-\infty}^{+\infty} \exp\left[-\left(\sum_{\alpha=1}^{O} \left\{ \prod_{i=1}^{\alpha} \sum_{\Theta_i=1}^{N} \frac{\Delta\lambda(\Theta_\alpha)}{\Delta\Theta_i} \Omega(\Theta_i)\Delta\Theta_i \right\} \right) \right] d\Omega . \qquad (4.196)$$

Ω is defined by

$$\Omega = \Omega(1), \Omega(2), \ldots \Omega(N) . \qquad (4.197)$$

By starting from these equations an extension can be derived which is valid for continuous problems.

In order to get the continuous case, the limit $\lim_{\substack{\Delta\Theta \to 0 \\ N \to \infty}}$ has to be used, i. e. if the continuous case is considered, the functions

$$\boxed{ \begin{aligned} \rho_{\text{func}}(\Omega) = \\ \lim_{\substack{\Delta\Theta \to 0 \\ N \to \infty}} Z_{\text{func}}^{-1} \exp\left[-\left(\sum_{\alpha=1}^{O} \left\{ \prod_{i=1}^{\alpha} \sum_{\Theta_i=1}^{N} \frac{\Delta\lambda(\Theta_\alpha)}{\Delta\Theta_i} \Omega(\Theta_i)\Delta\Theta_i \right\} \right) \right] \end{aligned} } \qquad (4.198)$$

and

$$Z_{\text{func}} =$$
$$\int_{-\infty}^{+\infty} \exp\left[-\left(\sum_{\alpha=1}^{O} \left\{ \prod_{i=1}^{\alpha} \sum_{\Theta_i=1}^{N} \frac{\Delta\lambda(\Theta_\alpha)}{\Delta\Theta_i} \Omega(\Theta_i)\Delta\Theta_i \right\} \right) \right] d\Omega \qquad (4.199)$$

have to be used. These limits represent functional expressions, in which case the partition function is a functional integral with real parameters. By using a customary notation the functional distribution function

$$\rho_{\text{func}}(\boldsymbol{\Omega}) = Z_{\text{func}}^{-1} \exp\left[-\left(\sum_{\alpha=1}^{O} \left\{ \prod_{i=1}^{\alpha} \int_{\Theta_i} \frac{d\lambda(\boldsymbol{\Theta}_\alpha)}{d\Theta_i} \Omega(\Theta_i)d\Theta_i \right\} \right) \right] \qquad (4.200)$$

has to be considered, with the functional partition function being

$$Z_{\text{func}} =$$
$$\int_{-\infty}^{+\infty} \exp\left[-\left(\sum_{\alpha=1}^{O} \left\{ \prod_{i=1}^{\alpha} \int_{\Theta_i} \frac{d\lambda(\boldsymbol{\Theta}_\alpha)}{d\Theta_i} \Omega(\Theta_i)d\Theta_i \right\} \right) \right] \prod_{c=1}^{\text{cont}} d\Omega(c) . \qquad (4.201)$$

(This notation is similar to the notation which is normally used if *Wiener* integrals are considered.) The product sign $\prod_{i=1}^{\alpha}$ generates expressions of the kind

$$\left[\prod_{i=1}^{\alpha} \int_{\Theta_i} \frac{d\lambda(\boldsymbol{\Theta}_\alpha)}{d\Theta_i} \Omega(\Theta_i)d\Theta_i \right\} =$$

$$\int_{\Theta_1} \int_{\Theta_2} \cdots \int_{\Theta_\alpha} \frac{d\lambda(\boldsymbol{\Theta}_\alpha)}{d\Theta_1 d\Theta_2 \ldots d\Theta_\alpha} \Omega(\Theta_1)\Omega(\Theta_2) \ldots \Omega(\Theta_\alpha) d\Theta_1 d\Theta_2 \ldots d\Theta_\alpha =$$

$$\int_{\Theta_1} \int_{\Theta_2} \cdots \int_{\Theta_\alpha} \frac{d\lambda(\boldsymbol{\Theta}_\alpha)}{(d\Theta)^\alpha} \Omega(\Theta_1)\Omega(\Theta_2) \ldots \Omega(\Theta_\alpha) d\Theta_1 d\Theta_2 \ldots d\Theta_\alpha , \qquad (4.202)$$

where the notation

$$\frac{d\lambda(\boldsymbol{\Theta}_\alpha)}{(d\Theta)^\alpha} = \lambda_{\text{dens}}(\boldsymbol{\Theta}_\alpha) \qquad (4.203)$$

shall be used. If numerical calculations have to be done, the limit value equations (4.198) and (4.199) have to be used. This fact has to be emphasized, because this means that the introduced calculation scheme can now be applied, too. In order to get the functional relations, the limit $\lim_{\substack{\Delta\Theta \to 0 \\ N \to \infty}}$ has to be used. Therefore, the functional hyper-surface equations can be found without any calculation, i. e. the self-similar power series function (4.178) represents the solution of the functional inversion problem if the now occuring notation is used. In a nutshell, the functional hyper-surface equations are given by

$$\lambda_{\text{dens}}(\boldsymbol{\Theta}_\alpha) = \lim_{\substack{\Delta\Theta \to 0 \\ N \to \infty}} \left[\left\{ \prod_{i=1}^{\alpha} \sqrt{\langle \Omega^2(\Theta_i) \rangle} \right\}^{-1} \right.$$

$$\left. \left\{ \prod_{\beta=1}^{O} \prod_{i=1}^{\beta} \prod_{\Xi_i=1}^{N} \sum_{\Gamma_{\Xi_\beta}=0}^{\infty} CONT_{\boldsymbol{\Theta}_\alpha}^{(\boldsymbol{\Gamma})} CORR_\beta^{\Gamma_{\Xi_\beta}} \right\} \right] , \tag{4.204}$$

with the standard quotients being defined by

$$CORR_\beta = \frac{\left\langle \left[\prod_{i=1}^{\beta} \Omega(\Xi_i) \right\} \right\rangle}{\left[\prod_{i=1}^{\beta} \sqrt{\langle \Omega^2(\Xi_i) \rangle} \right\}} . \tag{4.205}$$

(4.204) is a functional expression. If the underlying part $d\lambda(\boldsymbol{\Theta}_\alpha)$ (see (4.203)) represents an intensive quantity, it is useful to write the limit $\lim_{\substack{\Delta\Theta \to 0 \\ N \to \infty}}$ before the whole exponential expression (see (4.198)). In this functional case instead of the numbers $\Lambda_{\boldsymbol{\Theta}_\alpha}^{(\boldsymbol{\Gamma})}$ the densities

$$CONT_{\boldsymbol{\Theta}_\alpha}^{(\boldsymbol{\Gamma})} = \frac{\Lambda_{\boldsymbol{\Theta}_\alpha}^{(\boldsymbol{\Gamma})}}{(d\Theta)^\alpha} \tag{4.206}$$

have to be used. In the same way functional basic equation systems and explicit forms of functional partition functions can be derived. However, this shall not be done in this book. If such calculations would be done, it would be possible to show that functional hyper-surface equations can be written in the form

$$\lambda_{\text{dens}}(\boldsymbol{\Theta}_\alpha) = \left\{ \prod_{i=1}^{\alpha} \sqrt{\langle \Omega^2(\Theta_i) \rangle} \right\}^{-1}$$

$$\left\{ \prod_{\beta=1}^{O} \sum_{\Gamma_\beta=0}^{\infty} \left[\prod_{i=1}^{\beta} \int_{\Theta_i} CONT_{\boldsymbol{\Theta}_\alpha}^{(\boldsymbol{\Gamma})} CORR_\beta \, d\Theta_i \right\}^{\Gamma_\beta} \right\} \tag{4.207}$$

(numerical values are neglected). However, this shall neither be done.

Therefore, it is obvious that the concept of system analysis can be extended to physical systems which need a functional description. However, in this chapter only real parameters are considered. This makes sense if non-quantum mechanical problems are considered. In particular, the introduced concept can be used to solve functional integrals which represent solutions of a *Fokker-Planck* equation. Such integrals are special path integrals. However, in quantum theory complex path integrals occur which can be used to solve the quantum mechanical *Schrödinger* equation. These are the famous *Feynman* path integrals. Thus, the question arises whether it is possible to extend the concept of system analysis to quantum mechanical systems. This is actually possible and will be shown in chapter 7. Additionally, it shall be remarked that functional integrals are necessary to deal with statistical problems of superconductivity.

Statistical distribution functions fulfil suitable evolution equations such as the *Fokker-Planck* equation, i. e. they are solutions of suitable statistical evolution equations. If a specific evolution equation is known, such a distribution function can be derived by solving this evolution equation. Then no macroscopic determination process is necessary. In the following basic statistical evolution equations and interrelations of these equations shall be considered.

5   Statistical Evolution Equations

So far, the problem of determination of statistical distribution functions, the problem of derivation of relevant relations (like phase transition conditions) as well as the problem of determination of the form of a distribution function by an elementary and macroscopic principle were in the centre of interest. Another essential aspect of every statistical theory of multi-component systems is the aspect of evolution of a physical system, i. e. the problem of time-space evolution. In order to describe all evolution possibilities of a physical system in a uniform way, one needs evolution equations, in which case it very often makes sense to use differential evolution equations or equations which base on small differences of physical quantities (such equations shall also be called *differential evolution equations!*). Such evolution equations shall be discussed in this chapter. It shall be started with an elementary evolution equation, and it shall be shown how to get differential evolution equations by a gradual determination of all physically relevant terms (such as fluctuation terms and drift terms). In this way it will be shown that one can get many different classes of differential evolution equations by using only one basic evolution equation. Thus, in this chapter universal dynamic aspects of statistical phenomena shall play a crucial role. The equations which shall now be considered are of the *Fokker-Planck* type, of a conjugate *Fokker-Planck* type, of a *Schrödinger* type and of a conjugate complex *Schrödinger* type, with all equations emerging by a special choice of evolution equation parameters. The solutions of an equation of the *Fokker-Planck* type can be interpreted as probability densities. Complex parameters generate then equations from the *Schrödinger* type, in which case it has to be noted that solutions of an equation of the *Schrödinger* type cannot be interpreted as measurable probability densities. In this book solutions which describe a physical system only in an indirect way will be called *indirect state functions*. Additionally, it will be shown that in the above mentioned derivation an equation of the *Schrödinger* type can be found by using a special measure which determines the fluctuation activity. In this context a relation is needed which shows the form of *Heisenberg's* energy-time relation. In this context it will be shown that a complex impulse operator occurs, where this operator shows the form of the quantum mechanical impulse operator. The determination of such an evolution equation shall be carried out by using macroscopic quantities, because one of the essential facts of this work is to deal with determination procedures which start from a macroscopic level. However, it has to be remarked that the following considerations are also possible without considering mean values. Both the equations of the *Fokker-Planck* type and the equations of the *Schrödinger* type will again be used later, i. e. an equation of the *Fokker-Planck* type can be used to derive the relevant statistical distribution functions of laser physics (chapter 6), and an equation of the *Schrödinger* type occurs in quantum physics (chapter 7).

5.1   A Universal Evolution Equation

In the following an evolution equation will be derived, in which case a specification of basic parameters generates various basic evolution equations. These considerations will be of a general kind.

5.1.1   The Structure of the Signal

In the following dynamic signals will be considered. Such signals can be written in the form

$$\Omega_i(t) = \langle \Omega_i \rangle_t + F_i^{(1)}(t) \,, \tag{5.1}$$

with $\Omega_i(t)$ being an observable signal. For example, the variable $\Omega_i(t)$ can be an observable position coordinate of a particle or a measured electric field component of the brain surface. Such a signal can be decomposed into two parts, namely a mean value part and a part which gives the difference between the mean value part and the observable signal. $\langle \Omega_i \rangle_t$ represents the part which can be measured on a mean value level (short time measurement, ensemble measurement) at time t, and $F_i^{(1)}(t)$ is the difference between the observable signal $\Omega_i(t)$ and the mean signal $\langle \Omega_i \rangle_t$. Very often (5.1) describes a physical system which has to be described by using a deterministic physical component and a statistical physical component. In this case (5.1) represents a signal which consists of a deterministic part, which is identical with the mean value $\langle \Omega_i \rangle_t$, and a perturbation $F_i^{(1)}(t)$. Then $F_i^{(1)}(t)$ can be called a *fluctuation term*. However, sometimes it is useful to describe a totally deterministic problem by using statistical methods, because the dynamic behavior is a complicated one. In this case $F_i^{(1)}(t)$ in principle is a deterministic component, too. However, in this case $F_i^{(1)}(t)$ shall also be called a *fluctuation term*. Therefore, it has to be taken into account that the meaning of the used signals can completely be different.

If multi-component system are considered, many signals of the kind (5.1) have to be taken into account. In this case the signal $\Omega(t)$ has to be used, i. e. the relation

$$\Omega(t) = \Omega_1(t), \Omega_2(t), \ldots \Omega_N(t) \tag{5.2}$$

holds, with N representing the number of considered variables.

This signal structure will be taken as a basis. In order to describe such statistical signals, it makes sense to use distribution functions. In the following a correlated universal statistical evolution equation will be derived, which restricts the evolution possibilities of such a distribution function (in this context, compare with [30]).

5.1.2   The Basic Differential Equation

A totally general expression to describe a distribution function of the kind $\rho(\Omega, t)$ is represented by the integral expression

$$\rho(\Omega, t) = \int_{-\infty}^{+\infty} \rho_{\Omega(t)} \delta[\Omega - \Omega(t)]\, d\Omega(t) , \tag{5.3}$$

with Ω being a special signal value. δ represents *Dirac's* delta function (i. e. a distribution = a special continuous linear functional). The integral (5.1) connects the probability density of the signal value Ω, namely the function $\rho(\Omega, t)$, with the probability density of one special curve $\Omega(t)$, namely the function $\rho_{\Omega(t)}$. (The probability density $\rho(\Omega, t)$ is a density with respect to the space element $\Delta\Omega$, and the probability density $\rho_{\Omega(t)}$ is a density with respect to $\Delta\Omega(t)$.) The distribution function can be interpreted as a mean value with respect to the distribution function $\rho_{\Omega(t)}$, i. e. the notation

$$\rho(\Omega, t) = \left\langle \delta[\Omega - \Omega(t)] \right\rangle \tag{5.4}$$

makes sense. Considering the time evolution the difference of two consecutive distribution functions can be presented in the form

$$\Delta\rho(\Omega, t) = \left\langle \delta[\Omega - \Omega(t + \Delta t)] \right\rangle - \left\langle \delta[\Omega - \Omega(t)] \right\rangle . \tag{5.5}$$

Later only small time differences Δt will be used. Then the point $[\Omega - \Omega(t + \Delta t)]$ does not diverge very much from the point $[\Omega - \Omega(t)]$ so that a *Taylor* expansion can be used. Using the *Taylor* formula (3.31) one obtains the relation

$$\left\langle \delta[\Omega - \Omega(t + \Delta t)] \right\rangle = \left\langle \delta[\Omega - \Omega(t)] \right\rangle +$$

$$\sum_{\alpha=1}^{\infty} \left\langle \left\{ \prod_{i=1}^{\alpha} \sum_{\Theta_i=1}^{N} T_{\Theta_\alpha} \left[-\Delta\Omega_{\Theta_i}(t) \right] \right\} \right\rangle , \tag{5.6}$$

where the *Taylor* coefficients are defined by the functional expression

$$T_{\Theta_\alpha} = z_{\Theta_\alpha} \left\{ \prod_{i=1}^{\alpha} \left\{ \partial_{\Omega_{\Theta_i} - \Omega_{\Theta_i}(t+\Delta t)} \delta[\Omega - \Omega(t + \Delta t)] \right\} \Big|_{\Omega_{\Theta_i} - \Omega_{\Theta_i}(t)} \right\}$$

$$= z_{\Theta_\alpha} \left\{ \prod_{i=1}^{\alpha} \partial_{\Omega_{\Theta_i} - \Omega_{\Theta_i}(t)} \delta[\Omega - \Omega(t)] \right\}$$

$$= (-1)^\alpha z_{\Theta_\alpha} \left\{ \prod_{i=1}^{\alpha} \partial_{\Omega_{\Theta_i}(t)} \delta[\Omega - \Omega(t)] \right\}$$

$$= z_{\Theta_\alpha} \left\{ \prod_{i=1}^{\alpha} \partial_{\Omega_{\Theta_i}} \delta[\Omega - \Omega(t)] \right\} . \tag{5.7}$$

(The numbers z_{Θ_α} are defined by (3.32). In order to get the formula (5.6), the identification

$$\Xi, f(\Omega) \to \Omega - \Omega(t), \delta[\Omega - \Omega(t + \Delta t)] \tag{5.8}$$

has to be made.) $\Delta\Omega_{\Theta_i}(t)$ is defined by

$$\Delta\Omega_{\Theta_i}(t) = \Omega_{\Theta_i}(t + \Delta t) - \Omega_{\Theta_i}(t) \,, \tag{5.9}$$

where $\Delta\Omega_{\Theta_i}(t)$ can be decomposed into a deterministic and a fluctuation part (see subsection 5.1.4). Inserting (5.6) into the difference (5.5) one obtains the differential power series function

$$\Delta\rho(\mathbf{\Omega}, t) = \sum_{\alpha=1}^{\infty} (C_{\text{basic}})^\alpha \left\langle \left\{ \prod_{i=1}^{\alpha} \sum_{\Theta_i=1}^{N} z_{\Theta_\alpha} \Delta\Omega_{\Theta_i}(t) \partial_{\Omega_{\Theta_i}} \delta[\mathbf{\Omega} - \mathbf{\Omega}(t)] \right\} \right\rangle$$

$$(C_{\text{basic}} = -1) \,. \tag{5.10}$$

By using the functional relation

$$\left\{ \prod_{i=1}^{\alpha} \Delta\Omega_{\Theta_i}(t) \partial_{\Omega_{\Theta_i}} \delta[\mathbf{\Omega} - \mathbf{\Omega}(t)] \right\} = \left\{ \prod_{i=1}^{\alpha} \partial_{\Omega_{\Theta_i}} \Delta\Omega_{\Theta_i}^*(t) \delta[\mathbf{\Omega} - \mathbf{\Omega}(t)] \right\}$$

$$\tag{5.11}$$

the equation (5.10) takes on the form

$$\begin{aligned}
\Delta\rho(\mathbf{\Omega}, t) &= \\
&\sum_{\alpha=1}^{\infty} (C_{\text{basic}})^\alpha \left\langle \left\{ \prod_{i=1}^{\alpha} \sum_{\Theta_i=1}^{N} z_{\Theta_\alpha} \partial_{\Omega_{\Theta_i}} \Delta\Omega_{\Theta_i}^*(t) \delta[\mathbf{\Omega} - \mathbf{\Omega}(t)] \right\} \right\rangle \\
&= \\
&\sum_{\alpha=1}^{\infty} (C_{\text{basic}})^\alpha \left\langle \left\{ \prod_{i=1}^{\alpha} \sum_{\Theta_i=1}^{N} z_{\Theta_\alpha} \partial_{\Omega_{\Theta_i}} \Delta\Omega_{\Theta_i}^*(t) \right\} \right\rangle \left\langle \delta[\mathbf{\Omega} - \mathbf{\Omega}(t)] \right\rangle \\
&= \\
&\sum_{\alpha=1}^{\infty} (C_{\text{basic}})^\alpha \left\langle \left\{ \prod_{i=1}^{\alpha} \sum_{\Theta_i=1}^{N} z_{\Theta_\alpha} \partial_{\Omega_{\Theta_i}} \Delta\Omega_{\Theta_i}^*(t) \right\} \right\rangle \rho(\mathbf{\Omega}, t) \\
&(C_{\text{basic}} = -1) \,.
\end{aligned} \tag{5.12}$$

$(\Delta\Omega_{\Theta_i}^*(t)$ is assigned to a curve $\Omega_{\Theta_i}(t)$ in such a way that the deterministic part is a function of the position Ω_{Θ_i} and the time t. See subsection 5.1.4. In order to demonstrate the validity of the functional relation (5.11), the reader has to multiply both sides with an ordinary function and integrate both sides with respect to $\mathbf{\Omega}$. Then partial integration will show the validity of the functional relation (5.11). In (5.12) *Dirac's* delta function splits off, because the parts $\Delta\Omega_{\Theta_i}^*(t)$ and *Dirac's* function are statistical independent, i. e. $\delta[\mathbf{\Omega} - \mathbf{\Omega}(t)]$ represents the behavior up to the time t, and $\Delta\Omega_{\Theta_i}^*(t)$ represents the behavior after the time t, with being independent from the behavior before. Moreover, it has to be remarked that the partial differential operators $\partial_{\Omega_{\Theta_i}}$ have as well an effect on the term $\langle \delta[\mathbf{\Omega} - \mathbf{\Omega}(t)] \rangle = \rho(\mathbf{\Omega}, t)$.) The differential power series function (5.12) represents nothing but the general difference (5.5) if small time differences Δt are considered, i. e. this power series function is a general evolution equation to determine the evolution of probability densities $\rho(\Omega, t)$. This equation shall be the basic differential equation of the following considerations.

5.1.3 The Multi-System Equation

In order to extend the validity of the differential power series function (5.12), the coefficient C_{basic} shall be replaced by a coefficient $STRUC$, where $STRUC$ can be equal ± 1 or $\pm$i (i represents the imaginary unit). Then the differential power series function

$$\Delta\Phi(\Omega, t) = \sum_{\alpha=1}^{\infty} (STRUC)^{\alpha} \left\langle \left[\left. \right> \prod_{i=1}^{\alpha} \sum_{\Theta_i=1}^{N} z_{\Theta_\alpha} \partial_{\Omega_{\Theta_i}} \Delta\Omega_{\Theta_i}^{*}(t) \left< \right. \right] \right\rangle \Phi(\Omega, t)$$
$$(STRUC = \pm 1, \pm \text{i})$$
(5.13)

has to be considered. The coefficient $STRUC$ is the *structure coefficient* which determines the structure of the considered differential power series function. This more generalized form contains the derived equation (5.12). Furthermore, (5.13) contains a conjugate form ($STRUC = +1$), which can be derived by using the condition $\rho(\Omega, t + \Delta t) := \rho(\Omega + \Delta\Omega, t)$ instead of the expansion (5.6). A *Taylor* expansion of the function $\rho(\Omega + \Delta\Omega, t)$ afterwards generates then this conjugate form. Moreover, (5.13) contains a complex form ($STRUC = $ i) and a conjugate complex form ($STRUC = -$i). However, solutions of such equations cannot be interpreted as observable probability densities, because such solutions can be complex. If such a complex differential power series function is usable, the solutions need to be *indirect state functions*, i. e. they describe the physical system in an indirect way. The complex differential power series can be taken as an analytical continuation of the basic differential equation (5.12). Therefore, the equation (5.13) is a universal evolution equation, namely a *multi-system equation*. The symbol Φ instead of ρ was used in (5.13), i. e. only in the real cases the distribution function Φ is identical with a measureable probability density.

In the following the multi-system equation (5.13) shall be put in more concrete terms.

5.1.4 The Correlated Langevin Equation

If a time difference Δt is considered, the signal difference is of the form

$$\Delta\Omega_{\Theta_i}(t) = \left[\left\langle \frac{\Omega_{\Theta_i}(t + \Delta t) - \Omega_{\Theta_i}(t)}{\Delta t} \right\rangle_t + F_{\Theta_i}^{(2)}(t) \right] \Delta t$$
$$:= \{ \langle v_{\Theta_i} \rangle_t [\Omega_{\Theta_i}(t)] + F_{\Theta_i}^{(2)}(t) \} \Delta t \, ,$$
(5.14)

with $F_{\Theta_i}^{(2)}(t)$ describing the now occuring fluctuating forces. (5.12) is a differential equation of the *Langevin* type, i. e. an evolution equation of a variable, in which case the evolution is driven by a deterministic force and by a fluctuating force. In (5.13) instead of $\Delta\Omega_{\Theta_i}(t)$ the quantity $\Delta\Omega_{\Theta_i}^*(t)$ arises, i. e. a function with a deterministic part which depends on the position Ω_{Θ_i} and the time t arises. Therefore, the relation

$$\Delta\Omega_{\Theta_i}^*(t) = \left[\langle v_{\Theta_i}\rangle_t(\Omega_{\Theta_i}, t) + F_{\Theta_i}^{(2)}(t)\right]\Delta t \tag{5.15}$$

has to be considered.

5.1.5 The Power Series of the Time Difference

Inserting this *Langevin* equation into the multi-system equation (5.13) one obtains an ordinary power series with respect to the time difference Δt, i. e.

$$\Delta\Phi(\Omega, t) = \sum_{\alpha=1}^{\infty} (STRUC)^\alpha T_\alpha \Phi(\Omega, t)(\Delta t)^\alpha \tag{5.16}$$

holds, with T_α being

$$T_\alpha = \left\langle \left\{ \prod_{i=1}^{\alpha} \sum_{\Theta_i=1}^{N} z_{\Theta_\alpha} \partial_{\Omega_{\Theta_i}} \left[\langle v_{\Theta_i}\rangle_t(\Omega_{\Theta_i}, t) + F_{\Theta_i}^{(2)}(t)\right] \right\} \right\rangle . \tag{5.17}$$

The first two terms of (5.17) show the explicit structure

$$
\begin{aligned}
T_1 &= \sum_{\Theta_1=1}^{N} z_{\Theta_1} \partial_{\Omega_{\Theta_1}} \left\langle \left[\langle v_{\Theta_1}\rangle_t(\Omega_{\Theta_i}, t) + F_{\Theta_1}^{(2)}(t)\right]\right\rangle \\
&= \sum_{\Theta_1=1}^{N} z_{\Theta_1} \partial_{\Omega_{\Theta_1}} \left[\langle v_{\Theta_1}\rangle_t(\Omega_{\Theta_i}, t) + \langle F_{\Theta_1}^{(2)}(t)\rangle\right] ,
\end{aligned} \tag{5.18}
$$

$$
\begin{aligned}
T_2 &= \sum_{\Theta_1=1}^{N}\sum_{\Theta_2=1}^{N} z_{\Theta_1,\Theta_2} \partial_{\Omega_{\Theta_1}} \partial_{\Omega_{\Theta_2}} \left\langle \left[\langle v_{\Theta_1}\rangle_t(\Omega_{\Theta_i}, t)\langle v_{\Theta_2}\rangle_t(\Omega_{\Theta_i}, t) + \right.\right. \\
&\qquad\qquad\qquad\qquad \langle v_{\Theta_1}\rangle_t(\Omega_{\Theta_i}, t)F_{\Theta_2}^{(2)}(t) + \\
&\qquad\qquad\qquad\qquad \langle v_{\Theta_2}\rangle_t(\Omega_{\Theta_i}, t)F_{\Theta_1}^{(2)}(t) + \\
&\qquad\qquad\qquad\qquad \left.\left. F_{\Theta_1}^{(2)}(t)F_{\Theta_2}^{(2)}(t)\right]\right\rangle \\
&= \sum_{\Theta_1=1}^{N}\sum_{\Theta_2=1}^{N} z_{\Theta_1,\Theta_2} \partial_{\Omega_{\Theta_1}} \partial_{\Omega_{\Theta_2}} \left[\langle v_{\Theta_1}\rangle_t(\Omega_{\Theta_i}, t)\langle v_{\Theta_2}\rangle_t(\Omega_{\Theta_i}, t) + \right. \\
&\qquad\qquad\qquad\qquad \langle v_{\Theta_1}\rangle_t(\Omega_{\Theta_i}, t)\langle F_{\Theta_2}^{(2)}(t)\rangle + \\
&\qquad\qquad\qquad\qquad \langle v_{\Theta_2}\rangle_t(\Omega_{\Theta_i}, t)\langle F_{\Theta_1}^{(2)}(t)\rangle + \\
&\qquad\qquad\qquad\qquad \left. \langle F_{\Theta_1}^{(2)}(t)F_{\Theta_2}^{(2)}(t)\rangle\right] .
\end{aligned} \tag{5.19}
$$

($<>$ symbolizes an average with respect to $\Omega_{\Theta_i}(t)$. Therefore, only the fluctuation parts $F_{\Theta_i}^{(2)}(t)$ have to be considered in the context of such an average. The operators $\partial_{\Omega_{\Theta_i}}$ have an effect to the function $\Phi(\Omega, t)$, this has to be taken into account.) In order to put the mean values of the fluctuation forces into concrete terms, the following requirements shall be used.

5.1.6 The Mean Values of the Fluctuation Forces and the Action Factor

The fluctuation forces shall be put into concrete terms by using special requirements.

The Principle Requirements

Firstly, it shall be required that the mean values of the fluctuation forces, the mean values $\left\langle \left[F_{\Theta_1}^{(2)}(t)\right]^{\nu_1} \ldots \left[F_{\Theta_N}^{(2)}(t)\right]^{\nu_N} \right\rangle$, shall only be existent if $\nu_i = 2, 4, 6, \ldots \infty$ holds. Otherwise such mean values shall be equal zero. Mathematically expressed, this means the validity of the relation

$$\left\langle \left[F_{\Theta_1}^{(2)}(t)\right]^{\nu_1} \ldots \left[F_{\Theta_N}^{(2)}(t)\right]^{\nu_N} \right\rangle = \begin{cases} 0 \text{ if } \nu_i \neq 2, 4, 6, \ldots \infty \\ \text{special values if } \nu_i = 2, 4, 6, \ldots \infty \end{cases} \qquad (5.20)$$

Secondly, it shall be required that the various fluctuation forces $F_{\Theta_i}^{(2)}(t)$ have to be statistical independent. Therefore, the mean values $\left\langle \left[F_{\Theta_1}^{(2)}(t)\right]^{\nu_1} \ldots \left[F_{\Theta_N}^{(2)}(t)\right]^{\nu_N} \right\rangle$ decompose into parts which contain only a single fluctuation force, i. e. the relation

$$\left\langle \left[F_{\Theta_1}^{(2)}(t)\right]^{\nu_1} \ldots \left[F_{\Theta_N}^{(2)}(t)\right]^{\nu_N} \right\rangle = \left\langle \left[F_{\Theta_1}^{(2)}(t)\right]^{\nu_1} \right\rangle \ldots \left\langle \left[F_{\Theta_N}^{(2)}(t)\right]^{\nu_N} \right\rangle \qquad (5.21)$$

holds. Now the question arises in which way the existent mean values of the fluctuation forces have to be defined. In the following this question shall be answered.

The Action Factor

Within the power series (5.16) products of mean values of fluctuation forces and powers of the time difference occur, i. e. mean values of the form $\left\langle \left[F_{\Theta_1}^{(2)}(t)\right]^{\nu_\alpha} \right\rangle (\Delta t)^\alpha$ occur. (For example, the reader has to insert (5.19) into the power series (5.16). Then the product $\left\langle \left[F_{\Theta_1}^{(2)}(t)\right]^2 \right\rangle (\Delta t)^2$ occurs.) In order to define the existent mean values of the fluctuating forces, first the relation

$$\frac{SYS}{2} \left\langle \left[F_{\Theta_1}^{(2)}(t)\right]^2 \right\rangle \Delta t = w_{\text{sig}} \qquad (5.22)$$

shall be used, i.e. the existent quadratic correlation functions of the fluctuation forces shall be correlated with a special factor w_{sig}. If the coefficient SYS is choosen in such a way that the l. h. s. of (5.22) shows the dimension of an action (action=energy . time difference), the factor w_{sig} is the action factor of the considered signal. This shall be assumed. The correlation function $\left\langle \left[F_{\Theta_1}^{(2)}(t) \right]^2 \right\rangle$ represents a quantity which is proportional to the fluctuation intensity. Therefore, this correlation function shall be called *the fluctuation intensity*. The second relation shall be of the kind

$$\boxed{\left(\frac{SYS}{2} \right)^{\nu/2} \left\langle \left[F_{\Theta_1}^{(2)}(t) \right]^\nu \right\rangle (\Delta t)^{\nu/2} \approx w_{\text{sig}}^{\nu/2} \ (\nu = 2, 4, 6, \ldots \infty) ,} \tag{5.23}$$

i. e. the action factor shall be a sufficient quantity to define all existent correlation functions of the fluctuation forces. (5.22) at least allows three interpretations which make sense. These three possibilities shall be considered in the following.

Δt is the time difference between the two time points of observation t and $t + \Delta t$, where a special point Ω_{Θ_1} is taken as a basis, i. e. the measurement has to be carried out at the point Ω_{Θ_1}. Therefore, to describe a special physical system, in principle many different time differences Δt can be choosen, where every time difference corresponds to a special point Ω_{Θ_1}. One possible interpretation of the relation (5.22) is that these time differences Δt can be choosen in such a way that the correlated fluctuation intensities $\left\langle \left[F_{\Theta_1}^{(2)}(t) \right]^2 \right\rangle$ fulfill the relation (5.22), i. e. the physical system shows a time and position independent action of the fluctuation activity. An interpretation which is similar to this interpretation is that the r. h. s. of the relation (5.22) can be considered as a mean value, i. e. if one special time difference Δt is taken as a basis, various fluctuation intensities are measurable, in which case a mean value exists in such a way that the relation (5.22) holds. Another possible interpretation is an interpretation in the sense of *Heisenberg's* energy-time relation, i. e. the fluctuation intensity and the observation time difference cannot be measured at the same time in an exact way. However, such an uncertainty relation has to be formulated by using an inequality. Therefore, (5.22) has to be interpreted in such a way that effects of higher order are neglected. These three possibilities make sense. However, in this context very often another interpretation can be found, i. e. in this context the relation

$$\frac{SYS}{2} \left\langle \left[F_{\Theta_1}^{(2)}(t) \right]^2 \right\rangle \Delta t \overset{\Delta t \text{ small enough}}{=} \frac{SYS}{2} \int_t^{t+\Delta t} \left\langle \left[F_{\Theta_1}^{(2)}(t) \right]^2 \right\rangle dt =$$

$$\frac{SYS}{2} \int_t^{t+\Delta t} \left[\left\langle F_{\Theta_1}^{(2)}(t) F_{\Theta_1}^{(2)}(t^*) \right\rangle = w_{\text{sig}} \delta(t - t^*) \right] dt = w_{\text{sig}} \tag{5.24}$$

is used. Such a relation assumes delta correlated fluctuations. However, this means that the measureable fluctuation intensity $\left\langle \left[F_{\Theta_1}^{(2)}(t) \right]^2 \right\rangle$ has to show an infinit value which corresponds to the property of *Dirac's* delta function. Therefore, from a physical point of view this interpretation makes not very much sense, and shall not be assumed in this book.

Using the requirements (5.20)-(5.23) the power series (5.16) can be put into concrete terms. This power series shall be considered in the following.

5.1.7 A Basic Borderline Case

Using the requirements (5.20)-(5.22) the power series (5.16) takes on the form

$$\Delta\Phi(\Omega,t) = \left[\sum_{i=1}^{\infty} IF^{(i)}(\Delta t)^i\right]\Phi(\Omega,t)\,, \tag{5.25}$$

with $IF^{(i)}$ being the operator coefficients of the inner function. However, it has to be remarked here that the now occuring power series is not identical with the power series (5.16), because the requirements (5.22) and (5.23) change the powers of Δt. Then the first operator coefficient is defined by

$$IF^{(1)} = \sum_{\Theta_1=1}^{N}\left[STRUC\partial_{\Omega_{\Theta_1}}\langle v_{\Theta_1}\rangle_t(\Omega_{\Theta_1},t) + (STRUC)^2\frac{w_{\text{sig}}}{SYS}\partial^2_{\Omega_{\Theta_1}}\right]\,, \tag{5.26}$$

where $z_{\Theta_1} = 1$, $z_{\Theta_1,\Theta_1} = 1/2$ was used considered. Operator coefficients of higher order consist of additive terms which correspond to powers $(\Delta t)^l$ $(l > 1)$. Due to the fact that the used time difference Δt is small, such higher terms shall be neglected. In this case one obtains the differential evolution equation

$$\Delta\Phi(\Omega,t) = \sum_{\Theta_1=1}^{N}\left[STRUC\partial_{\Omega_{\Theta_1}}\langle v_{\Theta_1}\rangle_t(\Omega_{\Theta_1},t)\Phi(\Omega,t) + \right.$$
$$\left.(STRUC)^2\frac{w_{\text{sig}}}{SYS}\partial^2_{\Omega_{\Theta_1}}\Phi(\Omega,t)\right]\Delta t\,. \tag{5.27}$$

By using the *nabla*

$$\nabla_N = \partial_{\Omega_1},\partial_{\Omega_2},\ldots,\partial_{\Omega_N} \tag{5.28}$$

and the *Laplacian*

$$\triangle_N = \sum_{\Theta_1=1}^{N}\partial^2_{\Omega_{\Theta_1}}\,, \tag{5.29}$$

and using the notation

$$\langle v\rangle_t(\Omega,t) = \langle v_1\rangle_t(\Omega_1,t),\langle v_2\rangle_t(\Omega_2,t),\ldots,\langle v_N\rangle_t(\Omega_N,t) \tag{5.30}$$

one obtains the equation

$$\Delta\Phi(\Omega, t) = \left\{ STRUC\, \nabla_N \left[\langle v \rangle_t(\Omega, t)\Phi(\Omega, t) \right] + \right.$$
$$\left. (STRUC)^2\, \frac{w_{\mathrm{sig}}}{SYS}\, \triangle_N \Phi(\Omega, t) \right\}\Delta t$$
$$(STRUC = \pm 1, \pm \mathrm{i})\ .$$

$$\tag{5.31}$$

Due to the fact that the elements Δt are small, instead of (5.31) a partial differential equation can be used, i. e. the non-linear partial differential equation

$$\frac{\partial\Phi(\Omega, t)}{\partial t} = \left\{ STRUC\, \nabla_N \left[\langle v \rangle_t(\Omega, t)\Phi(\Omega, t) \right] + \right.$$
$$\left. (STRUC)^2\, \frac{w_{\mathrm{sig}}}{SYS}\, \triangle_N \Phi(\Omega, t) \right\}$$
$$(STRUC = \pm 1, \pm \mathrm{i})$$

$$\tag{5.32}$$

can be used. (5.32) is a basic borderline case of the considered multi-system equation. In order to determine the special physical system, the correlation function of first order, the function $\langle v \rangle_t(\Omega, t)$, has to be put into concrete terms. Such an equation makes sense if an additional constraint of the form (5.22) is used, i. e. if a condition of *Heisenberg's* type is used. The identification

$$\frac{w_{\mathrm{sig}}}{SYS} := \frac{Q}{2}\ ,\ STRUC := -1 \tag{5.33}$$

causes then an equation of the *Fokker-Planck* type. Equations of the *Fokker-Planck* type can be used in various field of physics. For example, in the context of laser theory such an equation allows to find the statistical distribution function of the statistics of the relevant laser modes. This example will be considered in chapter 6.

Between correlation functions of first and second order mathematical relations exist. Thus, it has to be possible to eliminate the correlation function of first order, i. e. the function $\langle v \rangle_t(\Omega, t)$, by using such a relation. Then differential equations occur which can be determined by using correlation functions of second order. In the following this shall be done. However, only the basic borderline case (5.32) shall be considered.

5.2   Differential Equations of the Kinetic Type

Between correlation functions of the type $\langle v^2 \rangle(\Omega, t)$ (correlation functions of second order) and of the type $\langle v \rangle_t(\Omega, t)$ (correlation functions of first order) exists a relation of the kind

$$\langle v^2 \rangle_t(\Omega, t) = \langle v \rangle_t^2(\Omega, t) + \mathrm{CORRECTION}\ . \tag{5.34}$$

Therefore, it has to be possible to replace correlation functions of first order by correlation functions of second order. In a following such a connection shall be considered. Then the basic borderline case (5.32) represents the equation which has to be taken into account.

5.2.1 The Correlation Function Replacement

The stationary case of (5.32) is defined by

$$\frac{\partial \Phi(\Omega, t)}{\partial t} = 0 \; . \tag{5.35}$$

If only one variable is considered, the stationary equation

$$(STRUC)^2 \frac{w_{\text{sig}}}{SYS} \frac{\partial^2}{\partial \Omega_1^{\,2}} \Phi(\Omega_1) = -STRUC \frac{\partial}{\partial \Omega_1} \Big[\langle v_1 \rangle_t (\Omega_1) \Phi(\Omega_1) \Big] \tag{5.36}$$

holds, which is equivalent to the equation

$$-\Phi(\Omega_1) \frac{\partial}{\partial \Omega_1} \left[\frac{1}{STRUC} \frac{SYS}{w_{\text{sig}}} \langle v_1 \rangle_t (\Omega_1) \right] =$$

$$\frac{\partial^2}{\partial \Omega_1^{\,2}} \Phi(\Omega_1) + \left[\frac{1}{STRUC} \frac{SYS}{w_{\text{sig}}} \langle v_1 \rangle_t (\Omega_1) \right] \frac{\partial}{\partial \Omega_1} \Phi(\Omega_1) \; . \tag{5.37}$$

Due to the fact that a function $\Phi(\Omega_1)$ has to fulfill the general rule

$$\Phi(\Omega_1) \frac{\partial}{\partial \Omega_1} \left[\Phi^{-1}(\Omega_1) \frac{\partial}{\partial \Omega_1} \Phi(\Omega_1) \right] =$$

$$\frac{\partial^2}{\partial \Omega_1^{\,2}} \Phi(\Omega_1) - \left[\Phi^{-1}(\Omega_1) \frac{\partial}{\partial \Omega_1} \Phi(\Omega_1) \right] \frac{\partial}{\partial \Omega_1} \Phi(\Omega_1) \; , \tag{5.38}$$

the relation

$$\boxed{\Phi^{-1}(\Omega_1) \left(-STRUC \frac{w_{\text{sig}}}{SYS} \right) \frac{\partial}{\partial \Omega_1} \Phi(\Omega_1) = \langle v_1 \rangle_t (\Omega_1)} \tag{5.39}$$

holds. This relation describes the connection of the statistical level with the mean value level, i. e. the connection between a microscopic and a macroscopic level. Such an equation is of a well-known type. The reader may compare with the basic equation systems of chapter 4. For example, the reader may have a look at (4.4).

A correlation function $\langle v_1 \rangle_t (\Omega_1)$ is connected with a distribution function of the kind $\Phi_{v_1}(\Omega_1, v_1)$, i. e.

$$\langle v_1 \rangle_t (\Omega_1) = \frac{\int \Phi_{v_1}(\Omega_1, v_1) v_1 \, dv_1}{\int \Phi_{v_1}(\Omega_1, v_1) \, dv_1} \tag{5.40}$$

holds. (This integral relation represents a general relation. I have to refer the reader to chapter 4. Such descriptions are needed to describe macroscopic quantities if suitable distribution functions are given.) On the other hand, a function $\Phi(\Omega_1)$ can be calculated by integration, i. e.

$$\Phi(\Omega_1) = \frac{\int \Phi_{v_1}(\Omega_1, v_1)\, dv_1}{\int \Phi_{v_1}(\Omega_1, v_1)\, dv_1 d\Omega_1} \tag{5.41}$$

holds, with $\int \Phi_{v_1}(\Omega_1, v_1)\, dv_1 d\Omega_1$ being the normalization coefficient. Due to the fact that the relation (5.39) holds, an application of the operator $-STRUC\dfrac{w_{\text{sig}}}{SYS}\dfrac{\partial}{\partial\Omega_1}$ to the relation (5.41) has to generate the product $\langle v_1\rangle_t(\Omega_1)\Phi(\Omega_1)$, and due to the fact that simultaneously the integral relation (5.40) holds, such a derivative shows only an additional part v within the integral, without changing the other functional parts. Therefore, higher derivatives have to generate parts v, too. Mathematically expressed, the relation

$$\boxed{\Phi^{-1}(\Omega_1)\left[(STRUC)^2\frac{w_{\text{sig}}^2}{(SYS)^2}\right]\frac{\partial^2}{\partial\Omega_1{}^2}\Phi(\Omega_1) = \langle v_1^2\rangle_t(\Omega_1)} \tag{5.42}$$

holds. If the reader compares with the stationary equation (5.36), it is obvious that (5.42) represents the same stationary equation, however, with a correlation function of second order as a central part. (Multiplication of (5.36) with $\Phi^{-1}(\Omega_1)w_{\text{sig}}/SYS$ leads to a l. h. s. which is identical with the l. h. s. of (5.42).)

If more variables are considered, instead of the operator $\partial^2/\partial\Omega_1{}^2$ the *Laplacian*

$$\triangle_N = \sum_{i=1}^{N}\partial_{\Omega_i}^2 \tag{5.43}$$

has to be used, and instead of one quadratic correlation function $\langle v_1^2\rangle_t(\Omega_1)$ the function

$$\langle v^2\rangle_t(\Omega) = \sum_{i=1}^{N}\langle v_i^2\rangle_t(\Omega) \tag{5.44}$$

has to be considered. In this case a stationary differential evolution equation of the kind

$$\boxed{\left[(STRUC)^2\frac{w_{\text{sig}}^2}{(SYS)^2}\right]\triangle_N\Phi(\Omega) - \langle v^2\rangle_t(\Omega)\Phi(\Omega) = 0} \tag{5.45}$$

holds. However, the derivation of this more-dimensional case shall not be considered.

5.2.2 The Kinetic Differential Equations

If a time-dependence shall be considered, in (5.45) an additional term of the kind $\frac{w_{\text{sig}}}{SYS}\frac{\partial\Phi(\Omega,t)}{\partial t}$ has to occur. (The factor $\frac{w_{\text{sig}}}{SYS}$ is necessary, because the first term of (5.45) diverges from the original term in (5.32). A multiplication with the factor $\frac{w_{\text{sig}}}{SYS}$ guarantees conformity.) In this case the stationary equation (5.45) has to be replaced by

$$\frac{w_{\text{sig}}}{SYS}\frac{\partial\Phi(\Omega,t)}{\partial t} = \left[(STRUC)^2\frac{w_{\text{sig}}^2}{(SYS)^2}\right]\triangle_N\Phi(\Omega,t) - \langle v^2\rangle_t(\Omega,t)\Phi(\Omega,t) .$$

$$(5.46)$$

A multiplication with the factor $SYS/2$ generates then the differential equations

$$\boxed{\begin{array}{l}\dfrac{w_{\text{sig}}}{2}\dfrac{\partial\Phi(\Omega,t)}{\partial t} = \left[(STRUC)^2\dfrac{w_{\text{sig}}^2}{2SYS}\right]\triangle_N\Phi(\Omega,t) - KIN(\Omega,t)\Phi(\Omega,t) \\[2mm] [STRUC = \pm 1, \pm i ; \ (STRUC)^2 = \pm 1] ,\end{array}} \quad (5.47)$$

with $KIN(\Omega,t)$ being a term of the kind

$$KIN(\Omega,t) = \frac{SYS}{2}\langle v^2\rangle_t(\Omega,t) . \tag{5.48}$$

Due to the fact that (5.48) represents a kinetic energy if SYS is identical with a specific mass m_0, the term (5.48) shall be called *kinetic term*. Therefore, (5.47) represents four underlying kinetic differential equations $(STRUC = \pm 1, \pm i)$, in which case only two equations $((STRUC)^2 = \pm 1)$ are necessary to describe these four underlying equations. In this sense (5.32) represents impulse forms. In contrary to the original equations (5.32), quadratic correlation functions occur in (5.47).

In the following a special case of (5.47) shall be considered.

5.2.3 Kinetic Differential Equations with Potentials. Kinetic Differential Equations of the Stationary Schrödinger Type

If $KIN(\Omega,t)$ represents a kinetic energy, the decomposition

$$KIN(\Omega,t) = E - V(\Omega,t) \tag{5.49}$$

is possible, with E being the total energy, and with $V(\Omega,t)$ being the specific potential. (5.49) represents the conservation law of energy. In the following it shall be assumed that such a decomposition is possible. If such a decomposition is used, and if the stationary case

$$\frac{w_{\text{sig}}}{2}\frac{\partial\Phi(\Omega,t)}{\partial t} = 0 \tag{5.50}$$

is considered, the equation

$$\left[(STRUC)^2\,\frac{w_{\text{sig}}^2}{2SYS}\right]\triangle_N\Phi(\Omega)+V(\Omega)\Phi(\Omega)=E\Phi(\Omega)$$

$$[STRUC=\pm1,\pm\text{i}\,;\;\;(STRUC)^2=\pm1]$$

$$(5.51)$$

holds. Such a kind of kinetic equation includes equations of the stationary *Schrödinger* type, i. e. in the case $(STRUC)^2=-1$ the equation (5.51) represents a differential evolution equation of *Schrödinger's* type ($STRUC=\text{i}$) and a differential evolution equation of the conjugate complex *Schrödinger* type ($STRUC=-\text{i}$). In the following only equations of the *Schrödinger* type shall be considered.

The fact that (5.51) includes equations of the *Schrödinger* type can easily be seen, i. e. if the additional identification

$$SYS, w_{\text{sig}}, \Omega \rightarrow m_0, \hbar, \boldsymbol{x} \tag{5.52}$$

is made, one obtains an equation which has the form of a stationary *Schrödinger* equation, i. e.

$$-\frac{\hbar^2}{2m_0}\triangle_3\Phi(\boldsymbol{x})+V(\boldsymbol{x})\Phi(\boldsymbol{x})=E\Phi(\boldsymbol{x})\,, \tag{5.53}$$

and the introduced operator $-STRUC\,w_{\text{sig}}\frac{\partial}{\partial\Omega_i}$ (compare with relation (5.39)) represents an operator which has the form of *Schrödinger's* impulse operator, i. e. the operator $-\text{i}\hbar\frac{\partial}{\partial x_i}$, in which case

$$-\text{i}\hbar\,\frac{\partial}{\partial x_i}\Phi(\boldsymbol{x})=\langle m_0v_i\rangle_i(\boldsymbol{x})\Phi(\boldsymbol{x}) \tag{5.54}$$

holds. Therefore, it is obvious that (5.51) in particular includes equations of *Schrödinger's* type. ($\boldsymbol{x}$ represents the position vector of a particle, x_i represents the components of the position vector. $\hbar$ is *Planck's* constant. Due to the fact that three position coordinates are necessary, the *Laplacian* $\triangle_3$ occurs. The product m_0v_i represents an impulse.) In this context it has to be mentioned that a necessary constraint is the relation of Heisenberg's type (compare with (5.22)). Therefore, if the identification (5.52) is made, the relation

$$\langle E_{\text{fluctuations}}\rangle\Delta t=\hbar \tag{5.55}$$

has to be considered, i. e. if the above given derivation is taken as a basis, an equation of *Schrödinger's* type and a relation of *Heisenberg's* type are mutually conditional ($E_{\text{fluctuations}}$ being the energy of the fluctuations). Moreover, it has to be mentioned that the necessity to use an impulse operator of the kind $-\text{i}\hbar\frac{\partial}{\partial x_i}$ represents nothing but a rule of *Jordan's* type.

Another point which has to be pointed out is that only the stationary case of the kinetic equation (5.47) leads to an equation of *Schrödinger's* type. The time-dependent differential evolution equation (5.47) does not lead to a time-dependent equation of *Schrödinger's* type. However, this is not very surprising, because so far only single statistical systems were considered and the time-dependent *Schrödinger* equation corresponds to ensembles of statistical systems. Therefore, the question arises whether it is possible to derive differential equations which can be used to describe ensembles of such statistical systems. In the following such equations shall be called *ensemble equations*. The way of derivation of such differential equations shall now be described.

5.3 Ensemble Equations

In this section it will be shown that it is possible to gain differential evolution equations which are valid for ensembles of statistical systems. The starting point of this consideration will be the kinetic equation (5.51). In this context it will be shown that these considerations lead straightforward to an equation of the time-dependent *Schrödinger* type. An ensemble equation shall now be derived by using three essential derivation steps.

5.3.1 The First Step: A Quasi-Time-Dependent Kinetic Equation

The values E of (5.51) are eigenvalues which depend on the considered solution so that it makes sense to use an additional index ν. Without any restrictions such eigenvalues can be written in the form

$$E^{(\nu)} = w_{\text{sig}}\omega^{(\nu)} \,, \tag{5.56}$$

with $\omega^{(\nu)}$ being a free parameter which shows the dimension of a frequency if $E^{(\nu)}$ represents the total energy. In this case the formulation

$$\left[(STRUC)^2 \frac{w_{\text{sig}}^2}{2SYS}\right] \triangle_N \Phi_t^{(\nu)}(\Omega) + V(\Omega)\Phi_t^{(\nu)}(\Omega) = STRUC\, w_{\text{sig}} \frac{\partial \Phi_t^{(\nu)}(\Omega)}{\partial t} \tag{5.57}$$

is possible, with (5.57) being equivalent to the kinetic equation (5.51), because

$$\Phi_t^{(\nu)}(\Omega) = \Phi^{(\nu)}(\Omega)\exp\left(INV\omega^{(\nu)}t\right) \tag{5.58}$$

holds. In this context non-degenerate sets of functions shall be considered, i. e. every eigenvalue has only one corresponding solution. Furthermore, the relation

$$STRUC = +1 \leftrightarrow INV = +1 \, ,$$
$$STRUC = -1 \leftrightarrow INV = -1 \, ,$$
$$STRUC = +\mathrm{i} \leftrightarrow INV = -\mathrm{i} \, ,$$
$$STRUC = -\mathrm{i} \leftrightarrow INV = +\mathrm{i} \tag{5.59}$$

has to be considered. Due to the fact that (5.57) is equivalent to the stationary equation (5.51), it makes sense to name (5.57) a *quasi-time-dependent kinetic equation*. Such an equation is valid for one single statistical system. In order to find an equation which is valid for an ensemble, this quasi-time-dependent equation can be taken as a basis.

5.3.2   The Second Step: Ensemble Functions

Not only (5.58) fulfills the differential equation (5.57), every linear combination of functions of the kind (5.58) fulfills (5.57) if the coefficients of such a linear combination are time-independent. If time-dependent coefficients are used, i. e. if a linear combination

$$\boxed{\Psi(\Omega, t) = \sum_{\nu=1}^{\nu_{\max}} c^{(\nu)}(t)\Phi_t^{(\nu)}(\Omega)} \tag{5.60}$$

is used, an additional term of the kind

$$STRUC\, w_{\mathrm{sig}} \sum_{\nu=1}^{\nu_{\max}} \Phi_t^{(\nu)}(\Omega)\frac{\partial c^{(\nu)}(t)}{\partial t} = INT(\Omega, t) \sum_{\nu=1}^{\nu_{\max}} c^{(\nu)}(t)\Phi_t^{(\nu)}(\Omega)$$
$$= INT(\Omega, t)\Psi(\Omega, t) \tag{5.61}$$

occurs on the r. h. s. of (5.57), with $INT(\Omega, t)$ being a function which guarantees the equivalence of the l. h. s. and the r. h. s. of (5.61). The functions $\Phi_t^{(\nu)}(\Omega)$ are state functions which correspond to special eigenvalues $E^{(\nu)}$. If the set of state functions $\Phi_t^{(\nu)}(\Omega)$ describes all observable (i. e. stationary) states (for example, this is possible if transition states exist only a relatively short time), and if the coefficients $c^{(\nu)}(t)$ are choosen in such a way that they represent the probabilities of the observable states at least in an implicit way, the linear combination $\Psi(\Omega, t)$ represents an ensemble of statistical systems. Due to the fact that a function $INT(\Omega, t)$ only is unequal zero if the coefficients of an ensemble function are time-dependent (i. e. non-equilibrium interactions between the systems of the ensemble are observable), the term $INT(\Omega, t)$ has to be an interaction term which is unequal zero if non-equilibrium interactions exist. In last consequence $INT(\Omega, t)$ determines the coefficients $c^{(\nu)}(t)$. Due to the fact that $INT(\Omega, t)$ *only* determines the coefficients $c^{(\nu)}(t)$ and does not change the eigenstates $\Phi_t^{(\nu)}(\Omega)$, the interaction has to be weak, i. e. $INT(\Omega, t)$ has to represent a *perturbation*. In the real cases ($STRUC = \pm 1$) the coefficients are real quantities, and in the complex cases ($STRUC = \pm \mathrm{i}$) these coefficients are normally complex. In both cases the linear combination (5.60) represents a function which describes a measurable ensemble. Therefore, it makes sense to name the function (5.60) an *ensemble function*. (5.61) represents a special kind of master equation. This equation will be considered later (see 5.3.4). In which way the coefficients $c^{(\nu)}(t)$ describe the probabilitiy of the existence of a state ν shall be discussed later, i. e. in subsection 5.4.3 the solution of the master equation (5.61) and some essential properties of the coefficients $c^{(\nu)}(t)$ shall be discussed, and in chapter 7, subsection 7.4.2 the physical meaning shall be considered.

5.3.3   The Third Step: Ensemble Equations

As the reader can see above, a linear combination with time-dependent coefficients fulfills an equation of the kind (5.57) if an additional term of the kind $INT(\Omega, t)\Psi(\Omega, t)$ is used. If such a term is taken into account, one obtains an equation of the form

$$
\left[(STRUC)^2 \frac{w_{\mathrm{sig}}^2}{2SYS} \right] \triangle_N \Psi(\Omega, t) + V(\Omega)\Psi(\Omega, t) + INT(\Omega, t)\Psi(\Omega, t) =
$$
$$
STRUC\, w_{\mathrm{sig}} \frac{\partial \Psi(\Omega, t)}{\partial t} \ .
$$

$$(5.62)$$

The partial differential equation (5.62) represents *ensemble equations*, in which case the equation (5.57) represents a borderline case which holds if non-equilibrium interactions can be neglected. Additionally, (5.62) includes the kinetic equation (5.51), i. e. (5.62) includes the borderline case for one single system.

The evolution of the coefficients $c^{(\nu)}(t)$ is given by an equation of the form (5.61). This equation shall now be considered.

5.3.4   Master Equations

(5.61) determines the evolution of the coefficients $c^{(\nu)}(t)$. From a physical point of view such an equation represents a balance equation which describes the balance of coefficients. Very often such balance equations are called *master equations*. Multiplying the master equation (5.61) with both a special conjugate complex function $\Phi_t^{*(\mu)}(\Omega)$ and a specific factor $\left(STRUC\, w_{\mathrm{sig}} \right)^{-1}$, and integrating over all values Ω, one obtains a more familiar form of such a master equation, i. e. one obtains the equation

$$\frac{\partial c^{(\mu)}(t)}{\partial t} =$$
$$\frac{1}{STRUCw_{\text{sig}}} \sum_{\nu=1}^{\nu_{\text{max}}} c^{(\nu)}(t) M_{\mu,\nu} \exp\left[\left(INV\omega^{(\nu)} - INV\omega^{(\mu)}\right)t\right] , \tag{5.63}$$

with $M_{\mu,\nu}$ representing matrix elements which are defined by

$$M_{\mu,\nu} = \int_{-\infty}^{+\infty} \Phi^{*(\mu)}(\Omega) INT(\Omega, t) \Phi^{(\nu)}(\Omega)\, d\Omega . \tag{5.64}$$

(Due to the fact that both real and complex cases are considered, it has to be mentioned that if the functions $\Phi^{(\mu)}(\Omega)$ are real functions, the conjugate complex functions $\Phi^{*(\mu)}(\Omega)$ are identical with the original functions $\Phi^{(\mu)}(\Omega)$.) $M_{\mu,\nu\neq\mu}$ has to represent a measure of the transition probability from the state ν into the state μ, and $M_{\mu,\mu}$ has to represent a measure of the transition probability from the state μ into any other state. In this context it has to be assumed that the set of functions $\Phi^{(\nu)}(\Omega)$ consists of orthonormal functions, i. e. the orthonormality relation

$$\int_{-\infty}^{+\infty} \Phi^{*(\mu)}(\Omega)\Phi^{(\nu)}(\Omega)\, d\Omega = \delta_{\mu,\nu} \tag{5.65}$$

holds, with $\delta_{\mu,\nu}$ being *Kronecker's* delta. (5.63) represents an explicit form of the master equations (5.61). In the real cases $STRUC = \pm 1$, $INV = \pm 1$ has to be considered, and in the complex cases $STRUC = \pm i$; $INV = \mp i$ has to be used.

In order to show the validity of the relation (5.65) it has to be considered that the functions $\Phi^{(\nu)}(\Omega)$ fulfill the equation

$$\left[(STRUC)^2 \frac{w_{\text{sig}}^2}{2SYS}\right]\triangle_N \Phi^{(\nu)}(\Omega) + V(\Omega)\Phi^{(\nu)}(\Omega) = E^{(\nu)}\Phi^{(\nu)}(\Omega) \tag{5.66}$$

(compare with (5.57) and (5.58)). Moreover, the integral relation

$$\int_{-\infty}^{+\infty} \Phi^{*(\mu)}(\Omega)\hat{H}\Phi^{(\nu)}(\Omega)\, d\Omega = \int_{-\infty}^{+\infty} \Phi^{(\nu)}(\Omega)\left[\hat{H}\Phi^{(\mu)}(\Omega)\right]^* d\Omega \tag{5.67}$$

has to be considered, which shows that the operator

$$\hat{H} = \left[(STRUC)^2 \frac{w_{\text{sig}}^2}{2SYS}\right]\triangle_N + V(\Omega) \tag{5.68}$$

is a self-adjoint operator. (The relation (5.67) can be calculated by using the method of partial integration.) Using (5.67) and (5.66) one obtains the relation

$$\left(E^{(\mu)} - E^{(\nu)}\right)\int_{-\infty}^{+\infty} \Phi^{*(\mu)}(\Omega)\Phi^{(\nu)}(\Omega)\, d\Omega = 0 , \tag{5.69}$$

which leads to the relation

$$\int_{-\infty}^{+\infty} \Phi^{*(\mu)}(\Omega)\Phi^{(\nu)}(\Omega)\, d\Omega = 0 \tag{5.70}$$

if the eigenvalues $E^{(\nu)}$ and $E^{(\mu)}$ are different, i. e. if non-degenerate sets of functions are considered. A suitable normalization of the functions $\Phi^{(\nu)}(\Omega)$ guarantees then the validity of the orthonormality relation (5.65).

The expression (5.62) includes an ensemble equation which bases on an equation of the stationary *Fokker-Planck* type and an ensemble equation which bases on an equation of the stationary *Schrödinger* type. These two special cases shall now be considered.

5.3.5 Ensemble Equations of the Fokker-Planck- and of the Schrödinger Type

Using the relation (5.33) one obtains the ensemble equation

$$\frac{SYS}{2}\frac{Q^2}{4}\triangle_N\Psi(\Omega,t) + V(\Omega)\rho(\Omega,t) + INT(\Omega,t)\Psi(\Omega,t) =$$
$$-SYS\frac{Q}{2}\frac{\partial\Psi(\Omega,t)}{\partial t} \tag{5.71}$$

instead of the general form (5.62). Due to the fact that (5.71) bases on an equation of the stationary *Fokker-Planck* type (see the stationary case (5.32) with the additional constraint (5.33)), (5.71) represents an ensemble equation of the *Fokker-Planck* type. In contrast to (5.71) the equation

$$-\frac{\hbar^2}{2m_0}\triangle_3\Psi(\boldsymbol{x},t) + V(\boldsymbol{x})\Psi(\boldsymbol{x},t) + INT(\boldsymbol{x},t)\Psi(\boldsymbol{x},t) = i\hbar\frac{\partial\Psi(\boldsymbol{x},t)}{\partial t} \tag{5.72}$$

represents an ensemble equation of the *Schrödinger* type, because this ensemble equation bases on an equation of the stationary *Schrödinger* type (see (5.51), the case $STRUC = i$, with the additional identification (5.52)). (5.71) and (5.72), respectively, are special cases of the general ensemble equation (5.62). The reader who is familiar with quantum theoretical equations can immediately see that the special case (5.72) represents an equation of the time-dependent *Schrödinger* type, i. e. only the transition to an ensemble equation leads to an equation of the time-dependent *Schrödinger* type.

So far four levels of differential evolution equations were considered. In order to discuss the meaning of the introduced equations, and to discuss the possibilities of application, it makes sense to consider solutions of such evolution equations. Now this shall be done.

5.4 Solutions

In this section some basic solutions of the introduced evolution equations shall be considered. However, it shall not be examined if the system functions $\langle v \rangle_t$, $\langle v^2 \rangle_t$ can be interpreted as macroscopic functions. This is possible, because the above considerations are still valid if the system functions are no macroscopic functions. This is an important fact and has to be borne in mind.

5.4.1 Solutions of Impulse Forms and the MIEP

Equations of the impulse type are equations which are defined by (5.32). In particular, this expression includes an equation of the *Fokker-Planck* type, i. e. the stationary equation of the *Fokker-Planck* type is defined by

$$-\nabla_N \left[\langle v \rangle_t (\Omega) \Phi_{\text{FPE}}^{(\text{IF})}(\Omega) \right] + \frac{Q}{2} \triangle_N \Phi_{\text{FPE}}^{(\text{IF})}(\Omega) = 0 \tag{5.73}$$

if the identification $w_{\text{sig}}/SYS = Q/2$ is made. Such a non-linear differential equation can be solved by using a function of the kind

$$\Phi_{\text{FPE}}^{(\text{IF})}(\Omega) = Z_{\text{FPE}}^{-1} \exp[CP_{\text{FPE}}(\Omega)] \,, \tag{5.74}$$

with Z_{FPE} being the partition function, and with $CP_{\text{FPE}}(\Omega)$ being a function which shall be called the *central potential*. The central potential is defined by

$$\frac{Q}{2} \frac{\partial}{\partial \Omega_i} CP_{\text{FPE}}(\Omega) = \langle v_i \rangle_t (\Omega_i) \,, \tag{5.75}$$

in which case very often the decomposition $CP_{\text{FPE}}(\Omega) = -\frac{2}{Q} U(\Omega)$ is used. (In order to show the validity of the exponential solution, insert (5.74) into (5.73) and use (5.75).)

Another special case of (5.32) is represented by the complex equation

$$i\nabla_3 \left[\langle v \rangle_t (x) \Phi_{\text{SE}}^{(\text{IF})}(x) \right] - \frac{\hbar}{m_0} \triangle_3 \Phi_{\text{SE}}^{(\text{IF})}(x) = 0 \,. \tag{5.76}$$

This expression can be considered as an impulse form of an equation of the stationary *Schrödinger* type. Then stationary solutions of the kind

$$\Phi_{\text{SE}}^{(\text{IF})}(x) = Z_{\text{SE}}^{-1} \exp[CP_{\text{SE}}(x)] \tag{5.77}$$

are possible, and the definition

$$-i \frac{\hbar}{m_0} \frac{\partial}{\partial x_k} CP_{\text{SE}}(x) = \langle v_k \rangle_t (x_k) \tag{5.78}$$

has to be used, where Z_{SE}^{-1} represents any constant. (Insert (5.77) and (5.78) into (5.76).)

In chapter 3 the maximum information entropy principle (MIEP) was discussed. As it was discussed, if correlation functions are taken as constraints, one obtains exponential functions of a special order by using the MIEP. Therefore, it has to be pointed out that functions which can be determined by using a macroscopic procedure as well can be found by using equations of the impulse type. However, not only exponential functions are solutions of equations of the impulse type. For example, the reader may consider a one-dimensional problem and may use as a formulation of $\Phi(\Omega)$ a non-exponential power series and may insert this power series into the equation (5.73). Then the occuring macroscopic function $\langle v_1 \rangle_t(\Omega_1)$ can be calculated. However, such a macroscopic function will be a power series, too.

Exponential functions and power series can be solutions of equations of the kinetic type. Solutions of such equations shall now be considered. In this context a short discussion of the used parabolic cylinder functions shall be given (for more information, see [2]).

5.4.2 Solutions of Kinetic Equations and Parabolic Cylinder Functions

Kinetic equations are defined by (5.47) or (5.51), respectively. (5.51) includes kinetic equations of the *Fokker-Planck* type as well as equations of the stationary *Schrödinger* type. These two types of equations shall be considered now. For sake of simplicity, only one-dimensional problems shall be discussed, i. e. only one variable Ω has to be used.

Mathematical Properties

Using the potential of a harmonic oscillator, i. e. using

$$V(\Omega) := \frac{SYS\omega^2}{2}\Omega^2 \,, \tag{5.79}$$

one obtains the two kinetic equations

$$\left(\frac{w_{\text{sig}}^2}{2SYS} \frac{\partial^2}{\partial\Omega^2} + \frac{SYS\omega^2}{2}\Omega^2 \right) \Phi_{\text{FPE}}^{(\text{KF},\nu)}(\Omega) = E^{(\nu)} \Phi_{\text{FPE}}^{(\text{KF},\nu)}(\Omega) \,, \tag{5.80}$$

$$\left(-\frac{w_{\text{sig}}^2}{2SYS} \frac{\partial^2}{\partial\Omega^2} + \frac{SYS\omega^2}{2}\Omega^2 \right) \Phi_{\text{SE}}^{(\text{KF},\nu)}(\Omega) = E^{(\nu)} \Phi_{\text{SE}}^{(\text{KF},\nu)}(\Omega) \,, \tag{5.81}$$

where (5.80) represents an equation of the *Fokker-Planck* type and (5.81) represents an equation of the *Schrödinger* type. Using the abbreviations

$$\Xi = \frac{\Omega}{\sqrt{w_{\text{sig}}/2\omega SYS}} \,, \quad a^{(\nu)} = \frac{E^{(\nu)}}{w_{\text{sig}}\omega} := \nu + \frac{1}{2} \tag{5.82}$$

and

$$\Phi_{\text{FPE}}^{(\text{KF},\nu)}(\Xi) = \Phi^{(1,\nu)}(\Xi), \quad \Phi_{\text{SE}}^{(\text{KF},\nu)}(\Xi) = \Phi^{(2,\nu)}(\Xi) \tag{5.83}$$

a formulation is possible which does not show any dimension, i. e. the dimensionless formulation

$$\left(\frac{\partial^2}{\partial \Xi^2} + \frac{1}{4}\Xi^2 \right) \Phi^{(1,\nu)}(\Xi) = a^{(\nu)} \Phi^{(1,\nu)}(\Xi) , \tag{5.84}$$

$$\left(-\frac{\partial^2}{\partial \Xi^2} + \frac{1}{4}\Xi^2 \right) \Phi^{(2,\nu)}(\Xi) = a^{(\nu)} \Phi^{(2,\nu)}(\Xi) \tag{5.85}$$

is possible. Permitting also negative values $a^{(\nu)}$ the system of equations (5.84)-(5.85) can be replaced by

$$\left(\frac{\partial^2}{\partial \Xi^2} + \frac{1}{4}\Xi^2 \right) \Phi^{(1,\nu)}(\Xi) = a^{(\nu)} \Phi^{(1,\nu)}(\Xi) , \tag{5.86}$$

$$\left(-\frac{\partial^2}{\partial \Xi^2} + \frac{1}{4}\Xi^2 \right) \Phi^{(3,\nu)}(\Xi) = -a^{(\nu)} \Phi^{(3,\nu)}(\Xi) , \tag{5.87}$$

with $\Phi^{(3,\nu)}(\Xi)$ being the now relevant solutions. (In the case (5.87), $-a^{(\nu)} = \frac{E^{(\nu)}}{w_{\text{sig}}\omega} :=$ $\nu + \frac{1}{2}$ holds!) Solutions of the equation system (5.86)-(5.87) are normally called *parabolic cylinder functions*, in which case such parabolic cylinder functions are defined by the linear combinations

$$\Phi^{(1,\nu)}(\Xi) = A_{1,1} \Phi_-^{(\text{even},\nu)}(\Xi) + A_{1,2} \Phi_-^{(\text{odd},\nu)}(\Xi) , \tag{5.88}$$

$$\Phi^{(3,\nu)}(\Xi) = A_{3,1} \Phi_+^{(\text{even},\nu)}(\Xi) + A_{3,2} \Phi_+^{(\text{odd},\nu)}(\Xi) . \tag{5.89}$$

The coefficients $A_{1,1}$, $A_{1,2}$, $A_{3,1}$, $A_{3,2}$ can be any coefficients, and the functions $\Phi_\mp^{(\text{even},\nu)}(\Xi)$, $\Phi_\mp^{(\text{odd},\nu)}(\Xi)$ are power series which are defined by

$$\Phi_\mp^{(\text{even},\nu)}(\Xi) = \sum_{n=0}^{\infty} b_{2n} \frac{1}{(2n)!} \Xi^{2n} , \tag{5.90}$$

$$\Phi_\mp^{(\text{odd},\nu)}(\Xi) = \sum_{n=0}^{\infty} b_{2n+1} \frac{1}{(2n+1)!} \Xi^{2n+1} . \tag{5.91}$$

In this context the recursive definition

$$b_{n+2} = a^{(\nu)} b_n \mp \frac{n(n-1)}{4} b_{n-2} \tag{5.92}$$

has to be taken into account. (Inserting this formulation into (5.86)-(5.87) the validity of this formulation can be shown.) Such recursive definitions of coefficients of power series are used very often in this book. I have to refer the reader to chapter 4, i. e. the coefficients of the introduced hyper-surface equations are defined by recursive formulae. However, here only ordinary power series are considered so that more easily recursive definitions are needed. Furthermore, in chapter 4 parabolic cylinder functions of the kind $D_{-1/2}(\Xi)$ were needed. Therefore, it has to be remarked that such parabolic cylinder functions are defined by the relation

$$D_{-a^{(\nu)}-1/2}(\Xi) = D_{3,1}\Phi_+^{(\text{even},\nu)}(\Xi) + D_{3,2}\Phi_+^{(\text{odd},\nu)}(\Xi) \tag{5.93}$$

with the coefficients

$$D_{3,1} = \cos\left[\pi\left(\frac{1}{4}+\frac{1}{2}a^{(\nu)}\right)\right]\frac{1}{2^{\frac{a^{(\nu)}}{2}+\frac{1}{4}}\sqrt{\pi}}\Gamma\left(\frac{1}{4}-\frac{1}{2}a^{(\nu)}\right),$$

$$D_{3,2} = -\sin\left[\pi\left(\frac{1}{4}+\frac{1}{2}a^{(\nu)}\right)\right]\frac{1}{2^{\frac{a^{(\nu)}}{2}-\frac{1}{4}}\sqrt{\pi}}\Gamma\left(\frac{3}{4}-\frac{1}{2}a^{(\nu)}\right), \tag{5.94}$$

where $\Gamma(\text{argument})$ represents *Euler's* gamma function. (5.93) is a special case of (5.89). These coefficients guarantee the alternating occurrence of the even part $\Phi_+^{(\text{even},\nu)}(\Xi)$ and the odd part $\Phi_+^{(\text{odd},\nu)}(\Xi)$. For example, in the case $\nu = 0$ only the even part is unequal zero and in the case $\nu = 1$ only the odd part is unequal zero.

In the following the solutions $\Phi_\mp^{(\text{even},\nu)}(\Xi)$, $\Phi_\mp^{(\text{odd},\nu)}(\Xi)$ shall be illustrated. A picture of the function $D_{-1/2}(\Xi) = D_{-1/2}(\lambda_1/\sqrt{2\lambda_2})$ the reader can find in chapter 4 (see figure 4.7). Pictures of exponential functions the reader can find in chapter 4 (see figures (4.1)-(4.3)).

Figures

The even solutions of the equation of *Schrödinger's* type, the solutions $\Phi_+^{(\text{even},\nu)}(\Xi)$, are illustrated in figure 5.1. The odd solutions of the equation of the *Schrödinger* type, the solutions $\Phi_+^{(\text{odd},\nu)}(\Xi)$, are illustrated in figure 5.2. Figure 5.3 shows the even solutions $\Phi_-^{(\text{even},\nu)}(\Xi)$ of the equation of the *Fokker-Planck* type, and figure 5.4 shows the combined odd solutions $\Phi_-^{(\text{odd},\nu)}(\Xi)$. As the reader can see, the functions $\Phi_+^{(\text{even},\nu)}(\Xi)$ are of a parabolic kind. The functions which correspond to the values $\nu = 0, 2$ are non-divergent functions, in which case an increase of the values ν causes an increase of the number of the extreme points. The same behavior would be observable for higher even natural numbers $\nu = 4, 6, \ldots \infty$. The functions between such natural numbers are divergent functions. Below the point $\nu = 0$ only parabolic functions occur which tend towards $+\infty$. All functions are symmetrical with respect to zero. The functions $\Phi_+^{(\text{odd},\nu)}(\Xi)$ are of a parabolic kind, too. However, these functions are anti-symmetrical functions. The function which corresponds to the value $\nu = 1$ is a non-divergent function. The same behavior would be observable for higher odd natural numbers $\nu = 3, 5, \ldots \infty$, in which case an increase of the values ν causes an increase of the number of the extreme points. The functions between such natural numbers are divergent functions, too. Below the point $\nu = 0$ only anti-symmetrical parabolic functions occur which tend towards $\pm\infty$. The functions of the figures 5.3, 5.4 are as well of a parabolic kind, however, with an overlapping oscillatory behavior. As the reader can see, an increase of the values ν causes a continuous change of the functions, i. e. no emphasized points are observable. The even solutions show a kind of bifurcation, i. e. an extreme point splits into two

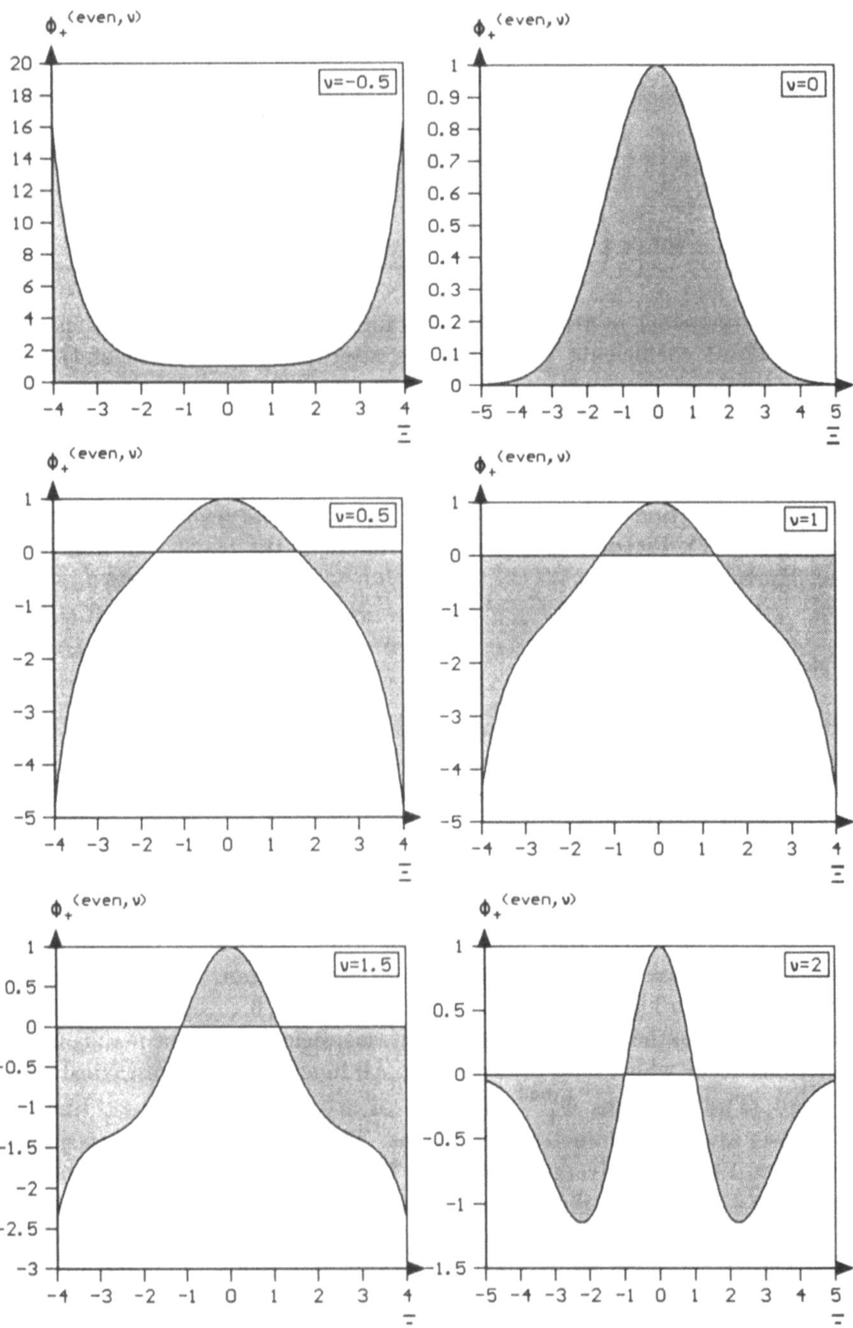

Figure 5.1 The even functions $\Phi_+^{(\text{even},\nu)}(\Xi)$

points. The functions $\Phi_+^{(\text{even},\nu)}(\Xi)$, $\Phi_+^{(\text{odd},\nu)}(\Xi)$ which correspond to natural numbers

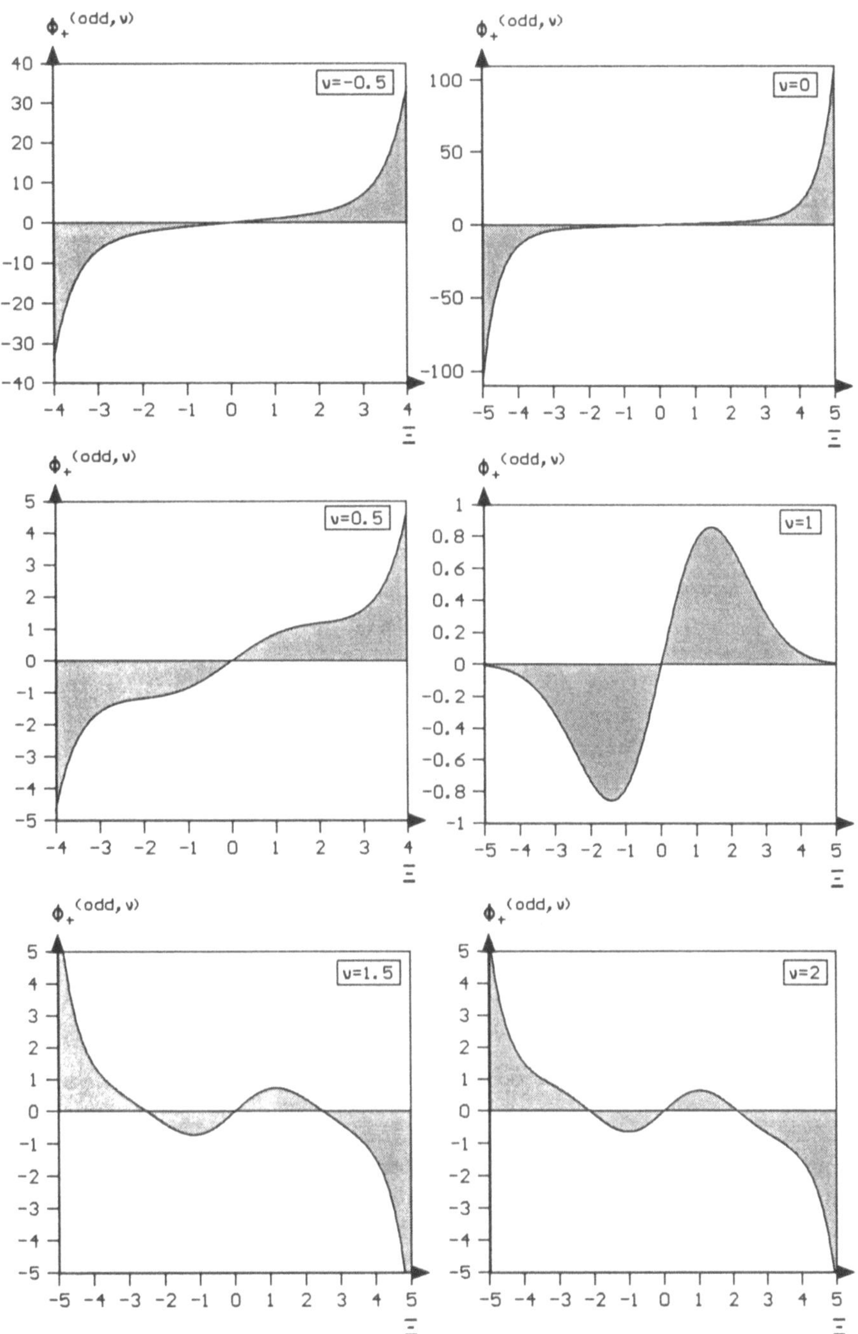

Figure 5.2   The odd functions $\Phi_+^{(odd,\nu)}(\Xi)$

$\nu := n = 0, 1, 2, 3, 4, \ldots \infty$ are normally well-known, i. e. these are the solutions of

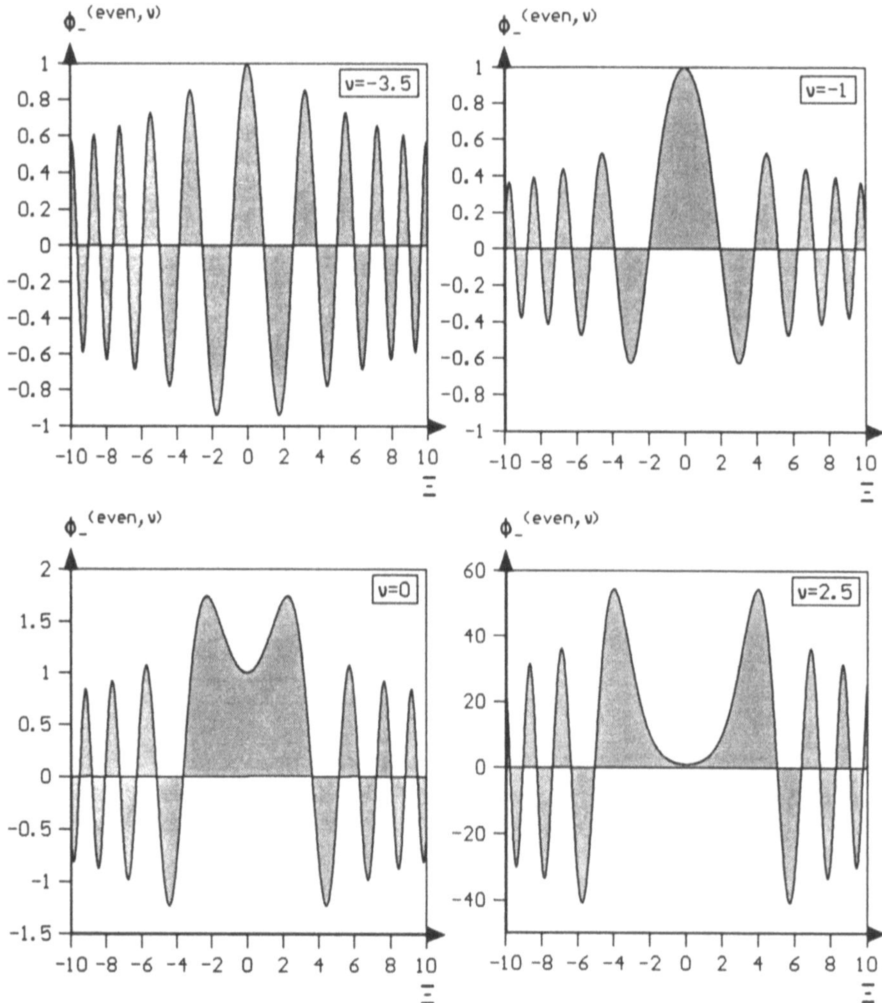

Figure 5.3   The even functions $\Phi_-^{(\text{even},\nu)}(\Xi)$

the quantum mechanical harmonic oscillator problem, however, without a suitable normalization factor. A uniform formulation of such solutions is possible by using the representation (5.93), i. e. introducing a normalization factor N_n these solutions can be written in the form (for more information, see [37])

$$\Phi^{(\text{OSZI},\nu:=n)}(\Xi) = N_n D_n(\Xi) \, . \tag{5.95}$$

However, in quantum theory the identification

$$\Xi := x\sqrt{2\epsilon} \, , \quad \epsilon = \frac{m_0\omega}{\hbar} \, . \tag{5.96}$$

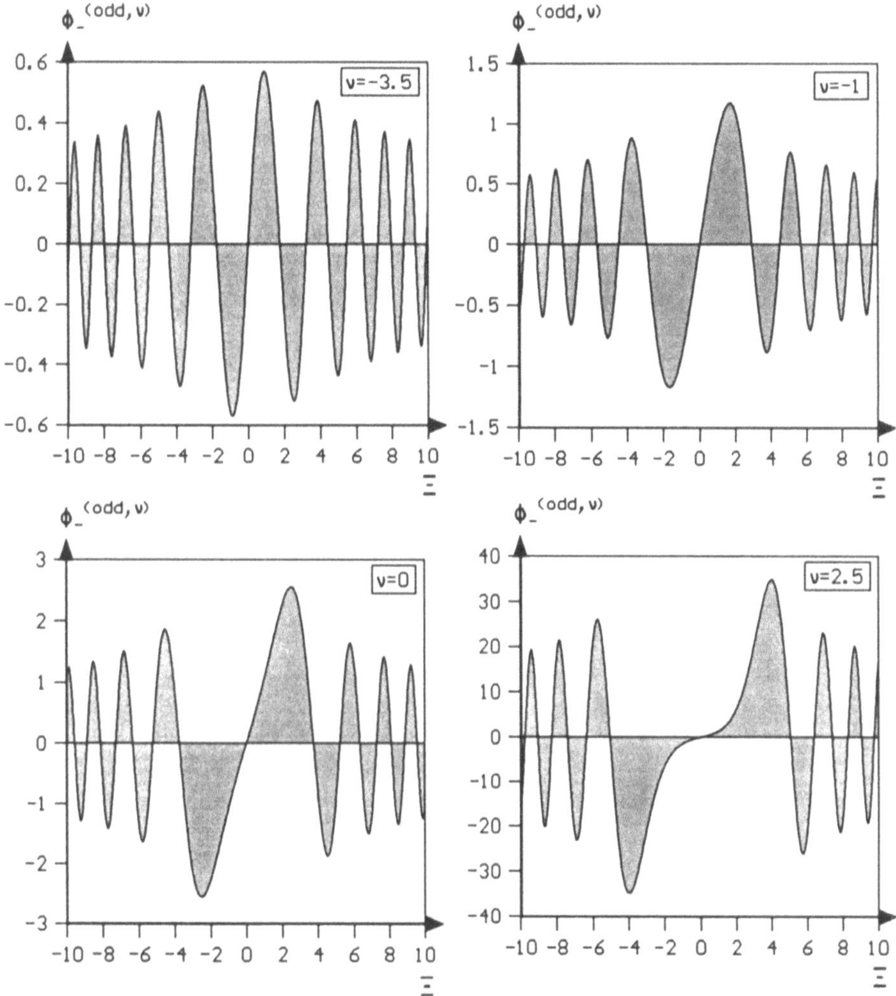

Figure 5.4 The odd functions $\Phi_-^{(\mathrm{odd},\nu)}(\Xi)$

has to be made. Therefore, a uniform formulation of the kind

$$\Phi^{(\mathrm{OSZI},\nu:=n)}(x) = \epsilon^{1/4} \exp\left(-\frac{\epsilon}{2}x^2\right) H_n(x\sqrt{\epsilon}) \tag{5.97}$$

is possible. $H_n(x\sqrt{\epsilon})$ represents *Hermite* polynomials, which are defined by

$$H_n(\xi) = (-1)^n \frac{1}{\sqrt{2^n n!}\sqrt{\sqrt{\pi}}} \exp\left(\xi^2\right) \frac{\partial^n \exp\left(-\xi^2\right)}{\partial \xi^n} . \tag{5.98}$$

Figure 5.5 shows some of these solutions. The increase of the extreme points can clearly be seen. However, it has to be remarked that a decomposition of parabolic

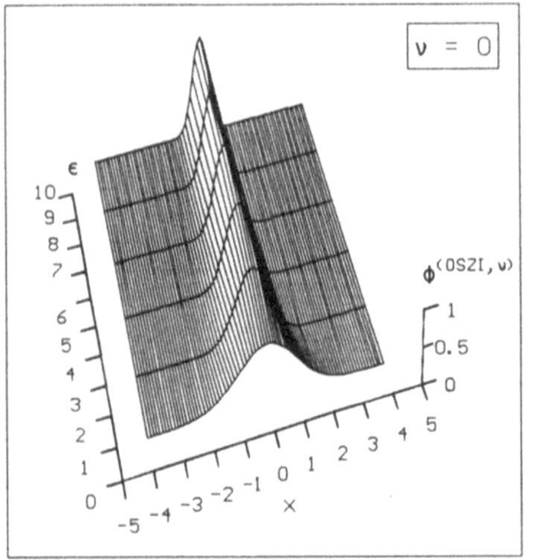

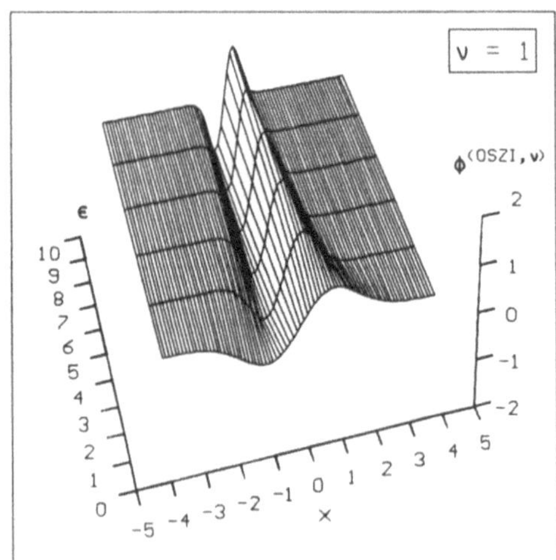

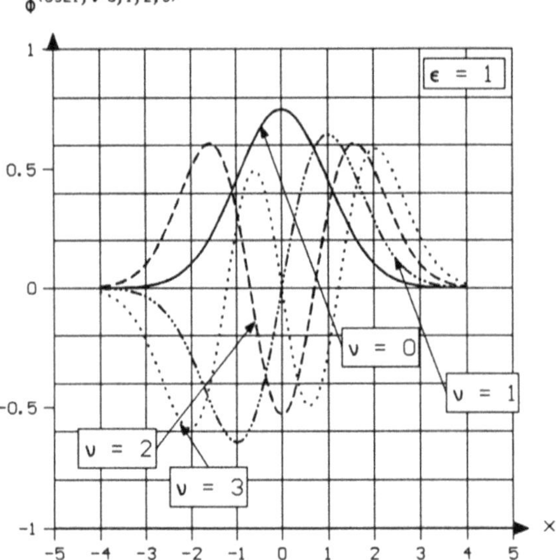

Figure 5.5
The solutions of the quantum mechanical harmonic oscillator problem

cylinder functions into *Hermite* polynomials and exponential functions only is possible if $\nu := 0, 1, 2, 3, 4, \ldots \infty$ holds.

Therefore, some solutions of the equations of the kinetic type were illustrated. In the following some interrelations between solutions of equations of the impulse type and solutions of the equations of the kinetic type shall be considered.

Interrelations Between Solutions of Equations of the Impulse Form and Solutions of Kinetic Equations

If the reader compares with the derivation of kinetic equations, it is obvious that an interrelation between solutions of the kinetic equations and solutions of the equations of the impulse form exists, i. e. solutions of one type of equations can be used to solve the other type of equation if suitable system functions are used. For example, if the exponential solution (5.74) is identical with a *Gaussian* function

$$\Phi_G^{(IF)}(\Omega) = Z_G^{-1} \exp\left(-\frac{2}{Q}\Omega^2\right), \tag{5.99}$$

the corresponding impulse form is a special case of (5.73), namely the equation

$$-\frac{\partial}{\partial\Omega}\left[-2\Omega\Phi_G^{(IF)}(\Omega)\right] + \frac{Q}{2}\frac{\partial^2}{\partial\Omega^2}\Phi_G^{(IF)}(\Omega) = 0 \tag{5.100}$$

holds. The same solution can be found if a suitable kinetic equation is used, i. e. if the equation

$$\frac{\partial^2}{\partial\Omega^2}\Phi_G^{(IF)}(\Omega) - \frac{2}{Q}\left(2 - \frac{8}{Q}\Omega^2\right)\Phi_G^{(IF)}(\Omega) = 0 \tag{5.101}$$

is used. (5.101) represents a special stationary case of the kinetic equation (5.47). Power series such as the above discussed parabolic cylinder functions as well can be used to solve equations of the impulse type. However, then system functions are needed which are power series themselves.

In the following solutions of the master equation (5.63) shall be considered.

5.4.3   The General Solution of the Master Equation

The master equation (5.63) can be solved. This shall now be done.

The General Solution

Integrating the master equation (5.63) with respect to the time t, one obtains the general solution

$$c^{(\mu)}(t) = c^{(\mu)}(0) +$$

$$\frac{1}{STRUCw_{\text{sig}}} \sum_{\nu=1}^{\nu_{\text{max}}} \int_0^t c^{(\nu)}(\tau) M_{\mu,\nu} \exp\left[-\left(INV\omega^{(\mu)} - INV\omega^{(\nu)}\right)\tau\right] d\tau . \tag{5.102}$$

In the following it shall be assumed that

$$c^{(\mu)}(0) = \delta_{\mu,\kappa} \tag{5.103}$$

holds (κ represents the state which is only observable at time $t = 0$, i. e. the initial state), the matrix elements shall be used in the form

$$M_{\mu,\nu} = \zeta M^{(1)}_{\mu,\nu} \,, \tag{5.104}$$

and the probability coefficients shall be written in the form

$$c^{(\mu)}(t) = \sum_{k=0}^{\infty} c^{(\mu)}_k(t)\zeta^k \,. \tag{5.105}$$

ζ can be any parameter. Inserting (5.105) into (5.102) and using (5.103)-(5.104) one obtains the power series

$$\left[c^{(\mu)}_0(t) - \delta_{\mu,\kappa} \right]\zeta^0 + \sum_{k=1}^{\infty} \left[c^{(\mu)}_k(t) - \frac{1}{STRUC w_{\text{sig}}} \sum_{\nu=1}^{\nu_{\text{max}}} I^{(\mu,\nu)}_{1,k}(t) \right]\zeta^k = 0 \,, \tag{5.106}$$

with $I^{(\mu,\nu)}_{1,k}(t)$ being defined by

$$I^{(\mu,\nu)}_{1,k}(t) = \int_0^t c^{(\nu)}_{k-1}(\tau) M^{(1)}_{\mu,\nu} \exp\left[-\left(INV\omega^{(\mu)} - INV\omega^{(\nu)} \right)\tau \right] d\tau \,. \tag{5.107}$$

(5.106) can be fulfilled by using a recursive relation, i. e. (5.106) can be solved by using the relation

$$c^{(\mu)}_0(t) = c^{(\mu)}(0) = \delta_{\mu,\kappa} \,, \quad c^{(\mu)}_{k>0}(t) = \frac{1}{STRUC w_{\text{sig}}} \sum_{\nu=1}^{\nu_{\text{max}}} I^{(\mu,\nu)}_{2,k}(t) \,, \tag{5.108}$$

in which case the second part of this relation is fulfilled by

$$c^{(\mu)}_{k>0}(t) = \left(\frac{1}{STRUC w_{\text{sig}}} \right)^k \left\{ \prod_{l=1}^{k} \sum_{\nu_l=1}^{\nu_{\text{max}}} I^{(\nu_{l-1},\nu_l)}_{k}(\tau_{l-1}) \right\}$$
$$(\nu_0 = \mu, \ \tau_0 = t) \,. \tag{5.109}$$

(In this formula product brackets are used.) $I^{(\nu_{l-1},\nu_l)}_{2,k}(\tau_{l-1})$ is defined by

$$I^{(\nu_{l-1},\nu_l)}_{2,k}(\tau_{l-1}) =$$
$$\int_{\tau_l=0}^{\tau_{l-1}} \delta_{\nu_k,\kappa} M^{(1)}_{\nu_{l-1},\nu_l} \exp\left[-\left(INV\omega^{(\nu_{l-1})} - INV\omega^{(\nu_l)} \right)\tau_l \right] d\tau_l \,. \tag{5.110}$$

(In order to show the validity of (5.109), insert (5.109) into (5.108).) Using the power series (5.105) and the definition (5.104) one obtains a general solution of the master equation (5.63), i. e. one obtains the relation

$$\boxed{\begin{aligned} c^{(\mu)}(t) &= \delta_{\mu,\kappa} + \sum_{k=1}^{\infty} \left(\frac{1}{STRUC w_{\text{sig}}} \right)^k \left\{ \prod_{l=1}^{k} \sum_{\nu_l=1}^{\nu_{\text{max}}} I^{(\nu_{l-1},\nu_l)}_{3,k}(\tau_{l-1}) \right\} \\ &(\nu_0 = \mu, \ \tau_0 = t) \,, \end{aligned}} \tag{5.111}$$

where $I_{3,k}^{(\nu_{l-1},\nu_l)}(\tau_{l-1})$ is defined by

$$
\boxed{
\begin{aligned}
&I_{3,k}^{(\nu_{l-1},\nu_l)}(\tau_{l-1}) = \\
&\int_{\tau_l=0}^{\tau_{l-1}} \delta_{\nu_k,\kappa} M_{\nu_{l-1},\nu_l} \exp\left[-\left(INV\omega^{(\nu_{l-1})} - INV\omega^{(\nu_l)}\right)\tau_l\right] d\tau_l \; .
\end{aligned}
}
\tag{5.112}
$$

In this context it has to be pointed out that a basic constraint is given by (5.103), i. e. at the time $t = 0$ only one state of the physical system is observable. The matrix elements which occur within (5.111) are measures of the transition probabilities from one state to another – this was already mentioned. However, in (5.111) products of matrix elements occur. This is obvious if the explicit form of (5.111) is considered, i. e. if

$$
\begin{aligned}
c^{(\mu)}(t) = \delta_{\mu,\kappa} + &S \int_{\tau_1=0}^{t} M_{\mu,\kappa} E_{\mu,\kappa}(\tau_1)\, d\tau_1 + \\
&S^2 \sum_{\nu_1=1}^{\nu_{\max}} \int_{\tau_1=0}^{t} \int_{\tau_2=0}^{\tau_1} M_{\mu,\nu_1} M_{\nu_1,\kappa} E_{\mu,\nu_1}(\tau_1) E_{\nu_1,\kappa}(\tau_2)\, d\tau_1 d\tau_2 +
\end{aligned}
$$

$$
\text{terms of higher order}
\tag{5.113}
$$

is considered, wherein

$$
E_{\alpha,\beta}(\tau_i) = \exp\left[-\left(INV\omega^{(\alpha)} - INV\omega^{(\beta)}\right)\tau_i\right] ,
\tag{5.114}
$$

$$
S = \frac{1}{STRUC w_{\text{sig}}}
\tag{5.115}
$$

holds. Due to the fact that one matrix element describes a direct transition between two states, such products have to describe transitions which need intermediate states. Therefore, if such a product is identical zero, a transition is not possible, i. e. such products describe *selection rules*. (5.111) is a general representation of the probability coefficients of the introduced ensemble functions (5.60). Some properties of the complex probability coefficients shall now be discussed. However, it has to be mentioned that the same analysis is possible if the real cases are considered.

Complex Probability Coefficients

If matrix elements does not depend on time, or if such a dependence is a weak one, the matrix elements and the integrations can be separated. If the *Schrödinger* case ($STRUC = i$, $INV = -i$, $w_{\text{sig}} = \hbar$, $S = -i/\hbar$) is considered, this means that (5.111) represents a sum over integrals which contain complex exponential functions. Then the integrations can be executed so that (5.111) is replaceable by

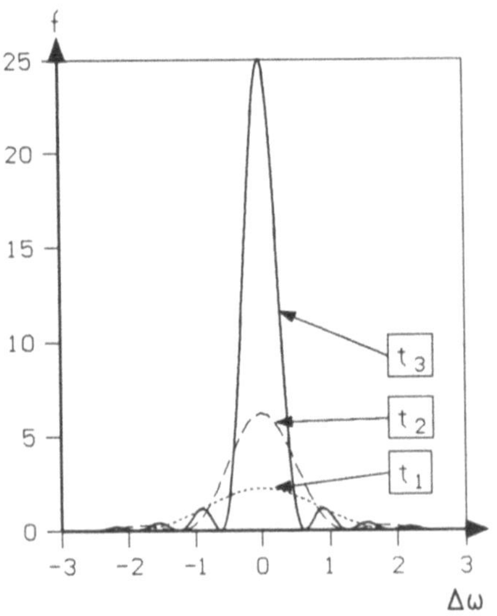
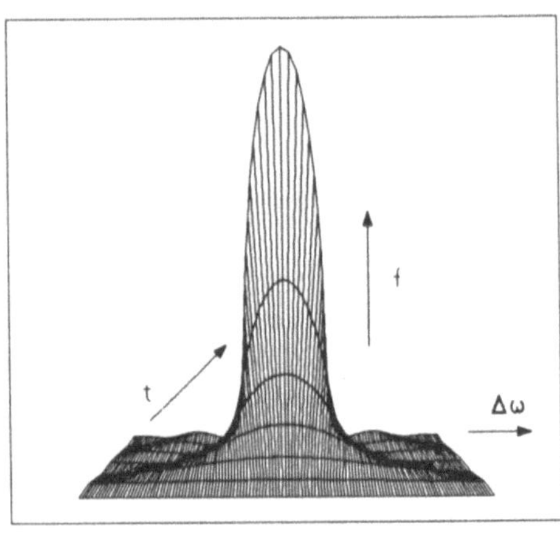

Figure 5.6   The function $f = \frac{\sin^2(\Delta\omega)}{(\Delta\omega)^2}$. $t_1 < t_2 < t_3$

$$
c_{\text{SE}}^{(\mu)}(t) = \delta_{\mu,\kappa} - \frac{1}{\hbar} M_{\mu,\kappa}^{(\text{SE,effective})} \frac{\exp\left[i\left(\omega^{(\mu)} - \omega^{(\kappa)}\right)t\right] - 1}{\omega^{(\mu)} - \omega^{(\kappa)}} , \tag{5.116}
$$

in which case an effective matrix element of the form

$$
M_{\mu,\kappa}^{(\text{SE,effective})} = M_{\mu,\kappa}^{(\text{SE})} + \frac{1}{\hbar} \sum_{\nu_1=1}^{\nu_{\max}} \frac{M_{\mu,\nu_1}^{(\text{SE})} M_{\nu_1,\kappa}^{(\text{SE})}}{\omega^{(\kappa)} - \omega^{(\nu_1)}} EXP_{\mu,\kappa,\nu_1} +
$$
$$
\text{terms of higher order} \tag{5.117}
$$

has to be used. The frequencies $\omega^{(\mu)}$ are correlated with the energies of the states (see (5.56)). (The index SE shows that now an equation of the time-dependent *Schrödinger* type is taken as a basis.)

EXP_{μ,κ,ν_1} is defined by

$$
EXP_{\mu,\kappa,\nu_1} = 1 - \frac{\left\{\exp\left[i\left(\omega^{(\mu)} - \omega^{(\nu_1)}\right)t\right] - 1\right\}\left(\omega^{(\mu)} - \omega^{(\kappa)}\right)}{\left\{\exp\left[i\left(\omega^{(\mu)} - \omega^{(\kappa)}\right)t\right] - 1\right\}\left(\omega^{(\mu)} - \omega^{(\nu_1)}\right)} . \tag{5.118}
$$

(5.118) guarantees that in the case $\kappa = \nu_1$ no divergent term of second order appears. (In this context it has to be pointed out that terms of higher order show the same behavior. However, this shall not be shown.)

If $\mu \neq \kappa$ holds, the square of the absolute amount of such a complex probability coefficient is given by

$$\left|c_{\text{SE}}^{(\mu)}(t)\right|^2 = \frac{4}{\hbar^2}\left|M_{\mu,\kappa}^{(\text{SE},\text{effective})}\right|^2 \frac{\sin^2\left[\left(\omega^{(\mu)}-\omega^{(\kappa)}\right)t/2\right]}{\left(\omega^{(\mu)}-\omega^{(\kappa)}\right)^2}\,, \tag{5.119}$$

and for large times t this expression can be replaced by

$$\left.\left|c_{\text{SE}}^{(\mu)}(t)\right|^2\right|_{t\to\infty}/t = \frac{2\pi}{\hbar^2}\left|M_{\mu,\kappa}^{(\text{SE},\text{effective})}\right|^2 \delta\left(\omega^{(\mu)}-\omega^{(\kappa)}\right)\,, \tag{5.120}$$

because the relation

$$\frac{2}{\pi t}\int_{-\infty}^{+\infty}\frac{\sin^2\left[\left(\omega^{(\mu)}-\omega^{(\kappa)}\right)t/2\right]}{\left(\omega^{(\mu)}-\omega^{(\kappa)}\right)^2}\,d\left(\omega^{(\mu)}-\omega^{(\kappa)}\right) =$$

$$\int_{-\infty}^{+\infty}\delta\left(\omega^{(\mu)}-\omega^{(\kappa)}\right)\,d\left(\omega^{(\mu)}-\omega^{(\kappa)}\right) \tag{5.121}$$

has to be considered (see figure 5.6), in which case the functional relation

$$\frac{2}{\pi t}\frac{\sin^2\left[\left(\omega^{(\mu)}-\omega^{(\kappa)}\right)t/2\right]}{\left(\omega^{(\mu)}-\omega^{(\kappa)}\right)^2} \stackrel{t\to\infty}{=} \delta\left(\omega^{(\mu)}-\omega^{(\kappa)}\right) \tag{5.122}$$

holds.

Such complex probability coefficients are not identical with the probability of a state, because these coefficients are normally complex. Due to the fact that the square of the absolute amount $\left|c_{\text{SE}}^{(\mu)}(t)\right|^2$ shows an oscillatory behavior even if a time-independent interaction function occurs (i. e. if a time-independent interaction is observable) the absolute amount $\left|c_{\text{SE}}^{(\mu)}(t)\right|^2$ cannot be interpreted as a probability of a state, because such a behavior cannot be observed, no matter which physical system is considered. Additionally, it has to be remarked that the relation (5.120) shows that for large times t the initial state and the final state have to be identical. However, this only makes sense if, for example, systems of particles are considered. In this case $\delta\left(\omega^{(\mu)}-\omega^{(\kappa)}\right)$ represents the conservation law of energy. Therefore, both complex probability coefficients and the squares of their absolute amounts have to be interpreted as indirect probability functions which allow to include the measurable interaction function $INT(\boldsymbol{x})$. However, if a physical system can be described by using an equation of the time-dependent Schrödinger type, the matrix elements $M_{\mu,\nu}^{(\text{SE})}$ have to represent selection rules. (This was above mentioned.) Therefore, it has to be possible to formulate a relation which describes the probability of a state by using the matrix elements $M_{\mu,\nu}^{(\text{SE})}$. This is indeed possible and shall now be discussed. Then an expression shall be used which is as similar to (5.120) as possible. Due to the fact that (5.120) shows the form of a transition probability, an expression which describes a transition probability shall be derived.

The Transition Probability of a Schrödinger System

Without any restrictions the change of a probability of a state κ can be written in the form

$$\boxed{dw_{\kappa\to\mu}/dt = \frac{2\pi}{\hbar^2} \int \rho \left| M_{\mu,\kappa}^{(\mathrm{SE,observ})} \right|^2 d\omega^{(\mu)}}$$

(5.123)

with

$$\boxed{\begin{array}{c} M_{\mu,\kappa}^{(\mathrm{SE,observ})} = M_{\mu,\kappa}^{(\mathrm{SE})} + \frac{1}{\hbar} \sum_{\nu_1=1}^{\nu_{\max}} \frac{M_{\mu,\nu_1}^{(\mathrm{SE})} M_{\nu_1,\kappa}^{(\mathrm{SE})}}{\omega^{(\kappa)} - \omega^{(\nu_1)}} K_{\mu,\kappa,\nu_1} + \\[2mm] \text{terms of higher order} \end{array}}$$

(5.124)

if a *Schrödinger* system is considered, i. e. if a system is considered which can be described by using an equation of the time-dependent *Schrödinger* type. $dw_{\kappa\to\mu}/dt$ represents the change of the probability of the state κ, i. e. the *transition probability*. Such a formulation is possible, because ρ can be any function (i. e. ρ represents a *mediation function*) and (5.123) includes the correct matrix elements. This expression is similar to (5.120). However, (5.123) uses an integral with respect to $\omega^{(\mu)}$, because if real statistical systems are considered, always distributions of a physical value are observable. This effect (5.120) does not describe. Furthermore, the factor EXP_{μ,κ,ν_1} is replaced by a new factor

$$K_{\mu,\kappa,\nu_1} = f(\text{specific quantities}) ,$$

(5.125)

because a time-dependence of the kind (5.118) is not observable. Additionally, (5.125) guarantees that no vanishing denominators occur. This factor has to be choosen in a suitable way. Such factors also occur if terms of higher order are considered. Additionally, these factors allow to determine all parts of the transition probability (5.123) in a correct way.

In this context it has to be pointed out that a matrix element $M_{\alpha,\beta}^{(\mathrm{SE})}$ can be written in the form

$$M_{\alpha,\beta}^{(\mathrm{SE})} = \left\langle \Phi^{(\alpha)}(x) \left| INT(x) \right| \Phi^{(\beta)}(x) \right\rangle .$$

(5.126)

So far, properties and a systematic way of derivation of several kinds of evolution equations were discussed. Some possibilities of application to physical systems were mentioned. For example, it has been mentioned that an equation of the *Fokker-Planck* type can be used in laser physics. In the following the meaning and the applicability of the introduced evolution equations shall be discussed in more detail.

5.5 Meaning and Applicability

In this chapter evolution equations were discussed. In particular, four levels of consideration were discussed. Table 5.1 shows the essential four levels and the general forms of the corresponding evolution equations. In this chapter it was shown that it is possible to calculate both equations of a *Fokker-Planck* type (a conjugate *Fokker-Planck* type, respectively) and a *Schrödinger* type (a conjugate complex *Schrödinger* type, respectively) by using only one basic evolution equation, namely the multi-system equation.

It was shown that in a first calculation step it is possible to derive impulse forms which are correlated to impulse functions by using an energy-time relation of *Heisenberg's* type. A special case of such an equation is represented by an equation of the *Fokker-Planck* type. Such equations can widely be used in statistical physics. An example was given in chapter 2. There it was discussed that distribution functions of thermodynamic systems in the context of *Landau's* theory can be found by solving a suitable equation of the *Fokker-Planck* type. As experience shows, statistical distribution functions in the context of laser theory also fulfill equations of the *Fokker-Planck* type. This will be discussed in the next chapter. Other examples are special biological systems. For example, in physiology the problem of coordination of finger movements is a widely discussed problem, because such a biological system represents a paradigma of a biological synergetic system. The fluctuating behavior of such finger movements can be described be using distribution functions which fulfill equations of the *Fokker-Planck* type. However, one point has to be stressed: The derivation of the evolution equations was a very specific one, i. e. special conditions and constraints were used. However, such equations can also be derived by using other conditions. An example was discussed, i. e. an equation of the *Fokker-Planck* type can also be derived by using delta-correlated fluctuating forces instead of a relation of *Heisenberg's* type. Therefore, if a special physical system is considered, strictly speaking it has to be examined whether all conditions are fulfilled. Then it can be decided whether or not the derivation of the evolution equations given in this chapter is the correct one. In this context it has to be pointed out that in the following such an examination will not be made, i. e. equations of the *Fokker-Planck* type will be used without examination of the physical background.

If kinetic functions instead of impulse functions shall be used, a derivation of kinetic equations is possible. The so-derived general equation includes an equation of the stationary *Schrödinger* type, i. e. a kind of equation which is used in quantum physics. In order to get equations of a time-dependent *Schrödinger* type, it was necessary to introduce an equation which corresponds to ensembles of physical systems. However, this is clear, because equations of the time-dependent *Schrödinger* type correspond to ensembles of physical systems, this shows the experience with quantum theoretical methods. Equations of the kinetic *Fokker-Planck* type are included, too. However, it has to be pointed out that solutions of such equations can show negative values. This shows the example in subsection 5.4.2. This means that solutions of real kinetic equations cannot always be interpreted as probability densities.

Table 5.1 Evolution equations: The four levels of consideration. $STRUC = \pm 1, \pm i$; $(STRUC)^2 = \pm 1$

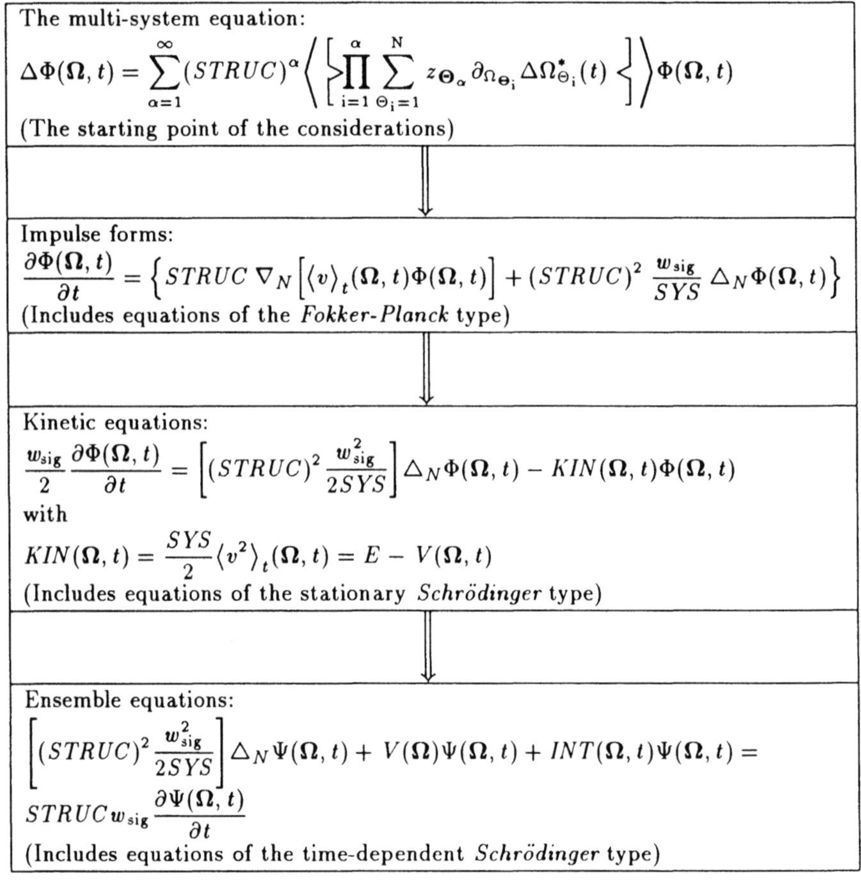

The multi-system equation:

$$\Delta\Phi(\mathbf{\Omega}, t) = \sum_{\alpha=1}^{\infty}(STRUC)^\alpha \left\langle \left[\prod_{i=1}^{\alpha} \sum_{\Theta_i=1}^{N} z_{\Theta_\alpha}\, \partial_{\Omega_{\Theta_i}} \Delta\Omega^*_{\Theta_i}(t) \right] \right\rangle \Phi(\mathbf{\Omega}, t)$$

(The starting point of the considerations)

Impulse forms:

$$\frac{\partial\Phi(\mathbf{\Omega}, t)}{\partial t} = \left\{ STRUC\, \nabla_N\left[\langle v\rangle_t(\mathbf{\Omega}, t)\Phi(\mathbf{\Omega}, t)\right] + (STRUC)^2\, \frac{w_{\mathrm{sig}}}{SYS}\, \triangle_N\Phi(\mathbf{\Omega}, t) \right\}$$

(Includes equations of the *Fokker-Planck* type)

Kinetic equations:

$$\frac{w_{\mathrm{sig}}}{2}\frac{\partial\Phi(\mathbf{\Omega}, t)}{\partial t} = \left[(STRUC)^2\, \frac{w_{\mathrm{sig}}^2}{2SYS}\right]\triangle_N\Phi(\mathbf{\Omega}, t) - KIN(\mathbf{\Omega}, t)\Phi(\mathbf{\Omega}, t)$$

with

$$KIN(\mathbf{\Omega}, t) = \frac{SYS}{2}\langle v^2\rangle_t(\mathbf{\Omega}, t) = E - V(\mathbf{\Omega}, t)$$

(Includes equations of the stationary *Schrödinger* type)

Ensemble equations:

$$\left[(STRUC)^2\, \frac{w_{\mathrm{sig}}^2}{2SYS}\right]\triangle_N\Psi(\mathbf{\Omega}, t) + V(\mathbf{\Omega})\Psi(\mathbf{\Omega}, t) + INT(\mathbf{\Omega}, t)\Psi(\mathbf{\Omega}, t) =$$

$$STRUC\, w_{\mathrm{sig}}\frac{\partial\Psi(\mathbf{\Omega}, t)}{\partial t}$$

(Includes equations of the time-dependent *Schrödinger* type)

This is in contrast to equations of the impulse type. Such equations always have exponential functions as stationary solutions, in which case such positive solutions can be interpreted as probability densities (see example 5.4.1).

Therefore, the statement can be made that it is possible to derive equations of types which are essential in statistical physics by using only one basic evolution equation, the multi-system equation, i. e. a uniform derivation is possible. If equations of the *Schrödinger* type are considered, it has to be pointed out that the complete quantum mechanical description scheme automatically arises. In order to find equations of the *Schrödinger* type, a basic complex power series function has to be used (see (5.13)). The use of such a complex function is equivalent to the requirement that solutions of the wave function kind have to be taken into account. In order to find an equation of the *Schrödinger* type, only three essential requirements have to be used. Firstly, it has to be used that the result of an impulse measurement is an uncertain one, in which case the strength of the uncertainty is defined by an energy-time relation of *Heisenberg's* type. (Principle of measurement uncertainty.) Secondly, it has to be used that the strength of the uncertainty (i. e. the fluctuation strength) is relatively small so that only power series terms of lowest order have to be taken into consideration. (Principle of simplicity.) Thirdly, it has to be used that solutions of equations of the time-dependent *Schrödinger* type have to correspond to ensembles. (Principle of ensemble interpretation.) Then the well-known quantum mechanical description scheme (*Jordan's* rule, *Hilbert* space, *Hermitian* operators, etc.) occurs. However, it has to be emphasized that without further ado it cannot be said that the given derivation describes the quantum mechanical background. This is the reason why the term *type* has always been used.

Moreover, it has to be noted that the structure of the real formalism and the complex formalism is totally similar, i. e. the complex (*Schrödinger*) formalism can be considered as an analytical continuation of the real (*Fokker-Planck*) formalism. For example, this means that an equation of the *Schrödinger* type can be considered as an analytical continuation of an equation of the kinetic *Fokker-Planck* type. Additionally, one point has to be stressed: The change to a complex equation causes an essential difference, i. e. a discrete structure of functions occurs which can be normalized. (In particular, see the discussion of 5.4.2. There the quantum mechanical harmonic oscillator was considered.) The correlated discrete energy eigenvalues are identical to measurable energy eigenvalues of quantum mechanical systems. Therefore, it has to be emphasized that the complex structure causes an essential difference. Another point which has to be stressed is that the close connection between equations of the *Fokker-Planck* and the *Schrödinger* type allows to use a solution of one type of equation to find a solution of the other type of equation, i. e. transitions from one type of solution to the other type are possible. In order to generate such a transition, a suitable replacement has to be carried out, i. e. real variables (parameters) have to be replaced by complex variables (parameters) and *vice versa*. This possibility is very interesting, because this possibility shows that special kinds of transformations allow to "switch" from one kind of physical equation to the other. However, this possibility shall not be discussed in this book.

As it was explained, the functions Φ of the real cases can normally directly be interpreted as probability densities, and it was explained that it is possible to gain expressions which describe the probability transition if an ensemble is considered. Moreover, it was shown that it is possible to gain energy eigenvalues of *Fokker-Planck* and *Schrödinger* systems by using a suitable differential equation. However, there are much more connections between the considered evolution equations and measurable quantities. This connection shall be discussed in chapter 7 (see section 7.4). Instead of this discussion, in the following an application of an equation of the *Fokker-Planck* type will be considered. In this context the introduced hyper-surface equations will be applied. The physical systems which will be considered are laser systems. (Such systems are often discussed systems. For example, see [30, 31, 32, 41, 64, 65].) Laser systems are typical synergetic systems. Therefore, in this context the phenomenon of synergetic behavior will be discussed.

6 The Laser, a Self-Organizing System

In the last few decades laser systems have frequently been discussed as multi-component systems (see figure 1.2 in chapter 1). A laser system is the paradigm of a self-organizing system, i. e. a multi-component system which is able to produce more or less complex patterns on a macroscopic level when control parameters cross critical values. Such systems are very often called *synergetic systems*, and this shall also be done here. In the context of laser theory *self-organization* means that only by changing the pump parameter, which is the control parameter, more or less complex field (light) patterns arise. For example, a lasing state occurs. But not only monochromatic oscillations, irregular motions like chaos are observable, too. Due to the fact that a light field can be decomposed by *Fourier's* method, a laser system is a multi-component system in the sense that it consists of many molecules and in the sense that a correlated light field very often can be decomposed into a large number of modes. Considering laser systems means that now systems far from thermodynamic equilibrium shall be considered. In this chapter a short summary of essential facts of the deterministic and statistical laser physics will be given. The far-reaching phenomenon of self-organization will then be discussed. In this context the slaving principle of synergetics shall also be considered. In particular, the connection with the contents of the chapters 5, 4 and 2 will be presented, i. e. the occurrence of a *Fokker-Planck* equation in the context of the statistical laser physics will be considered, the consequences of the existence of hyper-surface equations will be considered, and the maximum information entropy principle in the context of laser systems will be considered. In particular, it will be shown that the concept of hyper-surface equations evolved in this book is a concept of a universal meaning.

6.1 The Ensemble Level of the Laser Activity

In order to describe laser activity, both the level of the active laser molecules and the level of the whole system (ensemble level) can be taken as a basis. In order to describe the molecular activity, quantum mechanical methods are needed, and to describe the system level, laws have to be used which relate to the whole molecular ensemble. In this chapter such macroscopic laws shall be taken as a basis. Some remarks to quantum mechanical methods shall be given in chapter 7. First, some well-known basic facts shall be considered (for example, see [30, 31, 64, 65]).

6.1.1 Laser Equations of the Molecular Ensemble

An active laser medium is correlated with an electric field strength $\boldsymbol{E}(\boldsymbol{x}, t)$ which describes the electric field at time t and position $\boldsymbol{x}$. By using *Fourier's* decomposition such an electric field strength is of the form

$$E(\boldsymbol{x}, t) = \mathrm{i} \sum_w \left[N_w \, E_w^{\mathrm{c}}(t) \exp \left(\mathrm{i} \mathbf{k}_w \boldsymbol{x} \right) - \mathrm{c.c.} \right] \tag{6.1}$$

if only one polarization direction is taken as a basis, i. e. if only one component E of the vector $\boldsymbol{E}$ can be used. $E_w^{\mathrm{c}}(t)$ represents a time-dependent complex amplitude of one mode, in which case one mode shall be characterized by using the wavelength w. $\mathbf{k}_w$ represents the wave vector. N_w guarantees dimensionless complex amplitudes, and c.c. is identical with the conjugate complex part so that (6.1) represents the actually observable light field. These complex amplitudes fulfill the field equation

$$\boxed{\frac{dE_w^{\mathrm{c}}(t)}{dt} = \left(-\mathrm{i}\omega_w - \kappa_w \right) E_w^{\mathrm{c}}(t) - \mathrm{i} \sum_k g_{k,w}^{\mathrm{c}} \alpha_k^{\mathrm{c}}(t) + F_w^{(1)}(t) \, ,} \tag{6.2}$$

where (6.2) describes the change of the complex amplitude on grounds of free oscillations (which are represented by the angular frequencies ω_w), on grounds of damping (which is represented by the decay factors κ_w) and on grounds of oscillating molecular dipole moments $\alpha_k^{\mathrm{c}}(t)$ (this influence is represented by the terms $g_{k,w}^{\mathrm{c}} \alpha_k^{\mathrm{c}}(t)$, where the factors $g_{k,w}^{\mathrm{c}}$ are coupling factors). Additionally, fluctuation terms $F_w^{(1)}(t)$ are taken into account, which ascertain fluctuating effects (for example, the influence of non-ideal resonator end-plates or the process of spontaneous emission). k represents all active laser molecules. Such a complex amplitude is of the form

$$E_w^{\mathrm{c}}(t) = E_w(t) \exp \left[\mathrm{i}\phi_w(t) \right] \exp \left(-\mathrm{i}\omega_w t \right) , \tag{6.3}$$

with $E_w(t)$ being the real amplitude

$$E_w(t) = E_w^{(\mathrm{wd})}(t) \exp \left(-\kappa_w t \right) . \tag{6.4}$$

($E_w^{(\mathrm{wd})}(t)$ is the part without damping, $\phi_w(t)$ represents the phase.)

Additionally, the material equations

$$\boxed{\begin{aligned}
\frac{d\alpha_k^{\mathrm{c}}(t)}{dt} &= \left(-\mathrm{i}\nu - \gamma \right) \alpha_k^{\mathrm{c}}(t) + \mathrm{i} \sum_w g_{k,w}^{\mathrm{c.c.}} E_w^{\mathrm{c}}(t)\sigma_k(t) + F_k^{(2)}(t) \, , \\
\frac{d\sigma_k(t)}{dt} &= \gamma_{\mathrm{inv}} \left[d_0 - \sigma_k(t) \right] + 2\mathrm{i} \sum_w \left[g_{k,w}^{\mathrm{c}} \alpha_k^{\mathrm{c}}(t) E_w^{\mathrm{c.c.}}(t) - \mathrm{c.c.} \right] + F_k^{(3)}(t)
\end{aligned}} \tag{6.5}$$

have to be used, where the first equation describes the change of a molecular dipole moment on grounds of free oscillations (represented by the central frequency ν of a collective emission line), on grounds of damping (represented by a decay constant γ) and on grounds of coupling to the electric field (represented by the coupling term $g_{k,w}^{\text{c.c.}} E_w^{\text{c.}}(t)\sigma_k(t)$). The components $\sigma_k(t)$ guarantee the correct phase relation between the electric field and the molecular dipole moments. They fulfill the second equation of (6.5), and values $\sigma_k(t) < 0$ (absorption of field energy), $\sigma_k(t) > 0$ (emission of field energy) are possible. The sum

$$\sum_k \sigma_k(t) = D(t) = N_2 - N_1 \tag{6.6}$$

represents the inversion of the whole molecular ensemble, with N_2 and N_1 being the numbers of molecules in the upper and lower laser active energy states, i. e. D(t) represents the population inversion. Therefore, $\sigma_k(t)$ can be called the *molecular inversion*. Due to the fact that D_0 represents the equilibrium inversion which is the result of the pump process and incoherent decay processes of the molecular ensemble, $d_0 = D_0/N$ can be called *molecular equilibrium inversion*. (N represents the total number of active molecules.) The inverse relaxation time of a given inversion (relaxation into an equilibrium state) is represented by γ_{inv}.

The field equation (6.2) and the material equations (6.5) are evolution equations which determine the dynamics of the molecular system. Due to the fact that a laser system consists of many molecules (typically $> 10^{14}$ molecules), such a system of equations is a high-dimensional mathematical system. However, under certain conditions the degrees of freedom can be reduced and so-called order parameters occur. This effect shall be considered.

6.1.2 The Slaving Principle and Order Parameters

Under certain constraints a high-dimensional mathematical system can be reduced, i. e. a system of equations can be calculated which includes two parts. One part is a differential equation system which determines the behavior of normally few variables – the *order parameters* -, and the other part is a non-differential algebraic system which describes the connection between the order parameters and the remaining variables, in which case these remaning variables are functions of the order parameters. Therefore, it makes sense to call such variables *slaved variables*. If such a reduction is possible, the order parameters determine the physical system, i. e. the observable macroscopic level has to show a more or less complicated pattern. (In this book instead of the term *complicated* very often the term *complex* is used if patterns are characterized. However, then the term *complex* has not the mathematical meaning of *complex = real part + imaginary part*.) Such a reduction is only possible if the order parameters show small values, i. e. if critical points are considered. (A critical point characterizes a dramatic change of a physical state so that variables change their values in a dramatic way. Such variables naturally have small values. Examples of statistical systems which show critical points were already given, i. e. critical points are observable in the context of thermodynamic systems. Then equilibrium phase transitions are observable. In this chapter it will be shown that such critical points are also relevant in statistical laser theory. The patterns which occur during a phase transition are the patterns which can also be observed far from such a critical point, i. e. even if the mathematical conditions of reduction are not fulfilled any more. This effect shall be explained later.) Furthermore, some physical (and therefore: mathematical) constraints have to be fulfilled. In the following a possible mathematical procedure shall be considered which allow such a reduction of degrees of freedom.

A Mathematical Reduction Procedure

A homogeneous linear differential equation system of first order is defined by

$$\frac{dy(t)}{dt} = P(t)y(t) + Q(t) , \tag{6.7}$$

where an integrating factor of the form

$$\mu(t) = \exp\left[- \int P(t)\, dt \right] \tag{6.8}$$

exists which leads to the general solution

$$y(t) = \frac{\mu(0)y(0) + \int_0^t \mu(\tau)Q(\tau)\, d\tau}{\mu(t)} . \tag{6.9}$$

Without any restrictions the term $Q(t)$ can have the form

$$Q(t) = \sum_w Q_w(t) . \tag{6.10}$$

This formulation shall be considered. Due to the fact that the term $\mu(0)y(0)/\mu(t)$ represents switching effects, this term shall be neglected. In this case one obtains the result

$$y(t) = \int_0^t \exp\left[\int P(t)\, dt - \int P(\tau)\, d\tau \right] Q(\tau)\, d\tau . \tag{6.11}$$

Assuming a function $P(t)$ of the kind

$$P(t) = -\left(ip_1 + p_2\right) \tag{6.12}$$

one obtains a solution $y(t)$ of the kind

$$y(t) = \int_0^t \exp\left[-\left(ip_1 + p_2\right)(t - \tau) \right] Q(\tau)\, d\tau . \tag{6.13}$$

Without any restrictions the functions $Q_w(t)$ can be replaced by

$$Q_w(t) = Q_w^{(1)}(t) \exp\left[-\left(iq_{1,w} + q_{2,w}\right)t\right], \tag{6.14}$$

because $Q_w^{(1)}(t)$ can be any functions. Therefore, the equation

$$y(t) =$$
$$\exp\left[-\left(ip_1 + p_2\right)t\right] \sum_w \int_0^t \exp\left[\left(ip_1 - iq_{1,w} + p_2 - q_{2,w}\right)\tau\right] Q_w^{(1)}(\tau)\, d\tau .$$
$$\tag{6.15}$$

holds. If $Q_w^{(1)}(\tau)$ represents time-independent functions, or if the quantities $Q_w^{(1)}(\tau)$ show a weak dependence with respect to time, $Q_w^{(1)}(\tau)$ can be replaced by $Q_w^{(1)}(t)$ so that

$$y(t) =$$
$$\sum_w Q_w^{(1)}(t) \exp\left[-\left(ip_1 + p_2\right)t\right] \int_0^t \exp\left[\left(ip_1 - iq_{1,w} + p_2 - q_{2,w}\right)\tau\right] d\tau .$$
$$\tag{6.16}$$

holds. Then the integration can be executed so that one obtains the result

$$y(t) = \sum_w \frac{Q_w^{(1)}(t)}{ip_1 - iq_{1,w} + p_2 - q_{2,w}} \left\{ \exp\left[-\left(iq_{1,w} + q_{2,w}\right)t\right] - \right.$$
$$\left. \exp\left[-\left(ip_1 + p_2\right)t\right] \right\} . \tag{6.17}$$

If the decay factors $q_{2,w}$ are much smaller than the decay factor p_2, i. e. if

$$q_{2,w} \ll p_2 \tag{6.18}$$

holds, the second part of (6.17) can be neglected. Then the solution $y(t)$ is of the form

$$y(t) = \sum_w \frac{1}{ip_1 - iq_{1,w} + p_2 - q_{2,w}} Q_w^{(1)}(t) \exp\left[-\left(iq_{1,w} + q_{2,w}\right)t\right]$$
$$= \sum_w \frac{1}{ip_1 - iq_{1,w} + p_2 - q_{2,w}} Q_w(t) , \tag{6.19}$$

which can be replaced by

$$y(t) = \sum_w \frac{1}{ip_1 - iq_{1,w} + p_2} Q_w(t) , \tag{6.20}$$

because in the denominator $q_{2,w}$ can be negelected in comparison with p_2. Such an approximation can be called *adiabatic approximation*, because such a notation is normally used if one considers an approximation which leads to the behavior that one physical quantity instantaneously follows another physical quantity. (Other examples can be found in thermodynamics and quantum theory. For example, if one assumes that the inner energy U of a thermodynamic system instantaneously follows the change of the work A, such an assumption is called an *adiabatic approximation*. Another example represents the *Born-Oppenheimer approximation* of molecule physics (see [43]). There it is assumed that the electrons of a molecule instantaneously follow the corresponding nuclei.) Therefore, it was shown that under certain constraints a differential equation can be replaced by an algebraic equation in such a way that the solution of a differential equation (here: $y(t)$) is a function of given other functions (here: $Q_w(t)$). If such a procedure is possible within a set of equations, this system of equations can be reduced. If the determination functions (here: $Q_w(t)$) fulfill themselves differential equations of the considered system, the considered system splits into a part which consists of algebraic equations and a part which consists of differential equations. The solutions of these differential equations are the mentioned order parameters (here: $Q(t)$), and the algebraic equations determine the slaved variables (here: $y(t)$). The principle that only few measurement quantities can determine a whole set of other quantities is a universal principle. This is a consequence of a principle called *slaving principle* (see [30, 31]). The above introduced mathematical reduction procedure represents one possibility to put the slaving principle in concrete mathematical terms. However, other mathematical formulations are possible. One possibility represents the *centre manifold theorem*. However, these formulations shall not be discussed in this book. Furthermore, it has to be remarked that the approximation $(6.15) \to (6.16)$ normally makes sense if the values of $Q(t)$ are small, i. e. if the region of a critical point is considered. Outside this critical region one has to expect time-dependent functions $Q(t)$. Therefore, such an adiabatic approximation is normally valid near a critical point.

The equations of the molecular dipole moments exactly show the structure of the equation (6.7). Provided the required restrictions are valid, the above considerations can be applied to these equations. Then the original systems of evolution equations (6.2),(6.5) can be reduced so that order parameter equations and algebraic equations occur. The algebraic equations determine slaved variables. It has to be noticed that the results will show that the introduced adiabatic procedure in the context of laser theory is valid, but only if a critical point is considered. This shall now be discussed.

The Application to the Equations of the Dipole Moments

The evolution equations of the molecular dipole moments (see (6.5)) are of the type (6.7). Therefore, the equations of the molecular dipole moments are replaceable by algebraic equations of the type (6.20) if the required constraints hold. However, the fluctuation terms $F_k^{(2)}(t)$ shall not be taken into account, i. e. it is sufficient to introduce new fluctuation terms afterwards. In doing so the problem of dipole moments is identical with the above discussed pure mathematical problem if the representation

$$E_w^{\text{c.}}(t) = E_w^{(\text{wd})}(t) \exp\left[i\phi_w(t)\right] \exp\left[-\left(i\omega_w + \kappa_w\right)t\right] \tag{6.21}$$

(see (6.3)-(6.4)) is used and if the identification

$$
\begin{aligned}
&y(t) \leftrightarrow d\alpha_k^{\text{c.}}(t) \;, \quad P(t) \leftrightarrow -(i\nu + \gamma) \;, \quad p_1 \leftrightarrow \nu \;, \quad p_2 \leftrightarrow \gamma \;, \\
&Q_w(t) \leftrightarrow ig_{k,w}^{\text{c.c.}}\sigma_k(t)E_w^{\text{c.}}(t) \;, \\
&Q_w^{(1)}(t) \leftrightarrow ig_{k,w}^{\text{c.c.}}\sigma_k(t)E_w^{(\text{wd})}(t)\exp\left[i\phi_w(t)\right] \;, \\
&q_1 \leftrightarrow \omega_w \;, \quad q_2 \leftrightarrow \kappa_w
\end{aligned}
\tag{6.22}
$$

is used. By using this identification the necessary constraints are of the form

$$\kappa_w \ll \gamma \;, \quad ig_{k,w}^{\text{c.c.}}\sigma_k(t)E_w^{(\text{wd})}(t)\exp\left[i\phi_w(t)\right] = \text{weak time dependence} \;. \tag{6.23}$$

Provided these constraints hold, the dipole moments are functions of the complex electric field amplitudes and of the molecular inversions, i. e. the relation

$$\alpha_k^{\text{c.}}(t) = i\sum_w g_{k,w}^{\text{c.c.}} \frac{1}{i\nu - i\omega_w + \gamma} E_w^{\text{c.}}(t)\sigma_k(t) \tag{6.24}$$

holds.

Inserting the relation (6.24) into the second material equation of (6.5) the dipole moments can be eliminated. This shall now be shown.

The First Collective Slaved Variable: The Inversion

Inserting (6.24) into the second equation of (6.5) and using (6.6) (without considering fluctuations) one obtains the equation

$$\frac{dD(t)}{dt} = \gamma_{\text{inv}}\left[D_0 - D(t)\right] - 2\sum_{w_1,w_2} \Omega_{w_1} g_{w_2}^{\text{c.}} g_{w_1}^{\text{c.c.}} E_{w_1}^{\text{c.}}(t)E_{w_2}^{\text{c.c.}}(t) \;, \tag{6.25}$$

where

$$\Omega_{w_1} = \frac{2\gamma}{(\nu - \omega_{w_1})^2 + \gamma^2} \;, \quad g_{k,w}^{\text{c.}} = g_w^{\text{c.}} \tag{6.26}$$

has to be considered. Using (6.7)-(6.9) and neglecting again switching effects one obtains the solution

$$D(t) = -2\sum_{w_1,w_2} \Omega_{w_1} g_{w_2}^{\text{c.}} g_{w_1}^{\text{c.c.}} \int_0^t \exp\left[-\gamma_{\text{lin}}(t-\tau)\right] D(\tau)E_{w_1}^{\text{c.}}(\tau)E_{w_2}^{\text{c.c.}}(\tau)\,d\tau +$$

$$D_0\gamma_{\text{lin}} \int_0^t \exp\left[-\gamma_{\text{lin}}(t-\tau)\right]\,d\tau \;. \tag{6.27}$$

The first integral shall be solved by using the introduced adiabatic procedure. In the sense of this procedure the inversion $D(\tau)$ can be replaced by the critical inversion D_{c}. Additionally, the expression

$$\int_0^t \exp\left[-\gamma_{\mathrm{lin}}(t-\tau)\right] d\tau = \frac{1}{\gamma_{\mathrm{lin}}}\left[1 - \exp\left(-\gamma_{\mathrm{lin}}t\right)\right] \approx \frac{1}{\gamma_{\mathrm{lin}}} \tag{6.28}$$

shall be used. Then (6.27) has to be replaced by

$$D(t) = -2\frac{D_{\mathrm{c}}}{\gamma_{\mathrm{lin}}} \sum_{w_1,w_2} \Omega_{w_1} g^{\mathrm{c.}}_{w_2} g^{\mathrm{c.c.}}_{w_1} E^{\mathrm{c.}}_{w_1}(t) E^{\mathrm{c.c.}}_{w_2}(t) + D_0 + F_D(t) \ . \tag{6.29}$$

As the following calculations will show, the complex amplitudes are the order parameters of the system. Therefore, the equation (6.29) represents the connection between the order parameters and the collective slaved variable $D(t)$. (The fluctuating effects are now considered again.) Then the second slaved variable, the collective dipole moment, can be calculated.

The Second Collective Slaved Variable: The Dipole Moment

Considering (6.24), using the relation (6.26) and forming the sum with respect to all molecules k one obtains the collective dipole moment

$$S^{\mathrm{c.}}(t) = \mathrm{i}D(t) \sum_w g^{\mathrm{c.c.}}_w \frac{1}{\mathrm{i}\nu - \mathrm{i}\omega_w + \gamma} E^{\mathrm{c.}}_w(t) + F_S(t) \ , \tag{6.30}$$

which is determined by the order parameters $E^{\mathrm{c.}}_w(t)$, i. e. $S^{\mathrm{c.}}(t)$ represents the second collective slaved variable of the system. In this equation fluctuating effects are considered again. Then the field equation (6.2) can be rewritten.

The Order Parameter Equation

Inserting (6.24) into (6.2), formig the sum with respect to k, using the relation (6.26) and inserting (6.29) one obtains the differential equation

$$\frac{dE^{\mathrm{c.}}_{w_1}(t)}{dt} = \left(-\mathrm{i}\omega_{w_1} - \kappa_{w_1}\right) E^{\mathrm{c.}}_{w_1}(t) + D_0 \sum_{w_2} \Omega^{\mathrm{c.}}_{w_1,w_2} E^{\mathrm{c.}}_{w_2}(t) -$$
$$2\frac{D_{\mathrm{c}}}{\gamma_{\mathrm{lin}}} \sum_{w_2,w_3,w_4} \Omega^{\mathrm{c.}}_{w_1,w_2,w_3,w_4} E^{\mathrm{c.}}_{w_2}(t) E^{\mathrm{c.}}_{w_3}(t) E^{\mathrm{c.c.}}_{w_4}(t) + F_{w_1}(t) \ , \tag{6.31}$$

which represents evolution equations of the complex amplitudes $E^{\mathrm{c.}}_w(t)$. Due to the fact that these amplitudes determine the behavior of the remaining variables, such amplitudes are the order parameters of the system, i. e. (6.31) represents order parameter equations. ($F_{w_1}(t)$ represents the now necessary fluctuation terms.) In this context

$$\Omega^{c.}_{w_1,w_2} = g^{c.}_{w_1} g^{c.c.}_{w_2} \frac{1}{i\nu - i\omega_{w_2} + \gamma} \, ,$$

$$\Omega^{c.}_{w_1,w_2,w_3,w_4} = g^{c.}_{w_1} g^{c.c.}_{w_2} g^{c.c.}_{w_3} g^{c.}_{w_4} \Omega_{w_3} \frac{1}{i\nu - i\omega_{w_2} + \gamma} \qquad (6.32)$$

has to be considered.

Therefore, the original high-dimensional equation system can be replaced by a (very often) low-dimensional system of equations. Using the order parameter system (6.31) it can be shown that the observable physical situation near a critical point (a laser threshold) can be described. In particular, it can be shown that light patterns arise if the pump energy crosses a critical value. This shall not be considered in this book, because this book deals with the statistical behavior of physical systems. Deterministic behavior shall not be considered. However, it shall be emphasized that the spontaneous emergence of light patterns by changing a control parameter (the pump parameter) represents an example of self-organization, i. e. laser systems are self-organizing (synergetic) systems.

The fluctuating forces of the above equations cause additional random behavior, i. e. in particular the real amplitudes of the light field are statistical quantities. This statistical level shall be considered in the following section.

6.2 The Statistical Level of the Laser Activity

The order parameter equation (6.31) can be replaced by an equation of the *Langevin* type. Then it is possible to derive a statistical evolution equation of the *Fokker-Planck* type. As it will be shown, solutions of such an equation describe the observable statistical behavior in a correct way. First, an equation of the *Langevin* type will be considered.

6.2.1 The Langevin Level

Using (6.3), i. e. using

$$E^{c.}_{w_1}(t) = A^{c.}_{w_1}(t) \exp\left(-i\omega_{w_1} t\right) , \qquad (6.33)$$

the order parameter equation (6.31) can be rewritten, i. e. one obtains the complex equation

$$\frac{dA^{c.}_{w_1}(t)}{dt} = \sum_{w_2} \lambda^{c.}_{w_1,w_2}(t) A^{c.}_{w_2}(t) -$$

$$\sum_{w_2,w_3,w_4} \lambda^{c.}_{w_1,w_2,w_3,w_4}(t) A^{c.}_{w_2}(t) A^{c.}_{w_3}(t) A^{c.c.}_{w_4}(t) + F^{(4)}_{w_1}(t) , \qquad (6.34)$$

where the definition

$$\lambda^{\mathrm{c.}}_{w_1,w_2}(t) = \left(D_0\Omega^{\mathrm{c.}}_{w_1,w_2} - \kappa_{w_1}\delta_{w_1,w_2}\right)\exp\left[-\mathrm{i}(\omega_{w_2}-\omega_{w_1})t\right],$$

$$\lambda^{\mathrm{c.}}_{w_1,w_2,w_3,w_4}(t) = 2\frac{D_{\mathrm{c}}}{\gamma_{\mathrm{lin}}}\Omega^{\mathrm{c.}}_{w_1,w_2,w_3,w_4}\exp\left[-\mathrm{i}(\omega_{w_2}+\omega_{w_3}-\omega_{w_4}-\omega_{w_1})t\right] \quad (6.35)$$

has to be considered. Due to the fact that

$$A^{\mathrm{c.}}_{w_1}(t) = E_w(t)\exp\left[-\mathrm{i}\phi_w(t)\right] \tag{6.36}$$

holds (see (6.33)+(6.3)), it is possible to rewrite the equation (6.34), where the so-derived equation is a complex equation. Then the sum of this complex and the conjugate complex equation yields a real equation of the form

$$\boxed{\frac{dE_{w_1}(t)}{dt} = L_{w_1}[\boldsymbol{E}_w(t),t] + F^{(5)}_{w_1}(t),} \tag{6.37}$$

where

$$L_{w_1}[\boldsymbol{E}_w(t),t] = -\left[\sum_{w_2}\lambda^{(2)}_{w_1,w_2}(t)E_{w_2}(t) + \right.$$
$$\left. \sum_{w_2,w_3,w_4}\lambda^{(4)}_{w_1,w_2,w_3,w_4}(t)E_{w_2}(t)E_{w_3}(t)E_{w_4}(t)\right] \tag{6.38}$$

holds. (6.37) represents an equation of the *Langevin* type. The parameters are defined by

$$\lambda^{(2)}_{w_1,w_2}(t) = -\frac{\lambda^{\mathrm{c.}}_{w_1,w_2}(t)\exp\left[\mathrm{i}\phi_{w_2}(t)\right] + \mathrm{c.c.}}{2\cos\left[\phi_{w_1}(t)\right]},$$

$$\lambda^{(4)}_{w_1,w_2,w_3,w_4}(t) = \frac{\lambda^{\mathrm{c.}}_{w_1,w_2,w_3,w_4}(t)\exp\left\{\mathrm{i}\left[\phi_{w_2}(t)+\phi_{w_3}(t)-\phi_{w_4}(t)\right]\right\} + \mathrm{c.c.}}{2\cos\left[\phi_{w_1}(t)\right]}.$$

$$\tag{6.39}$$

In this sense the complex equation (6.31) represents an analytical continuation of an equation of the *Langevin* type.

As it was shown in chapter 5, an equation of the *Langevin* type can be taken as a basis to gain a connected equation of the *Fokker-Planck* type. This connection shall now be considered.

6.2.2 The Fokker-Planck Level

If an equation of the *Langevin* type is defined by (6.37), the connected *Fokker-Planck* equation is of the form

$$\frac{\partial \rho_l(\boldsymbol{E}_w,t)}{\partial t} = \left[-\sum_{w_1} \frac{\partial L_{w_1}(\boldsymbol{E}_w,t)}{\partial E_{w_1}} + \frac{Q_l}{2} \triangle_{\boldsymbol{E}_w} \right] \rho_l(\boldsymbol{E}_w,t) \tag{6.40}$$

(see (6.32) and (6.33)), where

$$L_{w_1}(\boldsymbol{E}_w,t) = -\left[\sum_{w_2} \lambda^{(2)}_{w_1,w_2}(t) E_{w_2} + \sum_{w_2,w_3,w_4} \lambda^{(4)}_{w_1,w_2,w_3,w_4}(t) E_{w_2} E_{w_3} E_{w_4} \right] \tag{6.41}$$

holds. Q_l defines the strength of the fluctuations. The *Laplacian* is defined by

$$\triangle_{\boldsymbol{E}_w} = \sum_{w_1} \frac{\partial^2}{\partial E_{w_1}{}^2} \; , \tag{6.42}$$

and the vector $\boldsymbol{E}_w$ includes all real amplitudes E_w. Solutions of this kind of *Fokker-Planck* equation are distribution functions $\rho_l(\boldsymbol{E}_w,t)$ which describe the statistical behavior of the real amplitudes. These solutions shall be considered in the following. However, only the stationary case

$$\boxed{\left[-\sum_{w_1} \frac{\partial L_{w_1}(\boldsymbol{E}_w)}{\partial E_{w_1}} + \frac{Q_l}{2} \triangle_{\boldsymbol{E}_w} \right] \rho_l(\boldsymbol{E}_w) = 0} \tag{6.43}$$

with

$$\boxed{L_{w_1}(\boldsymbol{E}_w) = -\left(\sum_{w_2} \lambda^{(2)}_{w_1,w_2} E_{w_2} + \sum_{w_2,w_3,w_4} \lambda^{(4)}_{w_1,w_2,w_3,w_4} E_{w_2} E_{w_3} E_{w_4} \right)} \tag{6.44}$$

shall be considered. Such a stationary represents a case which exists on a mean value level, i. e. the fluctuating phases $\phi_w(t)$ and the time-dependent parts $\lambda^{c}_{w_1,w_2}(t)$ and $\lambda^{c}_{w_1,w_2,w_3,w_4}(t)$ of the parameters (6.39) have to be replaced by mean values.

6.2.3 Stationary Solutions

Stationary solutions of an equation of the *Fokker-Planck* type are exponential functions of the kind (5.74) with a central potential which is defined by (5.75). This was shown in section 5.4. In the now considered case the equations (5.74) and (5.75) have to be replaced by

$$\boxed{\rho_l(\boldsymbol{E}_w) = Z_l^{-1} \exp\left[CP_l(\boldsymbol{E}_w) \right], \; Z_l = \int \exp\left[CP_l(\boldsymbol{E}_w) \right] d\boldsymbol{E}_w \; ,} \tag{6.45}$$

with the central potential $CP_l(\boldsymbol{E}_w)$ being defined by

$$\frac{Q_l}{2} \frac{\partial}{\partial E_{w_1}} CP_l(\boldsymbol{E}_w) = L_{w_1}(\boldsymbol{E}_w) \; . \tag{6.46}$$

Using the selection rules

$$\lambda^{(2)}_{w_1,w_2} = \delta_{w_1,w_2}\lambda^{(2)}_{w_1} \;, \;\; \lambda^{(4)}_{w_1,w_2,w_3,w_4} = \delta_{(w_1,w_4)(w_2,w_3)}\lambda^{(4)}_{w_2,w_3} \tag{6.47}$$

and the abbreviations

$$\frac{2}{Q_l}\frac{\lambda^{(2)}_{w_1}}{2} = \lambda^{(2,N)}_{w_1} \;, \;\; \frac{2}{Q_l}\frac{\lambda^{(4)}_{w_1,w_2}}{2} = \lambda^{(4,N)}_{w_1,w_2} \;, \;\; \frac{2}{Q_l}\frac{\lambda^{(4)}_{w_1,w_1}}{4} = \lambda^{(4,N)}_{w_1,w_1} \tag{6.48}$$

one obtains the central potential

$$CP_l(\boldsymbol{E}_w) = -\left(\sum_{w_1} \lambda^{(2,N)}_{w_1} E^2_{w_1} + \sum_{\substack{w_1,w_2 \\ w_1 \neq w_2}} \lambda^{(4,N)}_{w_1,w_2} E^2_{w_1} E^2_{w_2} + \sum_{w_1} \lambda^{(4,N)}_{w_1,w_1} E^4_{w_1} \right) \cdot$$

$$\tag{6.49}$$

The parameters of this central potential can be considered as *Lagrangian* multipliers. In the sense of this book they define a microscoipc level of consideration. The connection with a macroscopic level shall now be considered.

6.2.4 Micro- and Macro-States. Hyper-Surface Equations

The connection with a macroscopic level, the mean value level, is defined by suitable hyper-surface equations, which were introduced in chapter 4. In the following this connection in the context of laser theory shall be discussed.

The One-Mode Laser

The statistical distribution function of a one-mode laser is defined by (6.45) and (6.49) if only one real amplitude E_w is considered, i. e. a distribution function of a one-mode laser is defined by

$$\rho^{(1)}_l(E_w) = Z^{(1)^{-1}}_l \exp\left[-\left(\lambda^{(2,1)}_w E^2_w + \lambda^{(4,1)}_{w,w} E^4_w \right) \right] , \tag{6.50}$$

with the partition function $Z^{(1)}_l$ being defined by

$$Z^{(1)}_l = \int_{-\infty}^{+\infty} \exp\left[-\left(\lambda^{(2,1)}_w E^2_w + \lambda^{(4,1)}_{w,w} E^4_w \right) \right] dE_w \;. \tag{6.51}$$

The now occuring *Lagrangian* multipliers are of the form

$$\lambda^{(2,1)}_w = \frac{\left(\kappa_w - D_0\Omega^{c}_{w,w} \right) \exp\left(i\phi_{\mathrm{mean}} \right) + \mathrm{c.c.}}{2Q_l \cos\left(\phi_{\mathrm{mean}} \right)}$$

$$= \frac{\kappa_w}{Q_l} - D_0 |g^c_w|^2 \Omega^{(1)}_w \;, \tag{6.52}$$

$$\lambda_{w,w}^{(4,1)} = \frac{D_c}{\gamma_{\text{lin}}} \frac{\Omega_{w,w,w,w}^{c.} \exp\left(i\phi_{\text{mean}}\right) + \text{c.c.}}{2Q_l \cos\left(\phi_{\text{mean}}\right)}$$

$$= \frac{D_c}{\gamma_{\text{lin}}} \omega_w \left|g_w^{c.}\right|^4 \Omega_w^{(1)} \tag{6.53}$$

(see (6.49)+(6.48)+(6.39)+(6.35) and bear in mind that the fluctuating phases have to be replaced by mean values), where

$$\Omega_w^{(1)} = \frac{\left[\gamma + \left(\nu - \omega_w\right)\tan\left(\phi_{\text{mean}}\right)\right]}{Q_l\left[\gamma^2 + \left(\nu - \omega_w\right)^2\right]} \tag{6.54}$$

holds. However, instead of the notation (6.52)-(6.53), the notation

$$PH_w = -\lambda_w^{(2,1)} = D_0 \left|g_w^{c.}\right|^2 \Omega_w^{(1)} - \frac{\kappa_w}{Q_l} , \tag{6.55}$$

$$STAB_w = \sqrt{\frac{1}{2\lambda_{w,w}^{(4,1)}}} = \sqrt{\frac{1}{2\frac{D_c}{\gamma_{\text{lin}}} \omega_w \left|g_w^{c.}\right|^4 \Omega_w^{(1)}}} \tag{6.56}$$

shall be used in the following.

The exponential distribution function (6.50) is shown in figure 4.1, chapter 4. There a phase transition point is shown, i. e. a bifurcation occurs. On this statistical level of consideration, the phase transition is correlated with the change of the sign of the paramter $\lambda_w^{(2,1)}$, i. e. this parameter generates the phase transition. This is the reason way the notation (6.55) is used. Moreover, the parameter of fourth order, the parameter $\lambda_{w,w}^{(4,1)}$, guarantees that the distribution is a finite one, i. e. this parameter guarantees the stability of the distribution function. Therefore, the notation (6.56) is used. Due to the fact that this exponential function represents the statistical behavior of the real mode amplitude E_w, figure 4.1 shows that before the transition only one and after the transition two possible stable points exist. Before the transition this stable point has the value 0, i. e. no mean amplitude is observable. This case represents the non-lasing state of the laser system. After the transition two mean amplitudes are possible, i. e. laser activity is observable and a light pattern arises. This behavior is in fact observable. Due to the fact that far from such a critical point only one stable state can be observed (symmetry breaking) and only *Gaussian* fluctuations are possible, the distribution function (6.50) can only be valid near a critical point. This means that the used derivation scheme (in particular the adiabatic approximation) is only valid near a critical point.

The connection between this macroscopic (statistical) and the microscopic level can be described by using relations of chapter 4. This shall now be done.

Measurable quantities in particular are the first two moments of intensity $\langle I \rangle$, $\langle I^2 \rangle$, which are proportional to correlation functions of the real amplitudes, i. e.

$$\langle E_w^2 \rangle \sim \langle I \rangle, \ \langle E_w^4 \rangle \sim \langle I^2 \rangle \tag{6.57}$$

holds. These correlation functions define a macroscopic level of the laser system. The connection of these macroscopic quantities and the statistical variables can be found by using the relations (4.74) and (4.75) if the now necessary variables are used, i. e. if the identification

$$\Omega \leftrightarrow E_w \; ; \; \lambda_1, \sqrt{2\lambda_2} \leftrightarrow -PH_w, STAB_w \tag{6.58}$$

is made. Using (6.58) the one-mode laser case can be described, i. e. these correlation functions are power series of the statistical parameter $(STAB_w)(PH_w)$. These power series are defined by

$$
\begin{aligned}
\langle E_w^2 \rangle &= STAB_w \sum_{i=0}^{\infty} D_2^{(i)} (STAB_w)^i (PH_w)^i , \\
\langle E_w^4 \rangle &= (STAB_w)^2 \sum_{i=0}^{\infty} D_4^{(i)} (STAB_w)^i (PH_w)^i ,
\end{aligned}
\tag{6.59}
$$

where the numbers are defined by

$$D_2^{(0)} = -\frac{D^{(1)}}{D^{(0)}} \; , \quad D_2^{(1)} = \frac{1}{2} - D_{2,1}^{(1)} \; , \quad D_2^{(i>1)} = -D_{2,1}^{(i)} \; ,$$

$$D_{2,1}^{(i>0)} = (-1)^i (i+1) \frac{D^{(i+1)}}{D^{(0)}} - \sum_{k=0}^{i-1} (-1)^{i-k} D_{2,1}^{(k)} \frac{D^{(i-k)}}{D^{(0)}} \; , \tag{6.60}$$

$$D_4^{(0)} = \frac{1}{2} \; , \quad D_4^{(1)} = -\frac{1}{2} \frac{D^{(1)}}{D^{(0)}} \; , \quad D_4^{(2)} = \frac{1}{2} + \frac{1}{2} D_{4,1}^{(2)} \; ,$$

$$D_{4,1}^{(0)} = 0 \; , \quad D_{4,1}^{(1)} = -\frac{D^{(1)}}{D^{(0)}} \; ,$$

$$D_{4,1}^{(i>1)} = (-1)^i i \frac{D^{(i)}}{D^{(0)}} - \sum_{k=0}^{i-1} (-1)^{i-k} D_{4,1}^{(k)} \frac{D^{(i-k)}}{D^{(0)}} \; . \tag{6.61}$$

The numbers $D^{(k)}$ are the coefficients of the used parabolic cylinder function. These coefficients are defined by (4.51). The power series (6.59) are valid if the exponential function (6.50) is valid, i. e. the power series (6.59) are valid near a critical point. The narrow surrounding of a critical point is characterized by

$$|PH_w| \ll 1 \; . \tag{6.62}$$

Therefore, within the narrow surrounding of a critical point the power series (6.59) can be replaced by

$$
\begin{aligned}
\langle E_w^2 \rangle_{\text{narrow}} &= STAB_w \left[D_2^{(0)} + D_2^{(1)} (STAB_w)(PH_w) \right] , \\
\langle E_w^4 \rangle_{\text{narrow}} &= (STAB_w)^2 \left[D_4^{(0)} + D_4^{(1)} (STAB_w)(PH_w) \right] ,
\end{aligned}
\tag{6.63}
$$

and the critical point itself is characterized by the well-known relation

$$\left\langle E_w^2 \right\rangle_{\text{crit}} = D_2^{(0)} STAB_w = -\frac{D^{(1)}}{D^{(0)}} STAB_w \; ,$$

$$\left\langle E_w^4 \right\rangle_{\text{crit}} = D_4^{(0)} \left(STAB_w \right)^2 = \frac{1}{2} \left(STAB_w \right)^2 \; . \tag{6.64}$$

This critical point can be characterized by using a macroscopic phase transition condition, namely the condition (4.93). In the now used notation this means that the condition

$$\boxed{\left\langle E_w^4 \right\rangle_{\text{crit}} = C_{\text{crit}} \left\langle E_w^2 \right\rangle_{\text{crit}}^2 = \frac{D^{(0)^2}}{2 D^{(1)^2}} \left\langle E_w^2 \right\rangle_{\text{crit}}^2} \tag{6.65}$$

characterizes the laser threshold. Comparing the macroscopic condition (6.65) with the relations (6.64) it can be seen that the now derived relations (6.64) are equivalent to the macroscopic phase transition condition which was derived in chapter 4. Then it has to be noticed that at the critical point the second moment of intensity is proportional to the square of the first moment of intensity.

In the narrow surrounding of the critical point the *Lagrangian* multipliers are definable by critical hyper-surface equations. These equations shall now be considered.

In chapter 4 hyper-surface equations were derived which are valid within a narrow surrounding of a critical point (see (4.98) and (4.100)). Using the now necessary notation these critical hyper-surface equations are defined by

$$\boxed{\lambda_w^{(2,1)} = \frac{1}{\left\langle E_w^2 \right\rangle} \left[\frac{1}{2} - \frac{\left\langle E_w^4 \right\rangle}{\left\langle E_w^2 \right\rangle^2} \left(C_1 - \sqrt{C_1^2 - C_2 \frac{\left\langle E_w^4 \right\rangle}{\left\langle E_w^2 \right\rangle^2}} \right)^{-2} \right]} \tag{6.66}$$

$$\boxed{\lambda_{w,w}^{(4,1)} = \frac{1}{\left\langle E_w^2 \right\rangle^2} \left(\sqrt{2} C_1 - \sqrt{2} \sqrt{C_1^2 - C_2 \frac{\left\langle E_w^4 \right\rangle}{\left\langle E_w^2 \right\rangle^2}} \right)^{-2}} \tag{6.67}$$

(The coefficients C_1, C_2 are defined by (4.90).) Figure 6.1 illustrates these hyper-surface equations. Due to the fact that the *Lagrangian* parameters $\lambda_w^{(2,1)}$, $\lambda_{w,w}^{(4,1)}$ are functions of all relevant specific parameters such as the decay constants or the critical inversion, the hyper-surface equations determine these specific parameters by using moments of the mode intensity. As figure 6.1 shows, the parameter of second order $\lambda_w^{(2,1)}$ changes its sign at the critical point. This behavior cannot be observed if the picture of the parameter of fourth oder $\lambda_{w,w}^{(4,1)}$ is taken into consideration. Therefore, the behavior which is shown in figure 6.1 exactly represents the above discussed behavior.

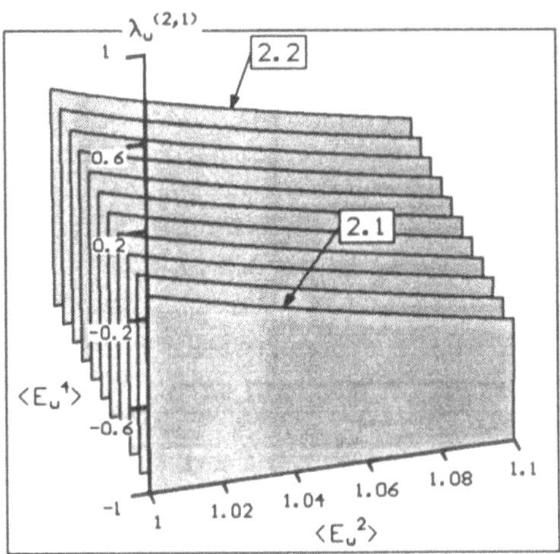
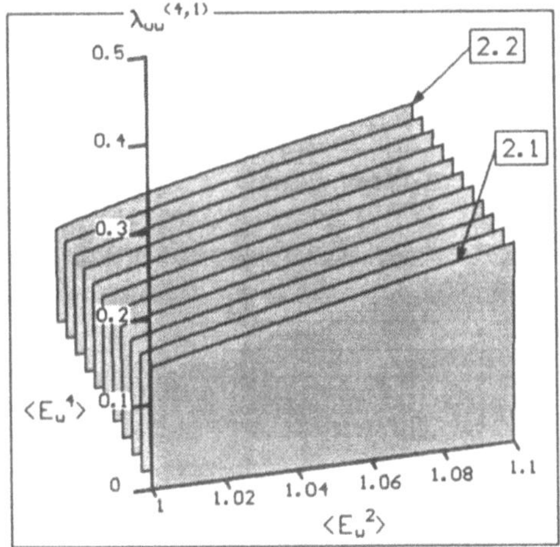

Figure 6.1   Critical Hyper-Surface Equations

The whole surrounding of the considered critical point has to be described by using hyper-surface equations of the kind (4.122). These equations shall now be discussed.

Using the identification (6.58) the hyper-surface equations (4.122) take on the form

$$
\lambda_w^{(2,1)} = \frac{1}{\langle E_w^2 \rangle} \sum_{i=0}^{\infty} \Lambda_1^{(i)} \left(\frac{\langle E_w^4 \rangle}{\langle E_w^2 \rangle^2} \right)^i ,
$$

$$
\lambda_{w,w}^{(4,1)} = \frac{1}{\langle E_w^2 \rangle^2} \sum_{i=0}^{\infty} \Lambda_2^{(i)} \left(\frac{\langle E_w^4 \rangle}{\langle E_w^2 \rangle^2} \right)^i ,
$$

$$(6.68)$$

where the coefficients are defined by a recursive scheme (see (4.124)ff.). As the reader can see, *Lagrangian* multipliers and moments of intensity are connected by ordinary power series. $\langle E_w^4 \rangle / \langle E_w^2 \rangle^2$ are dimensionless quotients and can be replaced by an ordinary number if a critical point is considered, because then the macroscopic phase transition condition (6.65) holds. In this case the well-known function

$$
\lambda_{w,w}^{(4,1)} \sim \frac{1}{\langle E_w^2 \rangle^2}
$$

$$(6.69)$$

replaces the second power series of (6.68). In order to show that the parameter $\lambda_w^{(2,1)}$ has to vanish in a self-consistent way, the coefficients of the first power series of (6.68) have to be calculated. However, this shall not be done in this book.

Using the above discussed hyper-surface equations a macroscopic determination of one-mode laser processes is possible. In the following the multi-mode laser shall be considered.

The Multi-Mode Laser

The statistical distribution function of the multi-mode laser is defined by

$$
\rho_l^{(N)}(\boldsymbol{E}_w) = Z_l^{(N)^{-1}} \exp\left[-\left(\sum_{w_1} \lambda_{w_1}^{(2,N)} E_{w_1}^2 + \sum_{\substack{w_1,w_2 \\ w_1 \neq w_2}} \lambda_{w_1,w_2}^{(4,N)} E_{w_1}^2 E_{w_2}^2 + \sum_{w_1} \lambda_{w_1,w_1}^{(4,N)} E_{w_1}^4 \right) \right] ,
\tag{6.70}
$$

with the partition function $Z_l^{(1)}$ being defined by

$$
Z_l^{(N)} = \int_{-\infty}^{+\infty} \exp\left[-\left(\sum_{w_1} \lambda_{w_1}^{(2,N)} E_{w_1}^2 + \sum_{\substack{w_1,w_2 \\ w_1 \neq w_2}} \lambda_{w_1,w_2}^{(4,N)} E_{w_1}^2 E_{w_2}^2 + \sum_{w_1} \lambda_{w_1,w_1}^{(4,N)} E_{w_1}^4 \right) \right] d\boldsymbol{E}_w .
\tag{6.71}
$$

Using the notation

$$
PH_{w_1} = -\lambda_{w_1}^{(2,N)} = D_0 \left| g_{w_1}^{\mathrm{c.}} \right|^2 \Omega_{w_1}^{(1)} - \frac{\kappa_{w_1}}{Q_l} ,
\tag{6.72}
$$

$$
STAB_{w_1} = \sqrt{\frac{1}{2\lambda_{w_1,w_1}^{(4,N)}}} = \sqrt{\frac{1}{2\frac{D_c}{\gamma_{\mathrm{lin}}}\omega_{w_1} \left| g_{w_1}^{\mathrm{c.}} \right|^4 \Omega_{w_1}^{(1)}}} ,
\tag{6.73}
$$

$$
\begin{aligned}
COUP_{w_1,w_2} &= \lambda_{w_1,w_2}^{(c,N)} = \lambda_{w_1,w_2}^{(4,N)} + \lambda_{w_2,w_1}^{(4,N)} \\
&= 2\frac{D_c}{\gamma_{\mathrm{lin}}} \left| g_{w_1}^{\mathrm{c.}} \right|^2 \left| g_{w_2}^{\mathrm{c.}} \right|^2 \left(\Omega_{w_1}^{(1)}\Omega_{w_2} + \Omega_{w_1}\Omega_{w_2}^{(1)} \right) ,
\end{aligned}
\tag{6.74}
$$

and using the explicit formulation (4.46) which is valid in the now considered case, one obtains the partitition function

$$
Z_l^{(N)} = \left\{ \prod_{\substack{w_1,w_2 \\ w_1 < w_2}} \sum_{\nu_{w_1,w_2}=0}^{\infty} \right\} \left[\prod_{w_1} k_{\mu_{w_1}}^{(w_1)} \right] \left\{ COUP_{w_1,w_2}^{\nu_{w_1,w_2}} \right\} \left[\left\{ \prod_{w_1} Z_{l,w_1}^{(1)} \right\} \right].
$$

$$(6.75)$$

(6.75) represents the explicit formulation of the integral expression (6.71). The notation (6.72) guarantees that the various parts of this partition function can directly be interpreted, i. e. the general partition function represents a product of elementary partition functions

$$
\begin{aligned}
Z_{l,w_1}^{(1)} &= \int_{-\infty}^{+\infty} \exp\left[-\left(\lambda_{w_1}^{(2,N)} E_{w_1}^2 + \lambda_{w_1,w_1}^{(4,N)} E_{w_1}^4 \right) \right] dE_{w_1} \\
&= \left(2\lambda_{w_1,w_1}^{(4,N)} \right)^{-1/4} \Gamma(1/2) \exp\left(\lambda_{w_1}^{(2,N)^2} / 8\lambda_{w_1,w_1}^{(4,N)} \right) \\
&\quad D_{-1/2}\left(\lambda_{w_1}^{(2,N)} / \sqrt{2\lambda_{w_1,w_1}^{(4,N)}} \right),
\end{aligned}
$$

$$(6.76)$$

where the copuling parameters $COUP_{w_1,w_2}$ generate the coupling of the elementary partition functions. The coefficients $k_{\mu_{w_1}}^{(w_1)}$ are the introduced real elementary coefficients (see (4.44) or (4.148), respectively), i. e. now the relation

$$
k_{\mu_{w_1}}^{(w_1)} = \sum_{\xi_{w_1}=0}^{\infty} k_{\mu_{w_1},\xi_{w_1}}^{(1)} (STAB_{w_1})^{\mu_{w_1}+\xi_{w_1}} (PH_{w_1})^{\xi_{w_1}}
$$

$$(6.77)$$

holds. The numbers of (6.77) are defined by

$$
k_{\mu_{w_1},\xi_{w_1}}^{(1)} = \left(-\sqrt{2} \right)^{\mu_{w_1}+\xi_{w_1}} g_{\mu_{w_1},\xi_{w_1}}^{(6)} \left\{ \prod_{\substack{w_2 \\ w_1 < w_2}} \frac{(-1)^{\nu_{w_1,w_2}}}{\nu_{w_1,w_2}!} \right\},
$$

$$
\mu_{w_1} = \sum_{\substack{w_2 \\ w_1 \neq w_2}} \nu_{w_1,w_2},
$$

$$(6.78)$$

wherein the numbers $g^{(6)}$ were introduced in chapter 4 (see (4.116)ff.). It has to be noted here that one gains the general partition function (6.75) by using partition functions which are connected with one single system, i. e. the elementary partition function is identical with the partition function of the one-mode laser (6.51). This mathematical result corresponds to the universal physical principle that physical systems of higher order are built up of elementary physical systems (see figure 1.1, in chapter 1).

Such a specific partition function can be taken as a basis to calculate macroscopic quantities. This possibility was shown in chapter 2. There thermodynamic systems were considered. However, such a partition function formalism is a universal formalism. For example, the correlation functions of the mode amplitudes can be calculated by using the partition function (6.75). This shall now be shown.

Using the partition function (6.75) the correlation functions $\langle E_{w_1}^2 \rangle$, $\langle E_{w_1}^4 \rangle$, $\langle E_{w_1}^2 E_{w_2}^2 \rangle$ can be represented by a system of differential equations, i. e. the differential system

$$
\boxed{
\begin{aligned}
\langle E_{w_1}^2 \rangle &= -Z_l^{(N)^{-1}} \frac{\partial Z_l^{(N)}}{\partial \lambda_{w_1}^{(2,N)}} \ , \\[2ex]
\langle E_{w_1}^4 \rangle &= Z_l^{(N)^{-1}} \frac{\partial^2 Z_l^{(N)}}{\partial \lambda_{w_1}^{(2,N)^2}} \ , \\[2ex]
\langle E_{w_1}^2 E_{w_2}^2 \rangle &= -Z_l^{(N)^{-1}} \frac{\partial Z_l^{(N)}}{\partial \lambda_{w_1,w_2}^{(c,N)}}
\end{aligned}
}
\tag{6.79}
$$

holds. Other macroscopic quantities can be calculated as well, because such a partition function includes the total statistical information of the physical system. Furthermore, this differential equation system represents nothing but a special case of a basic equation system (see chapter 4). Such a basic equation system can be used to calculate the correlated hyper-surface equations. This was shown in chapter 4. Due to the fact that the basic equation system (6.79) represents a special case of the basic equation system (4.144), the correlated hyper-surface equations are identical with (4.154) if a suitable identification is made. This shall now be discussed.

Using the notation

$$
\lambda_{w_1}^{(2,N)} := \lambda_{w_1,0}^{(\text{laser})} \ , \quad \lambda_{w_1,w_1}^{(4,N)} := \lambda_{w_1,w_1}^{(\text{laser})} \ , \quad \lambda_{w_1,w_2}^{(c,N)} := \lambda_{w_1,w_2}^{(\text{laser})} \ (w_1 \neq w_2) \ ,
\tag{6.80}
$$

and using the identification

$$
\lambda_{i,\gamma} \leftrightarrow \lambda_{w_1,w_2}^{(\text{laser})} \ , \quad \Omega_i \leftrightarrow E_{w_1} \ ,
\tag{6.81}
$$

the formula (4.154) can be used to describe the multi-mode laser system, i. e. in this case one obtains the hyper-surface equations

$$
\boxed{
\begin{aligned}
\lambda_{w_1,\gamma}^{(\text{laser})} &= \frac{1}{\langle E_{w_1}^2 \rangle \langle E_{\gamma}^2 \rangle} \left\{ \prod_{\substack{w_3,w_4 \\ w_3 \leq w_4}} \sum_{\epsilon_{w_3,w_4}=0}^{\infty} \Lambda_{w_1,\gamma}^{(\{\epsilon_{w_3,w_4}\})} \left(\frac{\langle E_{w_3}^2 E_{w_4}^2 \rangle}{\langle E_{w_3}^2 \rangle \langle E_{w_4}^2 \rangle} \right)^{\epsilon_{w_3,w_4}} \right\} \\[2ex]
&(\gamma = \{0, w_2\} \ , \ E_{\gamma=0} = 1) \ .
\end{aligned}
}
$$

$$
\tag{6.82}
$$

(The coefficients $\Lambda_{w_1,\gamma}^{(\{\epsilon_{w_3,w_4}\})}$ are defined by the recursive scheme (4.156)ff., chapter 4.) (6.83) represents a self-similar power series function of fourth order. All measurable real amplitudes E_{w_1} occur in these hyper-surface equations. Using (6.82) a totally analytical determination of the *Lagrangian* multipliers is possible. Then the access is a macroscopic one. Due to the fact that the MIEP can be taken as a basis to gain the distribution function (6.70), a totally macroscopic determination scheme is given. All measured details (for example, the dependence of the pump parameter) can be mapped, i. e. it can be calculated in which way statistical parameters depend on such details.

Far from a laser threshold a *Gaussian* distribution function is sufficient to describe the statistical behavior of the mode amplitudes. This was above mentioned. In this case hyper-surface equations exist, too. These equations shall now be considered.

Gaussian Hyper-Surface Equations

Far from a critical point a distribution function of a one-mode laser is of *Gaussian* form, i. e. in this case the function

$$\boxed{\rho_{l,G}^{(1)}(E_w) = Z_{l,G}^{(1)\,-1} \exp\left[-\left(\lambda_{w,G}^{(1,1)} E_w + \lambda_{w,G}^{(2,1)} E_w^2 \right) \right]} \tag{6.83}$$

is valid, with the partition function $Z_{l,G}^{(1)}$ being defined by

$$\boxed{Z_{l,G}^{(1)} = \int_{-\infty}^{+\infty} \exp\left[-\left(\lambda_{w,G}^{(1,1)} E_w + \lambda_{w,G}^{(2,1)} E_w^2 \right) \right] dE_w \; .} \tag{6.84}$$

Figure 6.2 shows such *Gaussian* distribution functions. (6.83) includes two kinds of distribution functions, namely functions with $\lambda_{w,G}^{(1,1)} = 0$ and with $\lambda_{w,G}^{(1,1)} \neq 0$. The first kind of distribution function is valid in the non-lasing state, i. e. before a phase transition occurs – far from the critical point. The second kind is valid in the lasing state, i. e. after the transition – far from the critical point. However, to describe the connection between the mean value level and the statistical level in an analytical way, suitable hyper-surface equations are needed. Due to the fact that the general formula (4.178) includes as well *Gaussian* problems, the now occuring statistical problem can be solved by using a special case of (4.178). This special case is defined by $O = 2$, $N = 1$, i. e. now the relation

$$\boxed{\begin{aligned} \lambda_{w,G}^{(1,1)} &= \frac{1}{\sqrt{\langle E_w^2 \rangle}} \sum_{i=0}^{\infty} \Lambda_{1,G}^{(i)} \left(\frac{\langle E_w \rangle}{\sqrt{\langle E_w^2 \rangle}} \right)^{i} \,, \\[2ex] \lambda_{w,G}^{(2,1)} &= \frac{1}{\langle E_w^2 \rangle} \sum_{i=0}^{\infty} \Lambda_{2,G}^{(i)} \left(\frac{\langle E_w \rangle}{\sqrt{\langle E_w^2 \rangle}} \right)^{i} \end{aligned}} \tag{6.85}$$

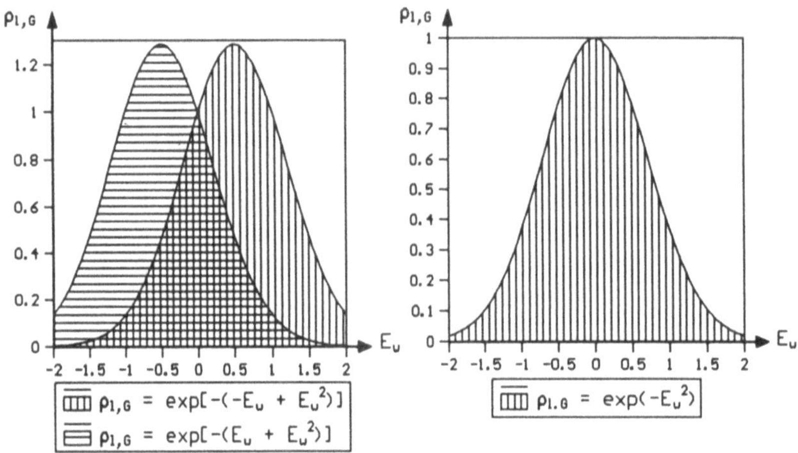

Figure 6.2   Gaussian distribution functions

holds, where the numbers $\Lambda^{(i)}_{1,G}$, $\Lambda^{(i)}_{2,G}$ of these ordinary power series are defined by a recursive scheme (see (4.181)ff., chapter 4). Due to the fact that in the non-lasing state $(\lambda^{(1,1)}_{w,G} = 0)$ the correlation function $\langle E_w \rangle$ vanishes, one obtains the well-known relation

$$\lambda^{(2,1)}_{w,G} \sim \frac{1}{\langle E_w^2 \rangle} \, . \tag{6.86}$$

Therefore, it has been shown that the concept of hyper-surface equations can be used to solve determination problems of statistical laser theory. Due to the fact that the maximum information entropy principle can be taken as a basis, a totally macroscopic determination strategy of statistical distribution functions has been introduced. Additionally, phase transition conditions and all relevant macroscopic quantities can be calculated. Therefore, a totally analytical concept of system theory has been introduced. Such a concept is a universal one, i. e. this concept can be used in various fields of statistical physics such as thermodynamics, laser physics and brain research.

In the following quantum systems shall be considered. Then an extension of methods of system theory shall be discussed.

7 Aspects of Quantum System Theory

Multi-component systems of the microscopic world, i. e. high-dimensional quantum systems, can be described by various formalisms, which are in the end equivalent. A formalism on the basis of a differential evolution equation, the *Schrödinger* equation, as well as a formalism on the basis of a path integral, the *Feynman* path integral, is possible, where this integral formalism represents nothing else but the method of Green's function. By some kinds of transformation one can get other possible formalisms like the *Heisenberg* formalism or the interaction formalism. In this chapter in particular I want to look at the method of *R. P. Feynman*, and I want to present the extension of the introduced method of integral calculation (see chapter 4) to quantum systems. In this context a statistical basic function will be introduced, which contains all statistically relevant information of both quantum systems and non-microscopic systems (like laser systems or thermodynamic systems). Thus, the universality of the introduced concept of system theory will become evident, and therefore a pathway to quantum systems is given. Furthermore, I will again attach importance to the problem of interrelation between equations of the *Schrödinger* and the *Fokker-Planck* type.

First, I would like to make some short remarks about my motivation to write this chapter.

7.1 Motivation

During 1989 and 1992 I participated in a project of the Deutsche Forschungsgemeinschaft (Sonderforschungsbereich 329, Physikalische und chemische Grundlagen der Molekularelektronik) which was founded to examine the possibility to construct molecular devices (see [78, 79]). In this context various organic molecules were examined. Figures 7.1 and 7.2 show some trivial examples of such molecules. However, it is not possible to construct devices by using such low-dimensional molecular structures, because stochastic effects such as spontaneous emission cause a fluctuating behavior which cannot be neglected. Due to the fact that technical devices normally have to be deterministic devices, it is normally not possible to use low-dimensional molecular structures. However, if high-dimensional molecular structures are used, a collective process signal is possible in such a way that the process signal and the overlaid stochastic signal are on different scales, in which case this behavior in last consequence defines the meaning of the term *high-dimensional*. Then a deterministic device can be constructed. An example of a very high-dimensional molecular structure which shows a process signal and a small overlaid stochastic signal is represented

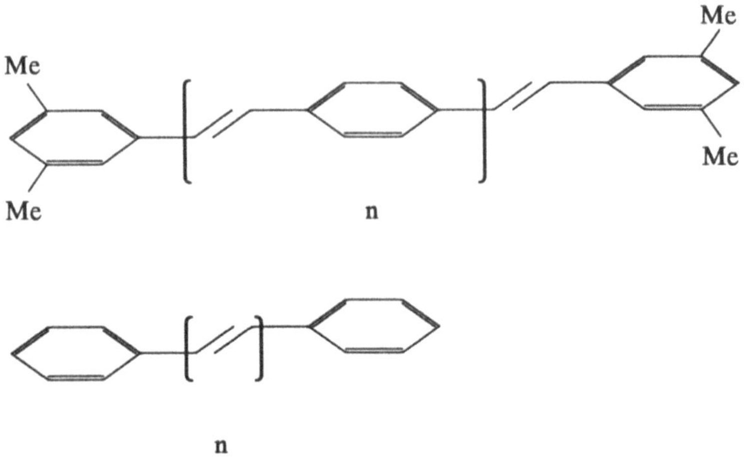

Figure 7.1 Above: The phenylene-vinylidene molecule. Below: The diphenylene-polyene molecule

by a laser system. Such a laser system represents a macroscopic device. However, it is not necessary to use macroscopic devices, microscopic - high-dimensional - molecular devices in principle are possible as well. Thus, it has to be emphasized that molecular devices normally have to be systems of basic molecules. Dealing with such a molecular system then means dealing with high-dimensional quantum sys-

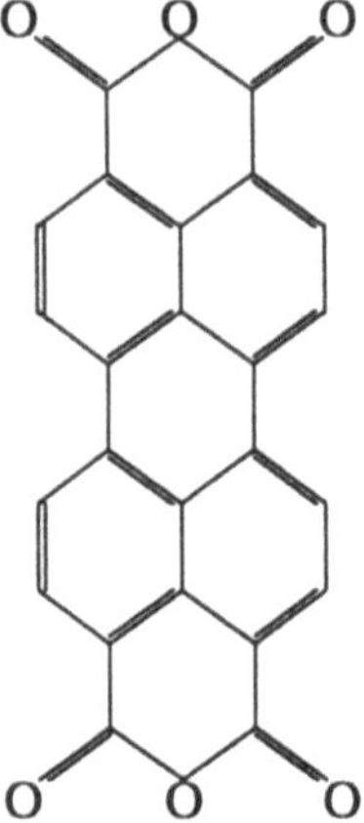

Figure 7.2
The perylenetetracarbonaciddianhydride
molecule (PTCDA)

tems, and this is what I have done. Due to the fact that I found many interesting aspects, I thought it would make sense to write some of these aspects down, i. e. this was motivation enough for me to write a chapter about quantum system theory. Due to the fact that this book is about *universality in statistical physics*, essential connections with *Fokker-Planck* systems shall be considered, too.

In the following the *Schrödinger* equation of quantum mechanics shall be considered. This shall be the starting point of the following considerations.

7.2 A Basic Evolution Equation

A basic evolution equation of quantum systems is represented by *Schrödinger's* time-dependent equation. If a multi-particle system is considered, this equation reads

$$\left[-\frac{\hbar^2}{2m_0} \sum_{l=1}^{N} \sum_{c=1}^{3} \frac{\partial^2}{\partial x_c^{(l)2}} + V(q) \right] \Psi(q,t) = i\hbar \frac{\partial}{\partial t} \Psi(q,t) \,, \qquad (7.1)$$

where it has been assumed that every particle has the non-relativistic mass m_0. N represents the number of the particles, $\hbar$ is *Planck's* constant. Additionally, it shall be assumed that the potential function is time-independent, i.e. $V(q)$ holds. $V(q)$ shall include interaction functions $INT(q)$. The vector q represents all *Cartesian* coordinates of all particles, i. e. the relation

$$q = \left(x_1^{(1)}, x_2^{(1)}, x_3^{(1)}, \ldots, x_1^{(N)}, x_2^{(N)}, x_3^{(N)} \right) \qquad (7.2)$$

holds. This equation actually represents a basic evolution equation of microscopic systems, i. e. the validity of (7.1) can be shown by many experiments. In particular, the time-independent *Schrödinger* equation of one particle, i. e. the equation

$$\left[-\frac{\hbar^2}{2m_0} \sum_{c=1}^{3} \frac{\partial^2}{\partial x_c^{(l)2}} + V\left(q^{(l)}\right) \right] \Phi^{(\nu)}\left(q^{(l)}\right) = E^{(\nu)}\Phi^{(\nu)}\left(q^{(l)}\right) \,, \qquad (7.3)$$

can be derived (in (7.3) a vector $q^{(l)}$ represents one special particle, to derive this equation the derivation scheme of chapter 5 can be used, i. e. insert (5.60) into (7.1) with $N = 1$ and use (5.58), (5.56) with $INV = -i$, $w_{\text{sig}} = \hbar$), and with this stationary *Schrödinger* equation actually measurable energy eigenvalues $E^{(\nu)}$ can be calculated. An example was given in chapter 5. In this chapter equations of Schrödinger's type were considered. Especially the quantum mechanical harmonic oscillator problem was explicitly solved. See figures 5.5 and 5.1, 5.2. There it was shown that if an equation of the stationary *Schrödinger* type is considered, only special eigenvalues are correlated with non-divergent solutions. As molecular experiments show, these eigenvalues actually represent measurable energy eigenvalues of bounded states. And the discrete structure of these eigenvalues represents actually the measurable discrete energy structure. Such a discrete structure is a characteristic property of quantum systems. Figure 7.3 shows an example. In figure 7.3 the

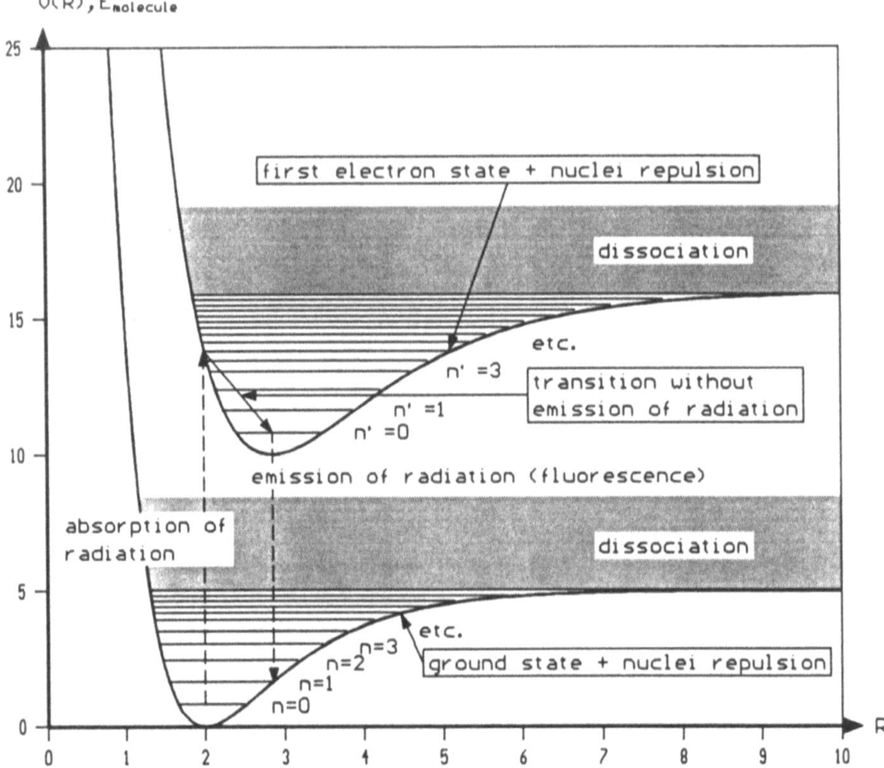

Figure 7.3 The energy scheme of the hydrogen molecule. R represents the space between the two nuclei. n represents a quantum number which characterizes the vibrational states of the nuclei. $V(R)$ represents a potential function which is the result of the nuclei repulsion and the attraction effect of the electrons between the nuclei

problem of a hydrogen molecule H_2 is presented. This picture represents the measurable energy states in a schematic way. As the reader can see, discrete electron states are observable, where additional discrete vibrational states $(n = 0, 1, \ldots \infty)$ of the nuclei can be observed. Absorption and emission of radiation (i. e. light) creates new states of the molecule. However, only special transitions are possible. Such transitions can be determined by using *selection rules* such as the *Franck-Condon principle*. These scheme can be calculated by using a *Schrödinger* equation. Other examples such as the well-known stationary hydrogen atom problem can also be calculated. Therefore, the time-dependent *Schrödinger* equation can be interpreted as a basic evolution equation of quantum systems.

In this context it has to be emphasized that it is not possible to gain a valid stationary *Schrödinger* equation by using the differential condition $\partial \Psi(q, t)/\partial t = 0$. For example, such conditions can be used if equations of the *Fokker-Planck* type are considered. In the context of time-dependent *Schrödinger* equations special formulations of the indirect state functions $\Psi(q,t)$ (which are normally called *wave functions*) have to be used. If the reader compares with the comments in chapter 5, the reason is obvious, i. e. there it was shown that equations of a time-dependent *Schrödinger* type are ensemble equations, in which case ensemble equations are connected with basic stationary equations by special formulations of the indirect state functions. However, there I did not consider multi-particle *Schrödinger* equations. This, however, does not change the principle statements.

Solutions of such a partial differential equation can be formulated in a general way by using a special functional integral, namely *Feynman's* path integral. In the following these integrals shall be considered.

7.3 Feynman's Path Integrals

Solutions of the *Schrödinger* equation (7.1) can be represented by using special functional integrals. These integrals are called *Feynman path integrals*. *Feynman* path integrals are integral representions of the indirect state functions $\Psi(q,t)$ by using *Green's* method. Using *Feynman's* path integrals then means using a special mathematical concept of quantum theory. In the following such functional integrals shall be discussed. In this context it shall be shown that the introduced concept of system theory (see chapters 3, 4) can be extended if an analytical continuation is used. Moreover, normally high-dimensional particle systems shall be taken as a basis. This is due to the fact that molecular devices have to be high-dimensional molecular systems. First, some basic remarks shall be given (for further information, see [17, 24]).

7.3.1 The Method of Green's Function

The operator of the l. h. s. of (7.1) represents a special *Hamiltonian*, where the notation

$$\hat{H}(q) = -\frac{\hbar^2}{2m_0} \sum_{l=1}^{N} \sum_{c=1}^{3} \frac{\partial^2}{\partial x_c^{(l)2}} + V(q) \tag{7.4}$$

shall be used. Then the time-dependent *Schrödinger* equation reads

$$\hat{H}(q)\Psi(q,t) = i\hbar\frac{\partial}{\partial t}\Psi(q,t) := g(q,t)\,, \tag{7.5}$$

where (7.5) can be replaced by

$$\int_{-\infty}^{+\infty} \hat{G}(q,\tilde{q})\hat{H}(\tilde{q})\Psi(\tilde{q},t)\,d\tilde{q} = \int_{-\infty}^{+\infty} \hat{G}(q,\tilde{q})g(\tilde{q},t)\,d\tilde{q} = \Psi(q,t) \tag{7.6}$$

if the operator

$$\hat{G}(\boldsymbol{q}, \tilde{\boldsymbol{q}}) = \delta(\boldsymbol{q} - \tilde{\boldsymbol{q}})\hat{H}^{-1}(\tilde{\boldsymbol{q}}) \tag{7.7}$$

is introduced. $\hat{H}^{-1}(\tilde{\boldsymbol{q}})$ represents the inverse operator of $\hat{H}(\tilde{\boldsymbol{q}})$. $\delta(\boldsymbol{q} - \tilde{\boldsymbol{q}})$ is *Dirac's* delta function. Using the representation

$$\Psi(\tilde{\boldsymbol{q}}, t) = \exp\left[-\frac{\mathrm{i}}{\hbar}\hat{H}(\tilde{\boldsymbol{q}})t\right]\Psi(\tilde{\boldsymbol{q}}, 0) \tag{7.8}$$

(to show the validity of (7.8), insert (7.8) into the *Schrödinger* equation (7.5)), and inserting (7.7) one obtains the relation

$$\int_{-\infty}^{+\infty} \hat{G}(\boldsymbol{q}, \tilde{\boldsymbol{q}}, t)\Psi(\tilde{\boldsymbol{q}}, 0)\, d\tilde{\boldsymbol{q}} = \Psi(\boldsymbol{q}, t)\ , \tag{7.9}$$

with $\hat{G}(\boldsymbol{q}, \tilde{\boldsymbol{q}}, t)$ being defined by

$$\hat{G}(\boldsymbol{q}, \tilde{\boldsymbol{q}}, t) = \delta(\boldsymbol{q} - \tilde{\boldsymbol{q}})\exp\left[-\frac{\mathrm{i}}{\hbar}\hat{H}(\tilde{\boldsymbol{q}})t\right]\ . \tag{7.10}$$

(7.9) is a special functional representation of the indirect state function $\Psi(\boldsymbol{q}, t)$. $\hat{G}(\boldsymbol{q}, \tilde{\boldsymbol{q}}, t)$ represents a time-dependent operator. Introducing a suitable function $G(\boldsymbol{q}, \tilde{\boldsymbol{q}}, t)$ the operator representation (7.9) can be replaced by

$$\boxed{\int_{-\infty}^{+\infty} G(\boldsymbol{q}, \tilde{\boldsymbol{q}}, t)\Psi(\tilde{\boldsymbol{q}}, 0)\, d\tilde{\boldsymbol{q}} = \Psi(\boldsymbol{q}, t)\ ,} \tag{7.11}$$

where $G(\boldsymbol{q}, \tilde{\boldsymbol{q}}, t)$ is nothing but a special kind of *Green's* function. Such a function very often is called *influence function, propagator* or *kernel*. Such a function is able to generate the indirect state function $\Psi(\boldsymbol{q}, t)$ if the initial function $\Psi(\boldsymbol{q}, 0)$ is given. The expression (7.11) represents a normalized function $\Psi(\boldsymbol{q}, t)$ if *Green's* function includes a normalization factor $\tilde{S}$. This shall be assumed. The explicit form of the kernel $G(\boldsymbol{q}, \tilde{\boldsymbol{q}}, t)$ shall now be considered.

7.3.2 The Feynman Kernel

The Kernel

If the kernel $G(\boldsymbol{q}, \tilde{\boldsymbol{q}}, t)$ is defined by (for example, see [15])

$$\boxed{\begin{aligned} &G(\boldsymbol{q}, \tilde{\boldsymbol{q}}, t) = \tilde{S}\int_{\Gamma} \exp\left[\frac{\mathrm{i}}{\hbar}\int_0^t \left\{\frac{m_0}{2}\left[\frac{dq(\tau)}{d\tau}\right]^2 - V[q(\tau)]\right\}d\tau\right] \prod_{0 \le \tau \le t}^{cont} dq(\tau) \\ &\Gamma = \{q(\tau) \in R^{3N}\ ,\ \boldsymbol{q}(0) := \boldsymbol{q}_0 = \tilde{\boldsymbol{q}}\ ,\ \boldsymbol{q}(t) := \boldsymbol{q}_P = \boldsymbol{q}\}\ , \end{aligned}}$$

$$\tag{7.12}$$

the integral relation (7.11) represents solutions of the basic evolution equation (7.1). R^{3N} represents the $3N$-dimensional *Euclidean position-time space*. $\tilde{q}$ is the vector of the coordiantes at time 0, and q is the vector at time t. Γ represents the set of all integration curves between the fixed points q and $\tilde{q}$. This kernel is a functional integral in the introduced sense (see chapter 4, subsection 4.3.8), i. e. to calculate such a kernel, a lattice representation of the kind

$$
G(q_P, q_0, \Delta t) =
$$
$$
S \int_{-\infty}^{+\infty} \exp\left\{ \frac{i}{\hbar} \sum_{k=0}^{P-1} \left[\frac{m_0}{2} \left(\frac{q_{k+1} - q_k}{\Delta t} \right)^2 - V(q_k) \right] \Delta t \right\} dq_{P-1} \ldots dq_1
$$

$$(7.13)$$

has to be used. The representation (7.12) (or (7.13), respectively) can be called *Feynman kernel*. In this book the *Feynman* kernel (7.13) sometimes is called a *profunction*. S represents the normalization factor of the lattice representation.

The Indirect State Function in the Discrete Case

In the discrete case (7.13) the indirecte state function has to be defined by

$$
\int_{-\infty}^{+\infty} \lim_{\substack{\Delta t \to 0 \\ P \to \infty}} G(q_P, q_0, \Delta t) \Psi(q_0, 0) \, dq_0 = \Psi(q_P, t) \, .
$$

$$(7.14)$$

In this formulation k represents the different time points, and the vectors q_k are defined by

$$
q_k = \left(x_{1,k}^{(1)}, x_{2,k}^{(1)}, x_{3,k}^{(1)}, \ldots, x_{1,k}^{(N)}, x_{2,k}^{(N)}, x_{3,k}^{(N)} \right) \, .
$$

$$(7.15)$$

The Action Function

The central part of such a *Feynman* kernel is nothing but the action function

$$
W = \begin{cases} \int_0^t L \, d\tau & \text{in the continuous case} \\ \sum_{k=0}^{P-1} L \, \Delta t & \text{in the discrete case} \end{cases} ,
$$

$$(7.16)$$

with L being the *Lagrangian* function

$$L = \text{kinetic energy} - \text{potential energy}$$

$$= \begin{cases} \dfrac{m_0}{2}\left[\dfrac{dq(\tau)}{d\tau}\right]^2 - V\big[q(\tau)\big] & \text{in the continuous case} \\[2ex] \dfrac{m_0}{2}\left(\dfrac{q_{k+1} - q_k}{\Delta t}\right)^2 - V\big(q_k\big) & \text{in the discrete case} \end{cases} \tag{7.17}$$

Due to the fact that the action function (7.15) describes deterministic particle motions, the expression (7.12) can be interpreted as a certain kind of mean value with respect to all possible deterministic action functions.

The calculation of the vector products $\left(q_{k+1}-q_k\right)^2$ generates complex exponential functions which are nothing but analytical continuations of the often used real exponential functions if the potential functions $V(q_k)$ represent power series or polynomials of the variables $x_{c,k}^{(l)}$. In this case pro-functions of the kind (7.13) can be interpreted as analytical continuations of the introduced partition functions. Therefore, a generalized partition function can be introduced. This shall be done.

7.3.3   The Statistical Basic Function

Introducing the integral representation

$$SBF = F \int_{-\infty}^{+\infty} \exp\left[-\left(\sum_{l=1}^{O}\left\{\prod_{k=1}^{l}\sum_{\Theta_k}\lambda_{\Theta_l}^{c.}\,\Omega_{\Theta_k}\right\}\right)\right]d\Omega \tag{7.18}$$

an expression is given which includes both the introduced real partition functions (for example, see (6.71) or (6.76)) and kernels of the type (7.13). F represents parts which do not depend on the integration variables Ω_{Θ_k}. O is the order of the mathematical problem. $>$ and $<$ represent the introduced product brackets (see subsection 3.3.2). Ω represents all statistical variables Ω_{Θ_k}. Θ_l represents all indices Θ_k. A suitable identifiaction of the parameters $\lambda_{\Theta_l}^{c.}$ and the integration variables Ω_{Θ_k} generates then the special case. Complex distribution function parameters $\lambda_{\Theta_l}^{c.}$ occur if (7.18) represents a pro-function, and real parameters $\lambda_{\Theta_l}^{c.}$ generate partition functions such as the partition function of the multi-mode laser (see (6.71)) or partition functions of thermodynamics (see (2.57)). In the case of laser theory, the statistical variables Ω_{Θ_k} are identical with the mode amplitudes E_w, and in the case of thermodynamics, Ω_{Θ_k} has to be replaced by a polarisation or a magnetization component. In these cases $F = 1$ holds. Due to the fact that such a partition function (or a pro-function, respectively) can be taken as a basis to calculate relevant statistical quantities of microscopic and macroscopic consideration levels, the function (7.18) shall be called a *statistical basic function*. Therefore, a statistical basic function represents a widely usable statistical function. Due to the fact that such functions in this book are called *universal functions*, the statistical basic function represents a universal statistical function.

The reader may remember that it is possible to gain a distribution function by using an extreme principle, the maximum information entropy principle (MIEP). The standardization factor of such a distribution function is nothing but a reciprocial partition function. This was discussed in chapter 3. However it is also possible to gain the integrand of a partition function, because such an integrand is nothing but the distribution function without the standardization factor. Therefore, it is possible to introduce an extreme principle to gain the integrands of partition functions. Due to the fact that *Feynman* kernels show the form of complex partition functions, it has to be possible to introduce a principle to gain integrands of partition functions as well as integrands of *Feynman* kernels. As the difference between such an integrand and special kinds of distribution functions is given by a normalization factor, such a principle can be taken to gain distribution functions, too. This shall be shown. Then the (discrete) statistical basic function (7.18) shall be taken as a basis, i. e. if *Feynman* kernels are considered, the discrete pro-function (7.13) has to be taken as a basis.

7.3.4   The Statistical Basic Principle

In the following the notation

$$\phi(\Omega) = F \exp\left[-\left(\sum_{l=1}^{O}\left\{\prod_{k=1}^{l}\sum_{\Theta_k}\lambda^{c.}_{\Theta_l}\Omega_{\Theta_k}\right\}\right)\right] \tag{7.19}$$

shall be used, and it shall be shown that it is possible to gain such a function by using a suitable extreme principle.

The starting point of the considerations shall be an expression of the form

$$\int_m \phi(\Omega)\ln[\phi(\Omega)/d]\,d\Omega = GI\,, \tag{7.20}$$

which can be called *generalized information*, because this expression can be considered as a generalized information expression (see (3.10)). $\int_m$ represents measurement integrals (see chapter 3, subsection 3.1.3), and the constant d guarantees a dimensionless argument of the natural logarithm if $\phi(\Omega)$ has a physical dimension. (In this case again measurement integrals have to be used, because the factor d can be equivalent to the element $1/d\Omega$. Therefore, the considerations introduced in 3.1.3 have to be considered.) Using the generalized information GI the extreme principle

$$\delta\int_m \phi(\Omega)\ln[\phi(\Omega)/d]\,d\Omega = 0\,,$$

$$\delta\int_m \phi(\Omega)\,d\Omega = 0\,,$$

$$\delta\int_m \phi(\Omega)\left\{\prod_{k=1}^{l}\Omega_{\Theta_k}\right\}d\Omega = \delta C_{\Theta_l} = 0 \tag{7.21}$$

can be introduced, where the quantities C_{Θ_l} represent various constraints. This variational principle requires an extreme point (maximum, minimum, point of inflexion) of the generalized information GI under consideration of some given constraints. In order to understand this principle much better, the reader may consider again section 3.2. There the MIEP was introduced. The variational principle (7.21) can be considered as a generalized form of the MIEP if special constraints are used. This generalized form shall be called the *statistical basic principle*. Using the method of *Lagrangian* multipliers (see section 3.2) (7.21) can be replaced by

$$\delta\left\{ \int_m \phi(\Omega)\ln[\phi(\Omega)/d]\,d\Omega + \lambda_0^c \int_m \phi(\Omega)\,d\Omega + \int_m \phi(\Omega)\sum_{l=1}^{O}\left\{\prod_{k=1}^{l}\sum_{\Theta_k}\lambda_{\Theta_l}^c\,\Omega_{\Theta_k}\right\}\,d\Omega \right\} = 0 \,, \tag{7.22}$$

where the parameters $\lambda_{\Theta_l}^c$ are the *Lagrangian* multipliers. In this case such multipliers can be complex. (7.22) can be considered as a generalized form of the expression (3.22) which describes the MIEP if special constraints are used. Moreover, (7.22) can be replaced by

$$\delta \int_m \phi(\Omega)D(\Omega)\,d\Omega = 0 \,, \tag{7.23}$$

with $D(\Omega)$ being defined by

$$D(\Omega) = \ln[\phi(\Omega)/d] + \lambda_0^c + \sum_{l=1}^{O}\left\{\prod_{k=1}^{l}\sum_{\Theta_k}\lambda_{\Theta_l}^c\,\Omega_{\Theta_k}\right\} \,, \tag{7.24}$$

and the variation can be executed so that one obtains the expression

$$\int_m [D(\Omega)\delta\phi(\Omega) + \phi(\Omega)\delta D(\Omega)]\,d\Omega = \int_m \phi(\Omega)[D(\Omega) + 1]\delta\phi(\Omega)\,d\Omega$$
$$= 0 \,. \tag{7.25}$$

Due to the fact that any elements $\delta\phi(\Omega)$ can be used, the equation (7.25) can be replaced by the requirement

$$D(\Omega) + 1 = 0 \,, \tag{7.26}$$

where (7.26) is equivalent to the relation

$$\ln[\phi(\Omega)/d] = -\left(1 + \lambda_0^c + \sum_{l=1}^{O}\left\{\prod_{k=1}^{l}\sum_{\Theta_k}\lambda_{\Theta_l}^c\,\Omega_{\Theta_k}\right\}\right) \,. \tag{7.27}$$

Considering an exponential operator, and using the condition

$$F = d \exp\left[-\left(1 + \lambda_0^{c.}\right)\right] \tag{7.28}$$

the expression (7.27) can be rewritten. In doing so one obtains the solution of the statistic basic principle, i. e. one obtains the exponential function

$$\phi(\Omega) = F \exp\left[-\left(\sum_{l=1}^{O}\left\{\prod_{k=1}^{l}\sum_{\Theta_k}\lambda_{\Theta_l}^{c.}\Omega_{\Theta_k}\right\}\right)\right]. \tag{7.29}$$

This exponential function is identical with (7.19), i. e. the statistical basic principle requires functions of the kind (7.19). Therefore, it has been shown that one gains solutions of the kind (7.19) if the the statistical basic principle is taken as a basis.

Due to the fact that exponential functions of the kind (7.29) represent both integrands of *Feynman* kernels and integrands of partition functions, an extreme principle is given which allows to determine the central parts of partition functions or *Feynman* kernels, respectively. Strictly speaking, this principle allows to gain the *form* of such a kernel, because the *Lagrangian* parameters are normally not known if only this principle is used. Therefore, the form of relevant statistical functions of many different classes of physical systems (for example, systems of thermodynamics, laser theory and quantum system theory) can be determined. In order to gain the statistical basic function, an integral operator has to be used, i. e. the expression

$$\int_{-\infty}^{+\infty} \phi(\Omega)\, d\Omega = SBF \tag{7.30}$$

has to be taken as a basis. However, if (7.29) is interpreted as a special kind of distribution function, the factor F has to be determined by using a suitable normalization condition. Therefore, special kinds of real and complex distribution functions, too, one can gain by using the statistical basic principle. (For example, if laser systems are considered, the function (7.29) has to be identical with the distribtuion function of the mode amplitudes (6.70).)

If a statistical basic function represents a real partition function, an explicit formulation can be found, i. e. in this case the integral (7.18) can be calculated. Therefore, the question arises whether it is possible to calculate pro-functions, too. This is indeed possible and shall now be discussed.

7.3.5 Feynman Kernels. A Basic Calculation Procedure

In the following it shall be shown that a *Feynman* kernel (7.13) can be calculated so that an explicit form is given. The procedure which will then be used can be interpreted as an analytical continuation of the introduced calculation procedure (see chapter 4, subsection 4.3.2). First, the explicit form of the exponential function shall be considered.

The Explicit Integral Form of the Feynman Kernel

After calculation of the vector products $\left(q_{k+1} - q_k\right)^2$ the kernel (7.13) can be rewritten. In doing so the formulation

$$G(q_P, q_0, \Delta t) = \tilde{F}_{\text{Fk}} \int_{-\infty}^{+\infty} \exp\left[-\frac{\mathrm{i}}{\hbar}\left(W_{\text{kin}} + W_{\text{pot}}\right)\right] dq_{P-1}\ldots dq_1 \qquad (7.31)$$

is possible, with W_{kin} and W_{pot} being defined by

$$W_{\text{kin}} = \sum_{c_1,k_1,l_1=1,1,1}^{3,P-1,N} \lambda^{(\text{kin})}_{c_1,k_1,l_1} x^{(l_1)}_{c_1,k_1} +$$
$$\sum_{c_1,k_1,l_1=1,1,1}^{3,P-1,N} \sum_{c_2,k_2,l_2=1,1,1}^{3,P-1,N} \lambda^{(\text{kin})}_{c_1,k_1,l_1,c_2,k_2,l_2} x^{(l_1)}_{c_1,k_1} x^{(l_2)}_{c_2,k_2} \qquad (7.32)$$

and

$$W_{\text{pot}} = \sum_{k=1}^{P-1} V\left(q_k\right)\Delta t \ . \qquad (7.33)$$

The factor of the *Feynman* kernel, the factor $\tilde{F}_{\text{Fk}}$, is defined by

$$\tilde{F}_{\text{Fk}} = S \exp\left\{\frac{\mathrm{i}}{\hbar}\left[\frac{m_0}{2\Delta t} \sum_{c_1,l_1=1,1}^{3,N} \left(x^{(l_1)\,2}_{c_1,0} + x^{(l_1)\,2}_{c_1,P}\right) - V(q_0)\Delta t\right]\right\}, \qquad (7.34)$$

and the parameters $\lambda^{(\text{kin})}_{c_1,k_1,l_1}$, $\lambda^{(\text{kin})}_{c_1,k_1,l_1,c_2,k_2,l_2}$ are defined by

$$\lambda^{(\text{kin})}_{c_1,k_1,l_1} = \frac{m_0}{\Delta t}\left(x^{(l_1)}_{c_1,0}\delta_{k_1,1} + x^{(l_1)}_{c_1,P}\delta_{k_1,P-1}\right),$$
$$\lambda^{(\text{kin})}_{c_1,k_1,l_1,c_2,k_2,l_2} = \frac{m_0}{\Delta t}\delta_{c_1,c_2}\delta_{l_1,l_2}\left(\delta_{k_1,k_2-1} - \delta_{k_1,k_2}\right) \ . \qquad (7.35)$$

The symbol δ represents *Kronecker's delta*. The *Kronecker* delta δ_{c_1,c_2} occurs, because the kinetic energies of different *Cartesian* coordinates are independent, the delta δ_{l_1,l_2} occurs, because the kinetic energies of different particles are independent, and the deltas δ_{k_1,k_2}, δ_{k_1,k_2-1} guarantee the necessary time ordering. In this case the formulation (7.31) is totally identical with the original *Feynman* kernel (7.13). However, in the following a special structure of the potential function W_{pot} shall be assumed, i. e. the relation

$$W_{\text{pot}} = \lambda^{(\text{pot})}_0 + \sum_{c_1,k_1,l_1=1,1,1}^{3,P-1,N} \lambda^{(\text{pot})}_{c_1,k_1,l_1} x^{(l_1)}_{c_1,k_1} +$$
$$\sum_{c_1,k_1,l_1=1,1,1}^{3,P-1,N} \sum_{c_2,k_2,l_2=1,1,1}^{3,P-1,N} \lambda^{(\text{pot})}_{c_1,k_1,l_1,c_2,k_2,l_2} x^{(l_1)}_{c_1,k_1} x^{(l_2)}_{c_2,k_2} + \text{tho} \qquad (7.36)$$

shall be used.

The relation (7.36) can be identical with an infinite power series or a finite polynomial. Such functions represent various kinds of non-linear molecular oscillator problems. For example, the non-linear oscillator problem

$$V(\boldsymbol{q}_k) = V_1 + V_2 \tag{7.37}$$

with

$$V_1 = \sum_{l_1=1}^{N} \left[\alpha_1^{(l_1)} \left(\boldsymbol{x}_k^{(l_1)} - \boldsymbol{R}^{(l_1)} \right)^2 + \alpha_2^{(l_1)} \left(\boldsymbol{x}_k^{(l_1)} - \boldsymbol{R}^{(l_1)} \right)^4 \right], \tag{7.38}$$

$$V_2 = \sum_{l_1,l_2=1,1}^{N,N} \left[\beta_1^{(l_1,l_2)} \left(\boldsymbol{x}_k^{(l_1)} - \boldsymbol{x}_k^{(l_2)} \right)^2 + \beta_2^{(l_1,l_2)} \left(\boldsymbol{x}_k^{(l_1)} - \boldsymbol{x}_k^{(l_2)} \right)^4 \right] \tag{7.39}$$

represents a special case of the assumed class of particle problems. (The vector $\boldsymbol{x}_k^{(l_1)} = \left(x_{c_1=1,k}^{(l_1)}, x_{c_1=2,k}^{(l_1)}, x_{c_1=3,k}^{(l_1)} \right)$ represents the position vectors of the single particles, and the vector $\boldsymbol{R}^{(l_1)} = \left(R_{c_1=1}^{(l_1)}, R_{c_1=2}^{(l_1)}, R_{c_1=3}^{(l_1)} \right)$ represents the vectors of the stable particle positions. The part V_1 shows that a system of decoupled non-linear oscillators is considered, with $\alpha_1^{(l_1)}$, $\alpha_2^{(l_1)}$ being the spring constants. The coupling of the system is guaranteed by the second term V_2. There coupling constants $\beta_1^{(l_1,l_2)}$, $\beta_2^{(l_1,l_2)}$ occur. Figure 7.4 illustrates this oscillator problem. In order to demonstrate that this special oscillator problem is a special case of (7.36), the various vector products of (7.37) have to be calculated.) Furthermore, functions of the kind (7.36) represent other molecular problems if suitable expansions instead of a potential function are used.

Using the abbreviation

$$F_{\mathrm{Fk}} = \tilde{F}_{\mathrm{Fk}} \exp\left(-\frac{\mathrm{i}}{\hbar} \lambda_0^{(\mathrm{pot})} \right), \tag{7.40}$$

and using the definition

$$\lambda_{c_1,k_1,l_1}^{(\mathrm{Feyn})} = \frac{\mathrm{i}}{\hbar} \left(\lambda_{c_1,k_1,l_1}^{(\mathrm{kin})} + \lambda_{c_1,k_1,l_1}^{(\mathrm{pot})} \right),$$

$$\lambda_{c_1,k_1,l_1,c_2,k_2,l_2}^{(\mathrm{Feyn})} = \frac{\mathrm{i}}{\hbar} \left(\lambda_{c_1,k_1,l_1,c_2,k_2,l_2}^{(\mathrm{kin})} + \lambda_{c_1,k_1,l_1,c_2,k_2,l_2}^{(\mathrm{pot})} \right),$$

$$\lambda_{c_1,k_1,l_1,c_2,k_2,l_2,\ldots\infty}^{(\mathrm{Feyn})} = \frac{\mathrm{i}}{\hbar} \lambda_{c_1,k_1,l_1,c_2,k_2,l_2,\ldots\infty}^{(\mathrm{pot})} \tag{7.41}$$

the *Feynman* kernel (7.31) can be replaced by

$$G(\boldsymbol{q}_P, \boldsymbol{q}_0, \Delta t) = F_{\mathrm{Fk}} \int_{-\infty}^{+\infty} \exp\left(-W_{\mathrm{Feyn}} \right) d\boldsymbol{q}_{P-1} \ldots d\boldsymbol{q}_1, \tag{7.42}$$

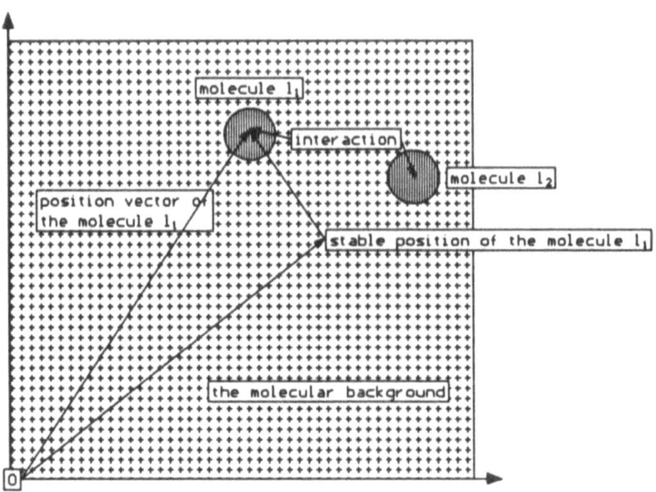

Figure 7.4 A system of non-linear oscillators

with W_{Feyn} being defined by

$$W_{\text{Feyn}} = \sum_{c_1,k_1,l_1=1,1,1}^{3,P-1,N} \lambda^{(\text{Feyn})}_{c_1,k_1,l_1} x^{(l_1)}_{c_1,k_1} +$$

$$\sum_{c_1,k_1,l_1=1,1,1}^{3,P-1,N} \sum_{c_2,k_2,l_2=1,1,1}^{3,P-1,N} \lambda^{(\text{Feyn})}_{c_1,k_1,l_1,c_2,k_2,l_2} x^{(l_1)}_{c_1,k_1} x^{(l_2)}_{c_2,k_2} + \text{tho} \tag{7.43}$$

The parameters of (7.43) can be considered as *Lagrangian* parameters. As the reader can see, *Lagrangian* parameters of first and second order include kinetic and potential parts, and *Lagrangian* parameters of higher order are proportional to the paramters of higher order of the potential function (7.36). Using a now familiar notation (7.43) can be replaced by

$$W_{\text{Feyn}} = \sum_{n=1}^{O} \left\{ \prod_{m=1}^{n} \sum_{c_m,k_m,l_m=1,1,1}^{3,P-1,N} \lambda^{(\text{Feyn})}_{\Theta_n} x^{(l_m)}_{c_m,k_m} \right\} , \tag{7.44}$$

where the index Θ_n includes the indices c_m, k_m, l_m, i. e.

$$\Theta_n = c_1, k_1, l_1, \ldots, c_m, k_m, l_m, \ldots, c_n, k_n, l_n ,$$

$$c_m = 1 \ldots 3 , \quad k_m = 1 \ldots P-1 , \quad l_m = 1 \ldots N ,$$

$$n = 1 \ldots O \tag{7.45}$$

holds. O represents the order of the considered problem, i. e. in particular $O = \infty$ generates a power series. As following calculations will show, the relation $O = 2, 4, 6, \ldots \infty$ has to be required.

If the reader compares (7.42) and (7.44) with the statistical basic function (7.18), it is obvious that *Feynman* kernels are special cases of a statistical basic function if a potential function of the kind (7.36) is taken as a basis. This fact already was mentioned, now it is obvious. (7.42) represents the explicit integral form of the considered *Feynman* kernel. This integral can exactly be calculated. In order to show this, power series representations of the occuring exponential functions have to be used. This shall now be shown.

The Analytical Continuation of the Gaussian Power Series Function

In order to calculate an integral of the kind (7.42), a formalism can be used which was discussed in chapter 4 (see (4.52)ff.). However, now an analytical continuation has to be used. In the following this way of calculation shall be considered. First, the exponential function of (7.42) shall be rewritten. Then self-similar power series functions will occur.

Using product brackets the exponential function of (7.42) can be written in the product form

$$
\exp\left(-W_{\text{Feyn}}\right) = \left\{ \prod_{\substack{n=1 \\ \Theta_O \neq \Theta_{conv}}}^{O} \prod_{\Theta_n} \exp\left(-\left\{\prod_{m=1}^{n} \lambda_{\Theta_n}^{(\text{Feyn})} x_{c_m,k_m}^{(l_m)}\right\}\right) \right\}
$$

$$
\left\{ \prod_{\Theta_{conv}} \exp\left[-\lambda_{\Theta_{conv}}^{(\text{Feyn})}\left(x_{c,k}^{(l)}\right)^O\right] \right\} , \tag{7.46}
$$

where the index Θ_{conv} represents all equal index sets which correspond to the order O of the problem, i. e. the relation

$$
\Theta_{conv} = c_1, k_1, l_1, \ldots, c_O, k_O, l_O \ ,
$$
$$
c_1 = \ldots = c_O = c \ , \quad k_1 = \ldots = k_O = k \ , \quad l_1 = \ldots = l_O = l \ ,
$$
$$
c = 1 \ldots 3 \ , \quad k = 1 \ldots P-1 \ , \quad l = 1 \ldots N \tag{7.47}
$$

holds so that the product rule

$$
\prod_{\Theta_{conv}} = \prod_{c,k,l=1,1,1}^{3,P-1,N} \tag{7.48}
$$

has to be considered. The separation of all exponential parts which correspond to equal index sets of highest order is an essential part of the calculation procedure. As the following will show, this will lead to infinite sums of easy integrable integrals. In this context the relation $\lambda_{\Theta_{conv}}^{(\text{Feyn})} \neq 0$ has to be required. The first part of (7.46) can be replaced by a power series function, because an exponential function can be defined by the power series

$$
\exp\left(-\left\{\prod_{m=1}^{n} \lambda_{\Theta_n}^{(\text{Feyn})} x_{c_m,k_m}^{(l_m)}\right\}\right) =
$$
$$
\sum_{\Lambda_{\Theta_n}=0}^{\infty} (-1)^{\Lambda_{\Theta_n}} \frac{1}{\Lambda_{\Theta_n}!} \left\{\prod_{m=1}^{n} x_{c_m,k_m}^{(l_m)}\right\}^{\Lambda_{\Theta_n}} \left(\lambda_{\Theta_n}^{(\text{Feyn})}\right)^{\Lambda_{\Theta_n}}. \tag{7.49}
$$

Inserting (7.49) into (7.46) one obtains a relation which shows the form of a power series function, i. e. one obtains the relation

$$
\exp\left(-W_{\text{Feyn}}\right) =
$$
$$
\left[\prod_{\substack{n=1 \\ \Theta_O \neq \Theta_{conv}}}^{O} \prod_{\Theta_n} \sum_{\Lambda_{\Theta_n}=0}^{\infty} z_{\Lambda}^{(3)} \left\{\prod_{\Theta_{conv}} \tilde{I}_{\Theta_{conv},\xi(\Lambda)}\right\} \left(\lambda_{\Theta_n}^{(\text{Feyn})}\right)^{\Lambda_{\Theta_n}}\right], \tag{7.50}
$$

where

$$
\tilde{I}_{\Theta_{conv},\xi(\Lambda)} = \left(x_{c,k}^{(l)}\right)^{\xi(\Lambda)} \exp\left[-\lambda_{\Theta_{conv}}^{(\text{Feyn})} \left(x_{c,k}^{(l)}\right)^{O}\right] \tag{7.51}
$$

holds. The numbers $z_{\Lambda}^{(3)}$ of (7.50) are defined by

$$
z_{\Lambda}^{(3)} = \left\{\prod_{\substack{n=1 \\ \Theta_O \neq \Theta_{conv}}}^{O} \prod_{\Theta_n} (-1)^{\Lambda_{\Theta_n}} \frac{1}{\Lambda_{\Theta_n}!}\right\}. \tag{7.52}
$$

Λ represents all indices Λ_{Θ_n}. The exponent $\xi(\Lambda)$ occurs, because the relation

$$
\left\{\prod_{\substack{n=1 \\ \Theta_O \neq \Theta_{conv}}}^{O} \prod_{\Theta_n} \left\{\prod_{m=1}^{n} x_{c_m,k_m}^{(l_m)}\right\}^{\Lambda_{\Theta_n}}\right\} = \left\{\prod_{\Theta_{conv}} \left(x_{c,k}^{(l)}\right)^{\xi(\Lambda)}\right\} \tag{7.53}
$$

has been used. (7.50) represents a power series function of the considered exponential function. Using this power series function the integral of the *Feynman* kernel (7.42) can be replaced.

Inserting (7.50) into (7.42) one obtains the power series function

$$\int_{-\infty}^{+\infty} \exp\left(-W_{\text{Feyn}}\right) dq_{P-1} \dots dq_1 =$$

$$\left\{ \prod_{\substack{n=1 \\ \Theta_O \neq \Theta_{conv}}}^{O} \prod_{\Theta_n} \sum_{\Lambda_{\Theta_n}=0}^{\infty} z_{\Lambda}^{(3)} \left[\prod_{\Theta_{conv}} I_{\Theta_{conv},\xi(\Lambda)} \right] \left(\lambda_{\Theta_n}^{(\text{Feyn})} \right)^{\Lambda_{\Theta_n}} \right\}, \qquad (7.54)$$

with $I_{\Theta_{conv},\xi(\Lambda)}$ representing integrals of the form

$$I_{\Theta_{conv},\xi(\Lambda)} = \int_{-\infty}^{+\infty} \left(x_{c,k}^{(l)} \right)^{\xi(\Lambda)} \exp\left[-\lambda_{\Theta_{conv}}^{(\text{Feyn})} \left(x_{c,k}^{(l)} \right)^{O} \right] dx_{c,k}^{(l)} . \qquad (7.55)$$

(7.55) represents elementary complex integrals of the *Fresnel* type. Using the transformation

$$x_{c,k}^{(l)} = \left(\tilde{x}_{c,k}^{(l)} \right)^{\frac{2}{O}} , \quad dx_{c,k}^{(l)} = \frac{2}{O} \left(\tilde{x}_{c,k}^{(l)} \right)^{\frac{2-O}{O}} d\tilde{x}_{c,k}^{(l)} \qquad (7.56)$$

one obtains *Fresnel* integrals of second order, i. e. if again the sign $x_{c,k}^{(l)}$ is used, (7.55) has to be replaced by

$$I_{\Theta_{conv},\xi(\Lambda)} = \frac{2}{O} \int_{-\infty}^{+\infty} \left(x_{c,k}^{(l)} \right)^{\frac{2\xi(\Lambda)+2-O}{O}} \exp\left[-\lambda_{\Theta_{conv}}^{(\text{Feyn})} \left(x_{c,k}^{(l)} \right)^{2} \right] dx_{c,k}^{(l)} . \qquad (7.57)$$

(Integrals of the *Fresnel* type are integrals with an imaginary unit i in the exponent of an exponential function. In particular, such *Fresnel* integrals are used in the theory of light diffraction. They can be considered as analytical continuations of *Gaussian* integrals.) Therefore, (7.54) represents a power series function of integrals of the *Fresnel* type. This power series function is an analytical continuation of the introduced real power series function (4.57). If such a real function is called *Gaussian power series function*, the now occuring function (7.54) represents an analytical continuation of a *Gaussian* power series function. However, the physical meaning of such an analytical continuation is another one, i. e. now *Feynman* kernels are considered, and in chapter 4 partition functions were considered. In chapter 4 the solution of an integral of the form (7.57) was already calculated (see (4.67)) if real parameters are taken as a basis. This result is also valid now if the identification $\lambda_{\Theta_{conv}} \leftrightarrow \lambda_{\Theta_{conv}}^{(\text{Feyn})}$ is made. In doing so one obtains the result

$$I_{\Theta_{conv},\xi(\Lambda)} = \delta_{\xi_{even}(\Lambda),\xi(\Lambda)} \frac{2}{O} \Gamma\left[\frac{\xi(\Lambda)+1}{O} \right] \left(\lambda_{\Theta_{conv}}^{(\text{Feyn})} \right)^{-\left[\frac{\xi(\Lambda)+1}{O} \right]}, \qquad (7.58)$$

where $\delta_{\xi_{even}(\Lambda),\xi(\Lambda)}$ represents *Kronecker's* delta, i. e. the relation

$$\delta_{\xi_{even}(\Lambda),\xi(\Lambda)} = \begin{cases} 0 & \text{if } \xi(\Lambda) = \xi_{odd}(\Lambda) \\ 1 & \text{if } \xi(\Lambda) = \xi_{even}(\Lambda) \end{cases} \qquad (7.59)$$

holds. Therefore, the explicit form of the *Feynman* kernel (7.42) can be calculated.

The Explicit Form of the Feynman Kernel

Using (7.54) and (7.58) the explicit form of the kernel (7.42) can be calculated, i. e. introducing the numbers

$$z_{\Lambda}^{(4)} = z_{\Lambda}^{(3)} \left(\frac{2}{O} \right)^{3(P-1)N} \left\{ \prod_{\Theta_{conv}} \delta_{\xi_{even}(\Lambda),\xi(\Lambda)} \Gamma \left[\frac{\xi(\Lambda) + 1}{O} \right] \right\} \tag{7.60}$$

one obtains the formulation

$$G(\boldsymbol{q}_P, \boldsymbol{q}_0, \Delta t) = F_{\mathrm{Fk}} \left\{ \prod_{\Theta_{conv}} \left(\lambda_{\Theta_{conv}}^{(\mathrm{Feyn})} \right)^{-\frac{1}{O}} \right\} \left\{ \prod_{\substack{n=1 \\ \Theta_O \neq \Theta_{conv}}}^{O} \prod_{\Theta_n} \prod_{\Theta_{conv}} \sum_{\Lambda_{\Theta_n}=0}^{\infty} z_{\Lambda}^{(4)} \frac{\left(\lambda_{\Theta_n}^{(\mathrm{Feyn})} \right)^{\Lambda_{\Theta_n}}}{\left(\lambda_{\Theta_{conv}}^{(\mathrm{Feyn})} \right)^{\frac{\xi(\Lambda)}{O}}} \right\}, \tag{7.61}$$

with F_{Fk} being defined by

$$F_{\mathrm{Fk}} = S \exp \left\{ -\frac{\mathrm{i}}{\hbar} \left[\lambda_0^{(\mathrm{pot})} - \frac{m_0}{2\Delta t} \sum_{c_1,l_1=1,1}^{3,N} \left(x_{c_1,0}^{(l_1)\,2} + x_{c_1,P}^{(l_1)\,2} \right) + V(\boldsymbol{q}_0)\Delta t \right] \right\} . \tag{7.62}$$

($\lambda_0^{(\mathrm{pot})}$ represents the first part of the assumed potential function (7.36).) Therefore, the explicit structure of the *Feynman* kernel is calculated. In particular, (7.61) represents a large class of solutions of non-linear oscillator problems. As the reader can see, such a *Feynman* kernel is represented by a power series function. Strictly speaking, a mathematical self-similar power series function (see subsection 4.3.5) has to be used to describe a *Feynman* kernel, in which case such a kernel can be considered as an analytical continuation of the partition function (4.70). In order to get such a *Feynman* kernel, potential functions of the kind (7.36) were assumed. However, other potential functions can be used as well, and the introduced concept of calculation can be taken as a basis. Therefore, other kinds of *Feynman* kernels can be calculated, in which case all calculation procedures are totally similar to procedures which one has to use to calculate non-quantum mechanical systems.

In the following a trivial example shall be considered. The starting point of the considerations shall be (7.13). Certainly, it would not be necessary to start with (7.13), because formula (7.61) includes this example, but I think it will deepen the understanding of the introduced mathematical concept.

7.3.6 The Eigenfunction Structure

Using the introduced concept of subsection 7.3.5 it is possible to calculate the explicit structure of the *Feynman* kernel of a *free particle*. The problem of a free particle is defined by the potential function $V(q_k) = 0$, i. e. no force influences the motion of the particle. Therefore, if an easy one-dimensional motion is considered, the general *Feynman* kernel (7.13) has to be replaced by

$$G^{(\mathrm{fp})}(x_P, x_0, \Delta t) =$$
$$S^{(\mathrm{fp})} \int_{-\infty}^{+\infty} \exp\left\{ \frac{\mathrm{i}}{\hbar} \sum_{k=0}^{P-1} \left[\frac{m_0}{2} \left(\frac{x_{k+1} - x_k}{\Delta t} \right)^2 \right] \Delta t \right\} dx_{P-1} \ldots dx_1 \ . \qquad (7.63)$$

Using (7.63) as a basis the explicit structure of the *Feynman* kernel of the free particle can systematically be calculated. This shall be shown.

(7.63) can be replaced by

$$G^{(\mathrm{fp})}(x_P, x_0, \Delta t) = F_{\mathrm{Fk}}^{(\mathrm{fp})} \int_{-\infty}^{+\infty} \exp\left(- W_{\mathrm{Feyn}} \right) dx_{P-1} \ldots dx_1 \ , \qquad (7.64)$$

with W_{Feyn} being defined by

$$W_{\mathrm{Feyn}} = \sum_{k_1=1}^{P-1} \lambda_{k_1}^{(\mathrm{fp})} x_{k_1} + \sum_{k_1=1}^{P-1} \sum_{k_2=1}^{P-1} \lambda_{k_1,k_2}^{(\mathrm{fp})} x_{k_1} x_{k_2} \ . \qquad (7.65)$$

(7.64) represents a special case of the general formulation (7.42), and (7.65) is a special case of the function (7.44). As the reader can see, the order of the problem is 2, i. e. $O = 2$ holds. In this case the factor $F_{\mathrm{Fk}}^{(\mathrm{fp})}$ is defined by

$$F_{\mathrm{Fk}}^{(\mathrm{fp})} = S^{(\mathrm{fp})} \exp\left\{ \frac{\mathrm{i}}{\hbar} \left[\frac{m_0}{2\Delta t} \left(x_P^2 + x_0^2 \right) \right] \right\} , \qquad (7.66)$$

and the parameters $\lambda_{k_1}^{(\mathrm{fp})}$, $\lambda_{k_1,k_2}^{(\mathrm{fp})}$ are defined by

$$\lambda_{k_1}^{(\mathrm{fp})} = \frac{\mathrm{i}}{\hbar} \frac{m_0}{\Delta t} \left(x_0 \delta_{k_1,1} + x_P \delta_{k_1, P-1} \right) \ ,$$
$$\lambda_{k_1,k_2}^{(\mathrm{fp})} = \frac{\mathrm{i}}{\hbar} \frac{m_0}{\Delta t} \left(\delta_{k_1,k_2-1} - \delta_{k_1,k_2} \right) \ . \qquad (7.67)$$

Furthermore, the exponential function of (7.64) can be replaced by a product of simple exponential functions, i. e. the relation

$$\exp\left(- W_{\mathrm{Feyn}} \right) = \left\{ \prod_{\substack{k_1,k_2=1 \\ k_1 \neq k_2}}^{P-1} \exp\left(- \lambda_{k_1}^{(\mathrm{fp})} x_{k_1} \right) \exp\left(\lambda_{k_1,k_2}^{(\mathrm{fp})} x_{k_1} x_{k_2} \right) \right\}$$
$$\left\{ \prod_{k=1}^{P-1} \exp\left(- \lambda_{k,k}^{(\mathrm{fp})} x_k^2 \right) \right\} \qquad (7.68)$$

holds, where (7.68) represents a special case of (7.46). As it was shown in subsection 7.3.5, the first exponential parts can be replaced by suitable power series, i. e.

$$\exp\left(-\lambda_{k_1}^{(\mathrm{fp})} x_{k_1}\right) = \sum_{\Lambda_{k_1}=0}^{\infty} (-1)^{\Lambda_{k_1}} \frac{1}{\Lambda_{k_1}!} \left(x_{k_1}\right)^{\Lambda_{k_1}} \left(\lambda_{k_1}^{(\mathrm{fp})}\right)^{\Lambda_{k_1}},$$

$$\exp\left(-\lambda_{k_1,k_2}^{(\mathrm{fp})} x_{k_1} x_{k_2}\right) =$$

$$\sum_{\Lambda_{k_1,k_2}=0}^{\infty} (-1)^{\Lambda_{k_1,k_2}} \frac{1}{\Lambda_{k_1,k_2}!} \left(x_{k_1} x_{k_2}\right)^{\Lambda_{k_1,k_2}} \left(\lambda_{k_1,k_2}^{(\mathrm{fp})}\right)^{\Lambda_{k_1,k_2}} \tag{7.69}$$

holds. Inserting (7.69) into (7.68), using the explicit formulation (7.67) and considering the integral formulation one obtains the power series function

$$\int_{-\infty}^{+\infty} \exp\left(-W_{\mathrm{Feyn}}\right) dx_{P-1} \dots dx_1 =$$

$$\sum_{\Lambda_1=0}^{\infty} \sum_{\Lambda_{P-1}=0}^{\infty} \sum_{\Lambda_{1,2}=0}^{\infty} \dots \sum_{\Lambda_{P-2,P-1}=0}^{\infty} I_\Lambda \left(x_0\right)^{\Lambda_1} \left(x_P\right)^{\Lambda_{P-1}}, \tag{7.70}$$

with I_Λ being defined by

$$I_\Lambda = c_\Lambda \int_{-\infty}^{+\infty} x_1^{\Lambda_1+\Lambda_{1,2}} \exp\left(-\frac{1}{i\hbar}\frac{m_0}{\Delta t} x_1^2\right) dx_1$$

$$\int_{-\infty}^{+\infty} x_2^{\Lambda_{1,2}+\Lambda_{2,3}} \exp\left(-\frac{1}{i\hbar}\frac{m_0}{\Delta t} x_2^2\right) dx_2$$

$$\cdots$$

$$\int_{-\infty}^{+\infty} x_{P-2}^{\Lambda_{P-3,P-2}+\Lambda_{P-2,P-1}} \exp\left(-\frac{1}{i\hbar}\frac{m_0}{\Delta t} x_{P-2}^2\right) dx_{P-2}$$

$$\int_{-\infty}^{+\infty} x_{P-1}^{\Lambda_{P-2,P-1}+\Lambda_{P-1}} \exp\left(-\frac{1}{i\hbar}\frac{m_0}{\Delta t} x_{P-1}^2\right) dx_{P-1}. \tag{7.71}$$

The quantities c_Λ are coefficients which contain the specific parameters. These coefficients are defined by

$$c_\Lambda = \frac{(-1)^{\Lambda_1+\Lambda_{P-1}+\Lambda_{1,2}+\dots+\Lambda_{P-2,P-1}}}{\Lambda_1!\Lambda_{P-1}!\Lambda_{1,2}!\dots\Lambda_{P-2,P-1}!} \left(\frac{i}{\hbar}\frac{m_0}{\Delta t}\right)^{\Lambda_1+\Lambda_{P-1}+\Lambda_{1,2}+\dots+\Lambda_{P-2,P-1}}.$$

$$\tag{7.72}$$

(7.70) represents a special case of (7.54). (7.70) can be rewritten, because an analysis of the integral parts of (7.71) shows that only in the case $\Lambda_1 = \Lambda_{P-1}$ the expression (7.71) does not vanish. Therefore, the integral (7.70) can be replaced by

$$\int_{-\infty}^{+\infty} \exp\left(-W_{\mathrm{Feyn}}\right) dx_{P-1} \dots dx_1 = \sum_{\Lambda_1=0}^{\infty} I_{\Lambda_1} \left(x_0 x_P\right)^{\Lambda_1}, \tag{7.73}$$

where the definition

$$I_{\Lambda_1} = \sum_{\Lambda_{P-1}=0}^{\infty} \sum_{\Lambda_{1,2}=0}^{\infty} \cdots \sum_{\Lambda_{P-2,P-1}=0}^{\infty} I_{\Lambda} \tag{7.74}$$

has to be used. A calculation of the various *Fresnel* integrals of (7.71) can easily be done. Using additionally (7.72) one obtains the result

$$I_{\Lambda_1} = I_P(-1)^{\Lambda_1} \frac{1}{\Lambda_1!} \left(\frac{\mathrm{i}}{\hbar} \frac{m_0}{\Delta t} \right)^{\Lambda_1}, \tag{7.75}$$

where I_P is proportional to $\mathrm{i}\hbar\Delta t/m_0$, i. e. the relation

$$I_P \sim \left(\frac{\mathrm{i}\hbar\Delta t}{m_0} \right)^{\frac{P-1}{2}} \tag{7.76}$$

holds. Therefore, (7.73) can be replaced by

$$\int_{-\infty}^{+\infty} \exp\left(-W_{\mathrm{Feyn}} \right) dx_{P-1} \ldots dx_1 =$$
$$I_P \sum_{\Lambda_1=0}^{\infty} (-1)^{\Lambda_1} \frac{1}{\Lambda_1!} \left(\frac{\mathrm{i}}{\hbar} \frac{m_0}{\Delta t} \right)^{\Lambda_1} (x_0 x_P)^{\Lambda_1}. \tag{7.77}$$

Due to the fact that this power series represents an ordinary exponential function, (7.77) can be replaced by

$$\int_{-\infty}^{+\infty} \exp\left(-W_{\mathrm{Feyn}} \right) dx_{P-1} \ldots dx_1 = I_P \exp\left(-\frac{\mathrm{i}}{\hbar} \frac{m_0}{\Delta t} x_0 x_P \right). \tag{7.78}$$

Inserting (7.77) into (7.64) one obtains the lattice representation of the *Feynman* kernel of the free particle, i. e. one obtains the relation

$$G^{(\mathrm{fp})}(x_P, x_0, \Delta t) = S_P^{(\mathrm{fp})} \exp\left\{ \frac{\mathrm{i}}{\hbar} \left[\frac{m_0}{2\Delta t} (x_P - x_0)^2 \right] \right\}, \tag{7.79}$$

where $S_P^{(\mathrm{fp})}$ includes the original normalization factor $S^{(\mathrm{fp})}$ as well as the factor I_P, i. e. $S_P^{(\mathrm{fp})} = S^{(\mathrm{fp})} I_P$ holds. $S_P^{(\mathrm{fp})}$ can directly be determined by using a suitable normalization condition. (7.79) describes a well-known result. Therefore, it exemplary has been shown that the introduced calculation procedure allows to calculate *Feynman* kernels. However, again it has to be emphasized tha highly complicated problems can be calculated by using the introduced procedure.

It is a well-known fact that *Feynman* kernels can be represented by using eigenfunctions of the stationary *Schrödinger* equation (see [24]). This can mathematically be expressed by using the formulation

$$\boxed{G(q_P, q_0, \Delta t) = \int_\nu \exp\left(-\frac{\mathrm{i}}{\hbar} E^{(\nu)}\Delta t \right) \Phi^{*(\nu)}(q_0)\Phi^{(\nu)}(q_P) \, d\nu.} \tag{7.80}$$

$E^{(\nu)}$ represents the energies of the eigenstates. The complex functions $\Phi^{(\nu)}(q_0)$ and $\Phi^{*(\nu)}(q_0)$ are eigenfunctions of the stationary *Schrödinger* equation, where the symbol $*$ denotes the conjugate complex eigenfunctions. Due to the fact that the above considered problem of a free particle represents a special case of this general equation, it has to be possible to calculate as well the eigenfunction structure of the kernel (7.79). In order to show this, it has to be used that (7.79) is equivalent to

$$G^{(\mathrm{fp})}(x_P, x_0, \Delta t) = \tilde{S}_P^{(\mathrm{fp})} \int_{-\infty}^{+\infty} \exp\left[ik(x_P - x_0) - i\Delta t \frac{\hbar k^2}{2m_0}\right] dk , \qquad (7.81)$$

where k is a wave vector, i. e. k represents the impulse coordinate of the considered particle. (The equivalence of (7.81) and (7.79) can be shown by using a quadratic completition. $\tilde{S}_P^{(\mathrm{fp})}$ includes the normalization factor of (7.79) and a correction factor which guarantees that the equations (7.81), (7.79) are identical.) Due to the fact that the eigenfunctions and the energy eigenvalues of a free particle can be defined by

$$\Phi^{(k)}(x) = \frac{1}{\sqrt{\tilde{S}_P^{(\mathrm{fp})}}} \exp(ikx) , \quad E^{(k)} = \frac{\hbar^2 k^2}{2m_0} , \qquad (7.82)$$

it is obvious that the *Feynman* kernel of the free particle can be written by using eigenfunctions and corresponding energy eigenvalues, i. e. it is obvious that the relation

$$G^{(\mathrm{fp})}(x_P, x_0, \Delta t) = \int_{-\infty}^{+\infty} \exp\left(-\frac{i}{\hbar} E^{(k)} \Delta t \right) \Phi^{*(k)}(x) \Phi^{(k)}(x) \, dk \qquad (7.83)$$

holds. Therefore, it exemplary has been shown that as well the eigenfunction structure of *Feynman* kernels can systematically be calculated. However, using the introduced procedure more complicated problems can be calculated, too.

As the reader can see, always the lattice representation has been considered. Therefore, it has to be remarked that in the borderline case of a continuous kernel instead of the time difference Δt the time $N\Delta t = t$ occurs. In this case Δt in (7.83) and (7.80) has to be replaced by t.

In this context some special points have to be stressed. Therefore, now some additional considerations.

7.3.7 The Principle of Coupling of Elementary Systems

The above calculations showed that a high-dimensional mathematical problem in the context of quantum system theory can be solved by using functions which correspond to elementary mathematical problems. So it was possible to reduce the problem of calculation of a *Feynman* kernel to the problem of calculation of elementary integrals of *Fresnel's* type, i. e. the solution of the total problem could be represented by a self-similar power series function which bases on such *Fresnel* integrals. This was described by (7.54). A similar mathematical construction principle could be used to solve the problem of calculation of partition functions in the context of laser theory and thermodynamics. In the context of partition function calculation this principle was already mentioned (see the remarks after the equation (4.71)). Therefore, it has to be emphasized that solutions of high-dimensional problems can be found by using solutions of elementary problems. The solution of the total problem is then represented by an equation which describes the coupling of the basic elementary solutions. This principle shall be called the *principle of coupling of elementary systems*. Due to the fact that such mathematical systems correspond to physical systems, this principle represents nothing but a universal principle of nature, namely the principle that complicated natural structures consist of elementary physical systems. Examples are represented by organic molecules which consist of elementary atoms. Other examples are represented by the human organs which are composed of cells. Therefore, the considerations set forth in this book show the close connection between this special aspect of natural behavior and mathematical calculation procedures.

Another point I want to stress is the problem of hyper-surface equations in the context of quantum system theory.

7.3.8 Hyper-Surface Equations in the Context of Quantum System Theory

As it was introduced, hyper-surface equations are analytical equations which connect distribution function parameters with certain measurement levels. In this book it was shown how to calculate hyper-surface equations (see chapter 4) if the basis of the calculation is represented by special self-similar power series functions. However, so far only real self-similar power series functions were considered. Due to the fact that *Feynman* kernels can be represented by using complex self-similar power series functions of the same kind, and due to the fact that such complex self-similar power series functions generate representations of measurable mean values which are self-similar functions themselves, it has to be emphasized that a calculation of hyper-surface equations is possible also now. The then necessary concept of derivation can be considered as an analytical continuation of the introduced concept (see section 4.3). Therefore, if quantum systems are considered, it is possible to calculate relations which describe the connection between a microscopic and a macroscopic level in a totally analytical way. However, such calculations shall not be presented in this book.

Another essential problem which shall be pointed out is the reference frame problem.

7.3.9 Physical Meaning and Reference Frame

In this book always *Cartesian* position coordinates were taken as a basis. Therefore, it has to be remarked that other reference frames, too, can be used to describe the position of a particle or the position of another physical quantity. However, this is not the point which shall be discussed. In particular, in the context of *Feynman* path integrals another point is much more relevant. In order to understand this problem, in the following some basic considerations about reference frame transformations follow.

As it was discussed, the self-similar power series function (7.61) represents a wealth of different physical systems. One example is the problem of the free particle. This problem was extensively discussed. However, the problem of the harmonic oscillator as well as the problem of non-harmonic oscillators and other problems are included. In order to get the solutions of special physical problems, one has to put the various *Lagrangian* multipliers of (7.61) in concrete terms. Due to the fact that these multipliers contain both position coordinates q_P, q_0 and physical quantities such as the non-relativistic mass m_0 and *Planck's* constant $\hbar$, this means to define the structure with respect to the position coordinates and with respect to all relevant physical quantities. In this context 0 represents the starting time and P the final time. (The position coordinates can be called *physical quantities*, too. In this context this shall not be done.) However, such results one can achieve by using special transformations of the kind

$$\lambda_k = \lambda_k(\tilde{\lambda}_1 \dots \tilde{\lambda}_N) \, , \tag{7.84}$$

where λ_k represents all *Lagrangian* multipliers $\lambda_1 \dots \lambda_N$ of (7.61). $\tilde{\lambda}_k$ represents the new multipliers. In other words, considering (7.61) as a general surface equation in which all possible *Lagrangian* multipliers occur, suitable transformations allow to find a wealth of different solutions, in which case the physical meaning of the solutions before and after transformation arises if the multipliers are put in concrete terms. Therefore, suitable transformations allow to "switch" from one physical problem to the other. In particular, fields (for example, electric and magnetic fields) can be created or annihilated by using a special reference frame, and different physical situations can be considered as one physical system which is observed in different reference frames. Such a procedure represents a kind of transformation which is well-known in mathematics, namely the *transformation of major axis*. Examples which show that it is possible to gain solutions of physical problems by using equations of other physical problems and using additional transformations are well-known. For example, it is well-known that it is possible to switch from the problem of a free particle to the problem of a harmonic oscillator. The above considerations explain this possibility in a clear way. Moreover, it is clear that this possibility is a wide-ranging possibility. Therefore, it has to be emphasized that two possibilities exist to generate a new physical situation. The first possibility is to change the physical conditions (i. e. to change the *Lagrangian* multipliers), and the second possibility is to change the reference frame (i. e. to use a special transformation). This can directly be compared with the facts given in the theory of relativity (see chapter 9). This theory requires that it is also possible to get a new physical situation by changing the inertial reference frame system. However, now a special class of reference frames has to be used. Due to the fact that it is possible to calculate a more general representation (in comparison with the representation (7.61)), it has to be emphasized that it is a wide-ranging possibility to find *Feynman* representations of new physical situations only by using suitable transformations. Furthermore, it has to be emphasized that the theory of *Feynman's* path integrals, too, shows that the change of physical situations is equivalent to the change of suitable reference frames.

General path integrals are general representations of solutions of differential equations. If a path integral includes all possible solutions of a differential equation, such a path integral represents a special representation of the basic physical system. So far, *Feynman* path integrals have been considered. Due to the fact that the method of path integrals is a general method, in the theory of *Fokker-Planck* systems, too, path integral solutions are possible. If *Fokker-Planck* equations of the ensemble type (see chapter 5) are considered, a direct correlation between path integrals of the *Feynman* type and the *Fokker-Planck* type exists, i. e. in this case path integrals of *Feynman's* type can be considered as analytical continuations of path integrals of the *Fokker-Planck* type. In particular, this means that if a *Fokker-Planck* path integral is used which corresponds to an ensemble *Fokker-Planck* equation, a *Feynman* path integral can be found by using suitable replacements, i. e. real variables (parameters) have to be replaced by complex variables (parameters). Therefore, it has to be emphasized that a close connection between *Feynman* path integrals and *Fokker-Planck* path integrals exists. Thus, here it makes sense to consider *Fokker-Planck* path integrals. Therefore, in the following some remarks on the problem of path integrals in the theory of *Fokker-Planck* systems shall be given. However, *Fokker-Planck* equations of the impulse type (see chapter 5) shall be taken as a basis.

7.3.10 Path Integrals of the Fokker-Planck Type

A *Fokker-Planck* equation of the impulse type can be defined by using the equation (5.32) if the additional constraint (5.33) is used. Solutions of such a differential evolution equation can be defined by using a path integral. If a lattice representation is used, this path integral has the well-known form (for example, see [40])

$$\int_{-\infty}^{+\infty} \lim_{\substack{\Delta t \to 0 \\ P \to \infty}} F(\Omega_P, \Omega_0, \Delta t)\Phi(\Omega_0, 0)\, d\Omega_0 = \Phi(\Omega_P, t)\,, \qquad (7.85)$$

with the *Fokker-Planck* kernel being defined by

$$F(\Omega_P, \Omega_0, \Delta t) =$$
$$S \int_{-\infty}^{+\infty} \exp\left\{ -\frac{1}{2Q} \sum_{k=0}^{P-1} \left[\frac{\Omega_{k+1} - \Omega_k}{\Delta t} - K(\Omega_k) \right]^2 \Delta t \right\} d\Omega_{P-1} \ldots d\Omega_1 \ .$$

$$(7.86)$$

Ω represents a set of statistical variables. Ω can be identical with q. However, it has to be emphasized that in such a case a path integral in the context of *Fokker-Planck* systems cannot be compared with path integrals in the context of *Schrödinger* systems, because in contrast to a *Fokker-Planck* equation of the impulse type a time-dependent *Schrödinger* equation represents an ensemble equation. This is a basic difference between a *Fokker-Planck* equation of the impulse type and an equation of the time-dependent *Schrödinger* type. In chapter 5 this was discussed. In contrast to solutions of an equation of the *Schrödinger* type, the solutions of an equation of the *Fokker-Planck* type can be interpreted as probability densities. Therefore, $\Phi(\Omega_P, t)$ represents a measurable distribution function. 0 and P characterize the time points of the initial state $\Phi(\Omega_0, 0)$ and the final state $\Phi(\Omega_P, t)$, where $\Omega_0 = \Omega(0)$, $\Omega_P = \Omega(t)$ holds. Q describes the stochastic effects, and S represents the normalization factor. The quantities $K(\Omega_k)$ represent the deterministic forces. Furthermore, *Fokker-Planck* path integrals can be calculated by using the mathematical concept introduced in chapter 4, i. e. an explicit formulation of (7.86) can be found. This shall now be shown.

If the notation and the physical background of chapter 5 is taken as a basis, the notation

$$K(\Omega_k) = \langle v \rangle_k (\Omega_k) \tag{7.87}$$

has to be used. If a component $K_l(\Omega_k)$ of the force $K(\Omega_k)$ can be written in the form

$$K_l(\Omega_k) = \sum_{l_1=1}^{N} \lambda_{l,k,l_1}^{(\text{force})} \Omega_k^{(l_1)} + \sum_{l_1=1}^{N} \sum_{l_2=1}^{N} \lambda_{l,k,l_1,k,l_2}^{(\text{force})} \Omega_k^{(l_1)} \Omega_k^{(l_2)} + \text{tho} \ , \tag{7.88}$$

the *Fokker-Planck* kernel (7.86) can be rewritten, i. e. in such a case the relation

$$F(\Omega_P, \Omega_0, \Delta t) = F_{\text{FP}} \int_{-\infty}^{+\infty} \exp\left(-W_{\text{FP}} \right) d\Omega_{P-1} \ldots d\Omega_1 \tag{7.89}$$

holds, where W_{FP} is defined by

$$W_{\text{FP}} = \sum_{n=1}^{O} \left\{ \prod_{m=1}^{n} \sum_{k_m, l_m = 1, 1}^{P-1, N} \lambda_{\Theta_n}^{(\text{FP})} \Omega_{k_m}^{(l_m)} \right\} , \tag{7.90}$$

and $F_{\rm FP}$ reads

$$F_{\rm FP} = S \exp\left\{ -\frac{1}{2Q}\left[-\frac{\Omega_0}{\Delta t} - K(\Omega_0)\right]^2 \Delta t - \frac{1}{2Q}\left(\frac{\Omega_P}{\Delta t}\right)^2 \Delta t \right\}. \tag{7.91}$$

(In order to get (7.90) one has to insert the polynomial (7.88) into (7.86). Then one has to separate all parts which only contain the vectors Ω_0 and Ω_P. The factor S and these separated parts can be combined so that (7.91) occurs. All other parts can be reformulated by introducing the parameters $\lambda_{\Theta_n}^{\rm (FP)}$.) Θ_n includes all indices k_m, l_m, i. e.

$$\Theta_n = k_1, l_1, \ldots, k_m, l_m, \ldots, k_n, l_n \tag{7.92}$$

holds. The parameters $\lambda_{\Theta_n}^{\rm (FP)}$ are the *Lagrangian* parameters of the considered *Fokker-Planck* system, and O represents the order of the system. The form of the now occuring *Fokker-Planck* kernel is identical with the form of the *Feynman* kernel (7.42). However, in contrast to a *Feynman* system the *Lagrangian* parameters are real. This means that such a *Fokker-Planck* kernel can be calculated in a similar way. However, instead of complex integrals of the *Fresnel* type, integrals of the *Gaussian* type occur. This calculation has not to be done, because the now occuring real calculation problem is identical with the calculation problem of a partition function, and such problems were solved in chapter 4. Due to the fact that the integral of (7.89) is identical with the partition function (4.52) if the identification

$$\Theta_i \leftrightarrow k_m, l_m \;,$$
$$\Theta_\alpha \leftrightarrow \Theta_n \tag{7.93}$$

is taken into account, the result (4.70) represents the explicit formulation of the integral in (7.89). Then (7.89) can be replaced by

$$F\left(\Omega_P, \Omega_0, \Delta t\right) = F_{\rm FP} \left[\prod_{\Theta_{conv}} \left(\lambda_{\Theta_{conv}}^{\rm (FP)}\right)^{-1/O} \right]$$
$$\left[\prod_{\substack{n=1 \\ \Theta_O \neq \Theta_{conv}}}^{O} \prod_{\Theta_n} \prod_{\Theta_{conv}} \sum_{\Lambda_{\Theta_n}=0}^{\infty} z_\Lambda^{(2)} \frac{\left(\lambda_{\Theta_n}^{\rm (FP)}\right)^{\Lambda_{\Theta_n}}}{\left(\lambda_{\Theta_{conv}}^{\rm (FP)}\right)^{\frac{\xi(\Lambda)}{O}}} \right], \tag{7.94}$$

where the product rule

$$\prod_{\Theta_{conv}} = \prod_{k,l=1,1}^{P-1,N} \tag{7.95}$$

has to be considered, i. e. Θ_{conv} is defined by

$$\Theta_{conv} = k_1, l_1, \ldots, k_O, l_O \ ,$$
$$k_1 = \ldots = k_O = k \ , \quad l_1 = \ldots l_O = l \ . \tag{7.96}$$

(The numbers $z_{\Lambda}^{(2)}$ are defined in chapter 4 (see (4.69)). ξ is defined in chapter 4 (see (4.60)).) (7.95) represents an explicit formulation of the considered *Fokker-Planck* kernel. The form of (7.95) is identical with the form of the *Feynman* kernel (7.61). These solutions are self-similar power series functions, too.

Therefore, it has been shown that path integrals as well in the context of *Fokker-Planck* systems can be calculated by using the concept introduced in this book. It has to be emphasized that also other forces as (7.88) can be used, i. e. the introduced concept allows to calculate as well path integrals with other basic force functions. Hyper-surface equations can be calculated, too, by using the introduced inversion scheme. However, this shall not be shown in this book. Furthermore, it has to be emphasized that as well in the context of *Fokker-Planck* path integrals reference frame transformations can be used to switch from one physical situation to the other.

In the last chapters differential evolution equations of the *Schrödinger* and the *Fokker-Planck* type were considered (in particular, see chapter 5). Solutions of such equations were considered (in particular, see chapter 5, section 5.4). The method of path integrals was discussed (see this section, i. e. see section 7.3). In this context it was explained that only solutions of equations of the *Fokker-Planck* type can directly be interpreted, i. e. only such solutions can be interpreted as probability densities. Therefore, such functions are *direct state functions*. By contrast, solutions of an equation of the *Schrödinger* type represent *indirect state functions*, i. e. they cannot directly be interpreted. Due to the fact that such indirect state functions are correlated with measurable quantities such as a potential energy and a total energy (this was discussed in chapter 5 and in 7.2), indirect state functions have to contain information about the basic physical system in an implicit way. This correlation is described by a differential equation of the *Schrödinger* type, i. e. a suitable differential operator generates a product of a measurable quantity and the considered indirect state function. In other words, a *Schrödinger* equation represents a rule to calculate measurable energy values. Therefore, the question arises whether it is possible to calculate other measurable quantities by taking an indirect state function as a basis. This is indeed possible and shall now be discussed. This means then that the connection between a microscopic (statistical) and a macroscopic level shall be considered. However, not only equations of the *Schrödinger* type shall be considered, equations of the *Fokker-Planck* type shall be discussed, too. This is useful to demonstrate again the close connection between *Schrödinger* and *Fokker-Planck* systems.

7.4 State Functions and Measurement

So far, equations of the *Fokker-Planck* and the *Schrödinger* type were considered, and correlated state functions were discussed. However, only few considerations were dedicated to the connection between the basic state functions (or differential equations) and measurable quantities. For example, sometimes it was remarked that a *Schrödinger* equation represents a basic evolution equation of quantum systems, but only the possibility to calculate the energy eigenvalues of stationary states was mentioned (see 5.5 and 7.2). Only one point was extensively discussed, namely the possibility to describe the probability of the occurrence of a stationary state within a *Schrödinger* ensemble by using a suitable equation which determines a transition probability (see subsection 5.4.3). Therefore, the connection between state functions and measurement quantities shall now be discussed.

7.4.1 Stationary Solutions and Measurement

In chapter 5 stationary functions of the kind $\Phi^{(\nu)}(\Omega)$ and $\Phi^{(\nu)}(\Omega, v)$ were considered, where Ω can represent any set of variables, and in which case v is the correlated set of impulse variables. (An example is represented by (5.41).) Due to the fact that now particles are considered, Ω has to be replaced by the position vector x of a particle, i. e. now stationary functions of the kind $\Phi^{(\nu)}(x)$ and $\Phi^{(\nu)}(x, v)$ have to be considered. The functions $\Phi^{(\nu)}(x)$ in particular are solutions of a kinetic equation of the type (5.51), i. e. if the vector x instead of Ω is used, the equation

$$\left[(STRUC)^2 \frac{w_{\mathrm{sig}}^2}{2SYS}\right] \triangle_3 \Phi^{(\nu)}(x) + V(x)\Phi^{(\nu)}(x) = E^{(\nu)}\Phi^{(\nu)}(x) \tag{7.97}$$

holds. (7.97) includes both an equation of the stationary *Schrödinger* type and a stationary kinetic equation of the *Fokker-Planck* type, where the identification SYS, w_{sig}, $STRUC \rightarrow m_0$, $\hbar$, i generates an equation of the *Schrödinger* type, and the identification SYS/w_{sig}, $STRUC \rightarrow Q/2$, -1 generates a kinetic equation of the *Fokker-Planck* type. Introducing a *Hamiltonian* of the type

$$\hat{H} = \left[(STRUC)^2 \frac{w_{\mathrm{sig}}^2}{2SYS}\right] \triangle_3 + V(x) \tag{7.98}$$

the differential equation (7.97) can be replaced by

$$\hat{H}\Phi^{(\nu)}(x) = E^{(\nu)}\Phi^{(\nu)}(x) , \tag{7.99}$$

where $\hat{H}$ is a self-adjoint operator (see (5.66)ff.). Multiplying with a conjugate complex function $\Phi^{*(\mu)}(x)$ (in the *Fokker-Planck* case such a conjugate complex function is identical with the original function $\Phi^{(\mu)}(x)$), and integrating with respect to x one obtains the relation

$$\int_{-\infty}^{+\infty} \Phi^{*(\mu)}(x)\hat{H}\Phi^{(\nu)}(x)\,dx = E^{(\nu)} \int_{-\infty}^{+\infty} \Phi^{*(\mu)}(x)\Phi^{(\nu)}(x)\,dx. \tag{7.100}$$

$\Phi^{(\nu)}(x)$) shall represent an orthonormal system of eigenfunctions (see (5.66)ff.), i. e. the orthonormality relation

$$\int_{-\infty}^{+\infty} \Phi^{*(\mu)}(x)\Phi^{(\nu)}(x)\,dx = \delta_{\mu,\nu} \tag{7.101}$$

can be used (compare with (5.65)) so that (7.100) can be replaced by

$$\int_{-\infty}^{+\infty} \Phi^{*(\nu)}(x)\hat{H}\Phi^{(\nu)}(x)\,dx = E^{(\nu)} \; . \tag{7.102}$$

(7.102) represents a special rule to calculate the measurement quantity $E^{(\nu)}$. Due to the fact that particles are considered, $E^{(\nu)}$ represents the total energy of the system. However, only if the *Hamiltonian* (7.98) is used, the scalar product of the l. h. s. of (7.102) generates a total energy. If any operator $\hat{M}$ is used, one obtains the relation

$$\boxed{\int_{-\infty}^{+\infty} \Phi^{*(\nu)}(x)\hat{M}\Phi^{(\nu)}(x)\,dx = M^{(\nu)}} \tag{7.103}$$

if the functions $\Phi^{(\nu)}(x)$ are as well eigenfunctions of the operator $\hat{M}$, i. e. if

$$\boxed{\hat{M}\Phi^{(\nu)}(x) = M^{(\nu)}\Phi^{(\nu)}(x)} \tag{7.104}$$

holds. This requirement is equivalent to the requirement that the occuring *commutator* $\left[\hat{M},\hat{H}\right]_{-}$ vanishes, i. e.

$$\left[\hat{M},\hat{H}\right]_{-} = \hat{M}\hat{H} - \hat{H}\hat{M} = 0 \tag{7.105}$$

holds. If (7.104) generates measurable eigenvalues $M^{(\nu)}$, (7.103) represents a rule to calculate measurable eigenvalues.

If ensembles are considered, the functions $\Phi^{(\nu)}(x)$ have to be replaced by ensemble functions $\Psi(x,t)$. This case shall now be considered.

7.4.2   Ensemble Functions and Measurement

An ensemble function

$$\Psi(x,t) = \sum_{\nu=1}^{\nu_{\max}} \epsilon^{(\nu)}(t)\Phi^{(\nu)}(x) \tag{7.106}$$

represents an ensemble of systems which obey an equation of the type (7.99) (see section 5.3). (7.106) represents a special case of the ensemble function (5.60), i. e.

$$\epsilon^{(\nu)}(t) = c^{(\nu)} \exp\left(INV\omega^{(\nu)} t \right) \tag{7.107}$$

holds (in the *Schrödinger* case $INV = -\mathrm{i}$ and in the *Fokker-Planck* case $INV = -1$ holds). Due to the fact that the stationary state functions $\Phi^{(\nu)}(\boldsymbol{x})$ are solutions of (7.99), and due to the fact that (7.104) holds, the relation

$$\boxed{\int_{-\infty}^{+\infty} \Psi^*(\boldsymbol{x},t)\hat{M}\Psi(\boldsymbol{x},t)\,d\boldsymbol{x} = \sum_{\nu=1}^{\nu_{\max}} \left|\epsilon^{(\nu)}(t)\right|^2 M^{(\nu)} := \langle M \rangle(t)} \tag{7.108}$$

is valid, with $\left|\epsilon^{(\nu)}(t)\right|^2$ being defined by

$$\left|\epsilon^{(\nu)}(t)\right|^2 = \begin{cases} \left|c_{\mathrm{SE}}^{(\nu)}(t)\right|^2 \ (Schrödinger \text{ case}) \\ \left|c_{\mathrm{FPE}}^{(\nu)}(t)\right|^2 exp\left(-2\omega^{(\nu)}t \right) \ (Fokker{-}Planck \text{ case}) \end{cases} . \tag{7.109}$$

If the quantities $\left|\epsilon^{(\nu)}(t)\right|^2$ represent the probability to find a measurement value $M^{(\nu)}$ within the considered ensemble, $\langle M \rangle(t)$ represents a time-dependent mean value which corresponds to the whole ensemble. Due to the fact that in subsection 5.4.3 was shown that it is possible to introduce relations which determine transition probabilities $dw/dt = d\left|\epsilon^{(\nu)}(t)\right|^2/dt$, it is possible to determine probabilities $\left|\epsilon^{(\nu)}(t)\right|^2$. In this case the result (7.108) represents the mean value of a measurement quantity M if a statistical ensemble is considered. (7.108) includes the case (7.103) so that in the following only ensemble functions have to be considered.

In conclusion it has to be noted that suitable measurement operators generate products of measurement values and specific stationary state functions $\Phi(\boldsymbol{x})$, i. e.

$$\hat{M}\Psi(\boldsymbol{x},t) = \sum_{\nu=1}^{\nu_{\max}} \epsilon^{(\nu)}(t) M^{(\nu)}\Phi^{(\nu)}(\boldsymbol{x}) \tag{7.110}$$

holds. This means that the time-dependent state function $\Psi(\boldsymbol{x},t)$ contains information of the considered physical system in an implicit way. In a nutshell, using a suitable state function various measurable quantities can be calculated. In this context it was assumed that (7.105) holds, i. e. the measurement operator $\hat{M}$ and the *Hamiltonian* commute so that the stationary functions $\Phi(\boldsymbol{x})$ simultaneously are eigenfunctions of $\hat{M}$ and $\hat{H}$. In this case $\Psi(\boldsymbol{x},t)$ contains information of the considered measurement. If this is not the case, i. e. if $\Psi(\boldsymbol{x},t)$ does not contain a specific measurement information, (7.110) has to be replaced by

$$\hat{M}\Psi(\boldsymbol{x},t) = \sum_{\nu=1}^{\nu_{\max}} \epsilon^{(\nu)}(t)\hat{M}\Phi^{(\nu)}(\boldsymbol{x}) . \tag{7.111}$$

Therefore, the connection between indirect state functions (or direct state functions, respectively) and correlated measurement quantities in the context of *Fokker-Planck* and quantum systems has been considered. As the reader can see, the dynamic behavior of the considered physical system is totally represented by the dynamic behavior of a suitable state function if an operator $\hat{M}$ does not represent a time-dependent quantity. However, to deal with high-dimensional quantum systems, it is as well possible to introduce a formalism which bases on time-dependent operators. This formalism shall now be considered.

7.5 Heisenberg's Formalism

It is possible to introduce a formalism which bases on time-dependent operators. In the following this formalism shall be considered. However, only the *Schrödinger* case shall be considered. In this case this formalism is nothing but the well-known *Heisenberg* formalism. In the following the basic formalism shall be discussed.

7.5.1 The Basic Formalism

Using the already introduced unitary transformation

$$\Psi(\boldsymbol{x}, t) = \exp\left[-\frac{i}{\hbar}\hat{H}(\boldsymbol{x})t\right]\Psi(\boldsymbol{x}, 0) \, ,$$

$$\hat{U}(t) = \exp\left[-\frac{i}{\hbar}\hat{H}(\boldsymbol{x})t\right] \, , \quad \hat{U}^{-1}(t) = \exp\left[\frac{i}{\hbar}\hat{H}(\boldsymbol{x})t\right] \tag{7.112}$$

(compare with (7.8)) and considering the *Schrödinger* case the relation (7.111) can be replaced by

$$\hat{M}_{\text{Hb}}(t)\Psi(\boldsymbol{x}, 0) = \sum_{\nu=1}^{\nu_{\max}} \epsilon^{(\nu)}(0)\hat{M}_{\text{Hb}}(t)\Phi^{(\nu)}(\boldsymbol{x}) \, , \tag{7.113}$$

where $\hat{M}_{\text{Hb}}(t)$ is defined by

$$\hat{M}_{\text{Hb}}(t) = \hat{U}^{-1}(t)\hat{M}_{\text{S}}\hat{U}(t) \, . \tag{7.114}$$

$\hat{M}_{\text{Hb}}(t)$ is a so-called *Heisenberg operator*, and $\hat{M}_{\text{S}}$ represents a *Schrödinger operator*. As the reader can see, instead of time-dependent state functions and time-independent operators, state functions at the time point $t = 0$ and time-dependent *Heisenberg* operators occur. (However, as it already was mentioned, sometimes a *Schrödinger* operator is a time-dependent operator itself. In such a case (7.114) connects two kinds of time-dependent operators.) Therefore, there is need of an evolution equation which determines the time behavior of such *Heisenberg* operators. In order to find such an evolution equation, the derivative of the relation (7.114) with respect to the time t has to be considered, i. e. the relation

$$\frac{d\hat{M}_{\text{Hb}}(t)}{dt} = \hat{U}^{-1}(t)\frac{d\hat{M}_{\text{S}}}{dt}\hat{U}(t) + \frac{d\hat{U}^{-1}(t)}{dt}\hat{M}_{\text{S}}\hat{U}(t) + \hat{U}^{-1}(t)\hat{M}_{\text{S}}\frac{d\hat{U}(t)}{dt} \tag{7.115}$$

has to be considered. Using the definition

$$\frac{\partial \hat{M}_{\mathrm{Hb}}(t)}{\partial t} = \hat{U}^{-1}(t)\frac{d\hat{M}_{\mathrm{S}}}{dt}\hat{U}(t) \tag{7.116}$$

the differential equation (7.115) can be replaced by

$$\frac{d\hat{M}_{\mathrm{Hb}}(t)}{dt} = \frac{\partial \hat{M}_{\mathrm{Hb}}(t)}{\partial t} - \frac{\mathrm{i}}{\hbar}\left[\hat{M}_{\mathrm{Hb}}(t), \hat{H}\right]_{-}, \tag{7.117}$$

with $\left[\hat{M}_{\mathrm{Hb}}(t), \hat{H}\right]_{-}$ being a special commutator. (The r. h. s. of (7.116) does not vanish if the operator M_{S} is a time-dependent operator.) This commutator is defined by

$$\left[\hat{M}_{\mathrm{Hb}}(t), \hat{H}\right]_{-} = \hat{M}_{\mathrm{Hb}}(t)\hat{H} - \hat{H}\hat{M}_{\mathrm{Hb}}(t) . \tag{7.118}$$

(7.117) represents the well-known *Heisenberg* equations. Such *Heisenberg* equations can be used instead of an equation of the *Schrödinger* type, i. e. they determine the dynamics of the physical system in an equivalent way. If a special *Heisenberg* measurement operator $\hat{M}_{\mathrm{Hb}}(t)$ is taken as a basis, and if $\hat{M}_{\mathrm{Hb}}(t)$ and $\hat{H}$ commute, the relation

$$\int_{-\infty}^{+\infty} \Psi^{*}(\boldsymbol{x}, 0)\hat{M}_{\mathrm{Hb}}(t)\Psi(\boldsymbol{x}, 0)\, d\boldsymbol{x} = \sum_{\nu=1}^{\nu_{\max}} \left|\epsilon^{(\nu)}(0)\right|^{2} M^{(\nu)} = \langle M\rangle(0) \tag{7.119}$$

holds (use (7.113), (7.114) and (7.104)), i. e. *Heisenberg* operators allow to calculate measurable quantities in the same way. Due to the fact that both the original operator representation (7.108) in the *Schrödinger* case and the *Heisenberg* representation (7.119) determine the same mean value if $t = 0$ is considered, the relation

$$\left\langle \Psi(\boldsymbol{x}, 0)\left|\hat{M}_{\mathrm{Hb}}(t)\right|\Psi(\boldsymbol{x}, 0)\right\rangle = \left\langle \Psi(\boldsymbol{x}, 0)\left|\hat{M}_{\mathrm{S}}\right|\Psi(\boldsymbol{x}, 0)\right\rangle \tag{7.120}$$

is valid, wherein *Dirac's* bra- and ket-notation has been used.

If physical systems are considered, very often physical quantities are observed which show only discrete states, in which case all states can be characterized by using natural numbers $n = 0, 1, 2, \ldots \infty$. In such a case *creation* and *annihilation operators* can be used. An example is represented by a laser system if a microscopic level is considered, because laser activity on a microscopic level means that interactions of a light field with electrons occur, in which case such an interaction means that a special number of light particles (photons) has to be absorbed or emitted. In order to describe laser activity on such a microscopic level, suitable *Heisenberg* equations can be taken as a basis. If creation and annihilation operators are used, *Heisenberg* equations can be taken as a basis, too. Due to the fact that in chapter 6 laser systems were discussed, it makes sense to show that such laser systems can be described as well by using *Heisenberg* equations and creation/annihilation operators. This problem shall now be considered. Then it will be shown that in this case the *Heisenberg* equations show the same form as the macroscopic equations of chapter 6. However, first the concept of creation/annihilation operators has to be introduced (for further information, see [45]). In order to introduce such operators, only experimental facts have to be taken into consideration, i. e. no basic theory is necessary to introduce such operators.

### 7.5.2	Creation and Annihilation Operators

If a physical quantity M is observed which shows discrete values M_ν with respect to a special measurement, all possible measurement results are desribed by

$$M = \sum_\nu n_\nu M_\nu , \qquad (7.121)$$

wherein n_ν represents the number which determines how often the measurement value M_ν has to be counted. ν normally represents a set of indices, in which case these indices characterize the possible states, i. e. if quantum systems are considered, these indices are special *quantum numbers*. For example, M_ν can represent the observable energy states of microscopic particles, and n_ν can represent the numbers of the particles in the various energy states. Moreover, macroscopic attractors can be characterized by using a suitable measurement value M_ν. Then n_ν represents the considered physical systems in the various attractor states. (7.121) includes the case $M = n_1 M_1$, too. If the reader compares with the discussion given in chapter 5, subsection 5.4.2, it is obvious that also problems such as the quantum mechanical harmonic oscillator are included. In such a case M_1 represents the basic part of the energy eigenvalues of the discrete bounded states, i. e. the relation $M_1 = \hbar\omega$ holds. Then n_1 represents a quantum number itself, i. e. in this case n_1 is no population number. In such a case the relation $M = E_n - \frac{\hbar\omega}{2}$ holds, with E_n representing the total energies of the bounded states. Very often the sum of (7.121) has to be replaced by an integral. However, this shall not be considered in this book. Introducing operators α_ν^+, α_ν^- which are defined by the relations

$$\left[\alpha_\nu^-, \alpha_\mu^+\right]_\mp = \delta_{\nu,\mu}, \left[\alpha_\nu^+, \alpha_\mu^+\right]_\mp = \left[\alpha_\nu^-, \alpha_\mu^-\right]_\mp = 0 \qquad (7.122)$$

and by the relation

$$\alpha_\nu^+ \alpha_\nu^- \Psi_\mathbf{n} = n_\nu \Psi_\mathbf{n} \qquad (7.123)$$

the eigenvalue equation

$$\left(\sum_\nu \alpha_\nu^+ \alpha_\nu^- M_\nu\right) \Psi_\mathbf{n} = M \Psi_\mathbf{n} \qquad (7.124)$$

can be introduced. The possible eigenvalues of (7.124) are identical with the measurable values (7.121). $\Psi_\mathbf{n}$ represents a special state, where $\mathbf{n}$ represents all natural numbers n_ν. The brackets $[]_-$ represent a so-called *commutator*, and the brackets $[]_+$ represent a so-called *anti-commutator*, where

$$[A, B]_{\mp} = AB \mp BA \tag{7.125}$$

holds. (7.124) represents a basic equation which describes both the measurable eigen-values M and the population of the various states. However, an equation of the form (7.124) does not contain more information.

If multi-particle systems of quantum system theory are taken as a basis, the state function Ψ_n has to represent bosons (for example, photons, phonons) and fermions (for example, electrons, positrons, protons, neutrons, neutrinos). Bosons are particles which can be found in one energy state several times. The spin of such particles is characterized by an even quantum number. If fermions are considered, only one fermion can be found in one energy state if interactions are neglected (if fermions with anti-parallel spins are considered, two fermions can be found in one energy state). Fermions are particles with an odd quantum number of spin. In the following such systems shall be considered. In this case the relation

$$\boxed{\Psi_n = \Psi_{n_B}\Psi_{n_F}} \tag{7.126}$$

holds, where Ψ_{n_B} denotes the state function of all bosons, and Ψ_{n_F} denotes the state functions of all fermions. In this case $\mathbf{n}$ represents population numbers, where $\mathbf{n}$ includes both the population numbers $n_{B,\nu}$ and the population numbers $n_{F,\nu}$. Such a state function has not to be defined in detail, and such a function has not to be interpreted as a function of position coordinates. One only has to know that the operators α_ν^+, α_ν^- generate new state functions in such a way that

$$\boxed{\begin{aligned}
\alpha_{B,\nu}^+ \Psi_{n \neq n_{B,\nu},n_{B,\nu}} &= \Psi_{n_B \neq n_{B,\nu},n_{B,\nu}+1}\,, \\
\alpha_{B,\nu}^- \Psi_{n \neq n_{B,\nu},n_{B,\nu}>0} &= \Psi_{n_B \neq n_{B,\nu},n_{B,\nu}-1}\,, \\
\alpha_{B,\nu}^- \Psi_{n \neq n_{B,\nu},n_{B,\nu}=0} &= 0
\end{aligned}} \tag{7.127}$$

and

$$\boxed{\begin{aligned}
\alpha_{F,\nu}^+ \Psi_{n \neq n_{F,\nu},n_{F,\nu}=0} &= \Psi_{n_F \neq n_{F,\nu},n_{F,\nu}=1}\,, \\
\alpha_{F,\nu}^+ \Psi_{n \neq n_{F,\nu},n_{F,\nu}=1} &= 0\,, \\
\alpha_{F,\nu}^- \Psi_{n \neq n_{F,\nu},n_{F,\nu}=1} &= \Psi_{n_F \neq n_{F,\nu},n_{F,\nu}=0}\,, \\
\alpha_{F,\nu}^- \Psi_{n \neq n_{F,\nu},n_{F,\nu}=0} &= 0
\end{aligned}} \tag{7.128}$$

holds, where $\alpha_{B,\nu}^\pm$ are *Bose-Einstein* operators and $\alpha_{F,\nu}^\pm$ are *Fermi-Dirac* operators. In particular, (7.127) includes the property that any population number makes sense if *Bose-Einstein* systems are considered, and (7.128) includes the property that only one particle can be found in on special state if *Fermi-Dirac* systems are considered. As the reader can see, the operators $\alpha_{B,\nu}^+$, $\alpha_{F,\nu}^+$ create a special population and the operators $\alpha_{B,\nu}^-$, $\alpha_{F,\nu}^-$ annihilate a special population. Therefore, the notation *creation* and *annihilation operators* makes sense. In this context it is useful to replace the eigenvalue equation (7.124) by

$$\boxed{\left(\sum_\nu \alpha_{B,\nu}^+ \alpha_{B,\nu}^- M_{B,\nu} + \sum_\nu \alpha_{F,\nu}^+ \alpha_{F,\nu}^- M_{F,\nu}\right)\Psi_n = M\Psi_n} \tag{7.129}$$

with M being defined by

$$M = \sum_\nu n_{\mathrm{B},\nu} M_{\mathrm{B},\nu} + \sum_\nu n_{\mathrm{F},\nu} M_{\mathrm{F},\nu} \ . \tag{7.130}$$

In order to "switch" from one state to the other, a creation or an annihilation operator has to be used, i.e. the relation

$$\alpha_\mu^\pm \left(\sum_\nu \alpha_{\mathrm{B},\nu}^+ \alpha_{\mathrm{B},\nu}^- M_{\mathrm{B},\nu} + \sum_\nu \alpha_{\mathrm{F},\nu}^+ \alpha_{\mathrm{F},\nu}^- M_{\mathrm{F},\nu} \right) \Psi_\mathbf{n} = M \alpha_\mu^\pm \Psi_\mathbf{n} \tag{7.131}$$

has to be used, where $\alpha_\mu^\pm$ represents *Bose-Einstein* operators as well as *Fermi-Dirac* operators. Both the *Bose-Einstein* and the *Fermi-Dirac* operators have to fulfill the relations (7.122) and (7.123), i. e. the identification $\nu \to \mathrm{B}, \nu, \ \mathrm{F}, \nu$ has to be used. However, if *Bose-Einstein* operators are considered, the commutator of (7.122) has to be used, and if *Fermi-Dirac* operators are considered, the anti-commutator of (7.122) has to be used. In this case (7.131) actually allows to switch from one *Fermi-Dirac* and *Bose-Einstein* eigenstate to the other. In order to show this, a short example shall now be considered.

If a *Bose-Einstein* creation operator is used to switch to another state, (7.131) has to be replaced by

$$\alpha_{\mathrm{B},\mu}^+ \left(\sum_\nu \alpha_{\mathrm{B},\nu}^+ \alpha_{\mathrm{B},\nu}^- M_{\mathrm{B},\nu} + \sum_\nu \alpha_{\mathrm{F},\nu}^+ \alpha_{\mathrm{F},\nu}^- M_{\mathrm{F},\nu} \right) \Psi_\mathbf{n} = M \alpha_{\mathrm{B},\mu}^+ \Psi_\mathbf{n} \ . \tag{7.132}$$

Using the commutator relations of (7.122) the relation (7.132) can be rewritten, i. e. one obtains the relation

$$\left(\sum_\nu \alpha_{\mathrm{B},\nu}^+ \alpha_{\mathrm{B},\nu}^- M_{\mathrm{B},\nu} + \sum_\nu \alpha_{\mathrm{F},\nu}^+ \alpha_{\mathrm{F},\nu}^- M_{\mathrm{F},\nu} \right) \alpha_{\mathrm{B},\mu}^+ \Psi_\mathbf{n} = (M + M_{\mathrm{B},\mu}) \alpha_{\mathrm{B},\mu}^+ \Psi_\mathbf{n} \ .$$
$$\tag{7.133}$$

Using (7.127) one obtains the relation

$$\alpha_{\mathrm{B},\mu}^+ \Psi_\mathbf{n} = \alpha_{\mathrm{B},\mu}^+ \Psi_{\mathbf{n} \ne n_{\mathrm{B},\mu}, n_{\mathrm{B},\mu}} = \Psi_{\mathbf{n}_\mathrm{B} \ne n_{\mathrm{B},\mu}, n_{\mathrm{B},\mu}+1} := \Psi_{\mathbf{n}_1} \ . \tag{7.134}$$

Additionally, the sum

$$\begin{aligned}
M + M_{\mathrm{B},\mu} &= \sum_\nu n_{\mathrm{B},\nu} M_{\mathrm{B},\nu} + \sum_\nu n_{\mathrm{F},\nu} M_{\mathrm{F},\nu} + M_{\mathrm{B},\mu} \\
&= \sum_{\nu \ne \mu} n_{\mathrm{B},\nu} M_{\mathrm{B},\nu} + \sum_\nu n_{\mathrm{F},\nu} M_{\mathrm{F},\nu} + (n_{\mathrm{B},\mu} + 1) M_{\mathrm{B},\mu} \\
&:= M_1
\end{aligned} \tag{7.135}$$

has to be considered. Inserting (7.135) and (7.134) into (7.133) one obtains the relation

$$\left(\sum_{\nu} \alpha_{\mathrm{B},\nu}^{+} \alpha_{\mathrm{B},\nu}^{-} M_{\mathrm{B},\nu} + \sum_{\nu} \alpha_{\mathrm{F},\nu}^{+} \alpha_{\mathrm{F},\nu}^{-} M_{\mathrm{F},\nu} \right) \Psi_{\mathbf{n}_1} = M_1 \Psi_{\mathbf{n}_1} , \tag{7.136}$$

which is noting but the original eigenvalue equation (7.131), however, now a new *Bose-Einstein* state with the eigenvalue M_1 and the state function $\Psi_{\mathbf{n}_1}$ has been created. In the same way it can be shown that *Fermi-Dirac* operators create new *Fermi-Dirac* states.

If such creation and annihilation operators are taken as a basis, relevant operators such as the *Hamiltonian* $\hat{H}$ or relevant measurement operators $\hat{M}$ can be reformulated. If such reformulated operators are used, a formulation can be introduced which only describes the possible eigenvalues (such as the total energy) and the population numbers of multi-particle systems. Using creation and annihilation operators the bounded states of one-particle systems can be described as well. If the original formulations (*Schrödinger*, *Feynman* or *Heisenberg* formulation) are used, one obtains solutions which describe non-bounded states, and if suitable creation/annihilation operators are introduced, it is possible to switch from one bounded state to the other and to neglect the non-bounded states between. In the following this concept shall be introduced, and it shall be made clear that this concept is a direct consequence of the experimental experience. However, only multi-particle systems shall be considered. In this context it has to be noted that very often the term *second quantization* is used.

7.5.3   The Second Quantization

The Quantization of the Basic Hamiltonian

In the *Schrödinger* case the *Hamiltonian* (7.98) has to be rewritten. In doing so one obtains the operator

$$\hat{H} = -\frac{\hbar^2}{2m_0} \Delta_3 + V(\boldsymbol{x}) . \tag{7.137}$$

Instead of (7.137) a formulation can be introduced which bases on creation and annihilation operators. This formulation shall now be derived.

If the *Hamiltonian* (7.137) and the ensemble function

$$\Psi(\boldsymbol{x}, t) = \sum_{\nu=1}^{\nu_{\max}} \alpha_\nu(t) \Phi^{(\nu)}(\boldsymbol{x}) \tag{7.138}$$

is considered, the correlated scalar product has to be defined by

$$\begin{aligned} E(t) &= \int_{-\infty}^{+\infty} \Psi^*(\boldsymbol{x}, t) \hat{H} \Psi(\boldsymbol{x}, t) \, d\boldsymbol{x} \\ &= \sum_{\nu=1}^{\nu_{\max}} \alpha_\nu^*(t) \alpha_\nu(t) E_\nu = \sum_{\nu=1}^{\nu_{\max}} |\alpha_\nu(t)|^2 E_\nu , \end{aligned} \tag{7.139}$$

where E_ν represents the energies of the possible bounded states, and where $\Phi^{(\nu)}(x)$ represents the eigenfunctions of the bounded states. This scalar product represents the mean energy $E(t)$ of a quantum system if $|\alpha_\nu(t)|^2$ represents the probability to find a state ν. If $|\alpha_\nu(t)|^2$ represents population numbers $n_\nu(t)$, $E(t)$ represents the total energy of the quantum system, and (7.139) can be replaced by

$$E(t) = \sum_{\nu=1}^{\nu_{\max}} n_\nu(t) E_\nu \ . \tag{7.140}$$

If the functions $\alpha_\nu^*(t)$, $\alpha_\nu(t)$ are replaced by operators α_ν^+, α_ν^-, and if these operators fulfill the relations (7.122) and (7.123), the quantity (7.140) can be replaced by the operator

$$\boxed{\hat{E} = \sum_{\nu=1}^{\nu_{\max}} \alpha_\nu^+ \alpha_\nu^- E_\nu \ ,} \tag{7.141}$$

where the eigenvalue equation

$$\boxed{\hat{E}\Psi_{\mathbf{n}} = E\Psi_{\mathbf{n}}} \tag{7.142}$$

holds, with E being defined by

$$\boxed{E = \sum_{\nu=1}^{\nu_{\max}} n_\nu E_\nu \ .} \tag{7.143}$$

This concept is due to experience. Very often (7.142) is used to describe electron states. In this case the anti-commutator relation of (7.122) has to be taken as a basis. (7.142) replaces an ordinary *Schrödinger* equation if creation and annihilation operators are used. In such a case (7.141) represents the correlated operator. The functions $\Psi_{\mathbf{n}}$ have to fulfill the relations (7.127) and (7.128). In this context it has to be remarked that the population numbers n_ν in principle can be time-dependent. However, the creation and annihilation operators α_ν^+, α_ν^- are time-independent so that they only create new populations, i. e. they do not determine the time evolution of a considered system. Therefore, (7.142) only determines the possible energy values E and the population numbers n_ν (if the energy eigenvalues E_ν are given).

If the functions $\Psi^*(x,t)$, $\Psi(x,t)$ instead of $\alpha_\nu^*(t)$, $\alpha_\nu(t)$ are replaced by operators, another operative representation is possible. This representation shall not be used in this book. However, it shall be remarked that in such a case the operator

$$\hat{E} = \int_{-\infty}^{+\infty} \Psi^+(x)\hat{H}\Psi^-(x)\,dx \tag{7.144}$$

instead of (7.141) has to be used. In this case the relation (7.122) has to be replaced by

$$\left[\Psi^-(\boldsymbol{x}), \Psi^+(\tilde{\boldsymbol{x}})\right]_{\mp} = \delta(\boldsymbol{x} - \tilde{\boldsymbol{x}}) \;,$$

$$\left[\Psi^-(\boldsymbol{x}), \Psi^-(\tilde{\boldsymbol{x}})\right]_{\mp} = \left[\Psi^+(\boldsymbol{x}), \Psi^+(\tilde{\boldsymbol{x}})\right]_{\mp} = 0 \;. \tag{7.145}$$

However, this shall only be remarked.

Not only the basic *Hamiltonian* (and thus the *Schrödinger* equation) can be reformulated by using creation and annihilation operators, macroscopic energy functions of fields, too, can be reformulated in this way. Such a formulation is directly correlated with the particle aspect of fields. Due to the fact that in last consequence laser problems shall be considered, only the quantization of a light field shall be discussed.

The Quantization of a Light Field

The macroscopic energy function of a light field is defined by

$$\begin{aligned}
E_{\text{light}}(t) &= \frac{1}{8\pi} \int_{-\infty}^{+\infty} \left(\boldsymbol{E}^2 + \boldsymbol{B}^2\right) \, d\boldsymbol{x} \\
&= \frac{1}{8\pi} \int_{-\infty}^{+\infty} \left[(4\pi c)^2 \left|\boldsymbol{\Pi}\right|^2 + \left|\nabla \times \boldsymbol{A}\right|^2\right] \, d\boldsymbol{x}
\end{aligned} \tag{7.146}$$

if the *CGS-system* is taken as a basis. ($\boldsymbol{E}$ represents the *electric field strength*, $\boldsymbol{B}$ represents the *magnetic induction*, $\boldsymbol{\Pi}$ is the *canonical conjugate potential* of the electric field, and $\boldsymbol{A}$ represents the *vector potential* of the magnetic field. $\times$ represents the vector product, and ∇ is the well-known *nabla*.) Due to the fact that the change of the energy of a field only is possible by exchange of energy quantums, it has to be possible to derive a formulation which is directly correlated with this quantum structure. This is indeed possible and can be done by starting from (7.146) and using a formal procedure. This procedure shall be considered.

Using *Fourier's* decomposition the vector potential reads

$$\boldsymbol{A} = \sum_{\mathbf{k},\sigma} \boldsymbol{e}_{\mathbf{k},\sigma} \left[\alpha_{\mathbf{k},\sigma}(t) \exp\left(\mathrm{i}\mathbf{k}\boldsymbol{x}\right) + \alpha^*_{\mathbf{k},\sigma}(t) \exp\left(-\mathrm{i}\mathbf{k}\boldsymbol{x}\right)\right] \;, \tag{7.147}$$

with $\alpha_{\mathbf{k},\sigma}(t)$ being defined by

$$\alpha_{\mathbf{k},\sigma}(t) = \alpha_{\mathbf{k},\sigma} \exp\left(-\mathrm{i}\omega_k t\right) \;. \tag{7.148}$$

Due to the fact that the vector potential $\boldsymbol{A}$ and the canonical congugate potential $\boldsymbol{\Pi}$ of the electic field is connected by

$$\boldsymbol{\Pi} = \frac{1}{4\pi c^2} \frac{\partial \boldsymbol{A}}{\partial t} \;, \tag{7.149}$$

the relation

$$\boldsymbol{\Pi} = \frac{1}{4\pi c^2} \sum_{\mathbf{k},\sigma} \boldsymbol{e}_{\mathbf{k},\sigma} \left[-\mathrm{i}\omega_k \alpha_{\mathbf{k},\sigma}(t) \exp\left(\mathrm{i}\mathbf{k}\boldsymbol{x}\right) + \mathrm{i}\omega_k \alpha^*_{\mathbf{k},\sigma}(t) \exp\left(-\mathrm{i}\mathbf{k}\boldsymbol{x}\right)\right] \tag{7.150}$$

holds. ($\mathbf{k}$ represents wave vectors with the amount k, σ represents polarization planes, c represents the light velocity, and $e_{\mathbf{k},\sigma}$ is the unit vector of the considered vector potential.) Inserting (7.150) and (7.147) into the macroscopic energy function (7.146) this energy function can be rewritten. If additionally the functional orthonormality relation

$$\int_{-\infty}^{+\infty} \exp\left[i\left(\mathbf{k} - \tilde{\mathbf{k}}\right)\boldsymbol{x}\right] = V\delta\left(\mathbf{k} - \tilde{\mathbf{k}}\right) \tag{7.151}$$

is used (δ represents *Dirac's* delta function), and if

$$e_{\mathbf{k},\sigma}e_{\mathbf{k},\tilde{\sigma}} = \delta_{\sigma,\tilde{\sigma}} \tag{7.152}$$

is used, one obtains the energy function

$$E_{\text{light}}(t) = \sum_{\mathbf{k},\sigma} \tilde{\alpha}_{\mathbf{k},\sigma}^{*}(t)\tilde{\alpha}_{\mathbf{k},\sigma}(t)\hbar\omega_k \tag{7.153}$$

if the choice

$$\alpha_{\mathbf{k},\sigma}(t) = \tilde{\alpha}_{\mathbf{k},\sigma}(t)\sqrt{\frac{2\pi\hbar c^2}{\omega_k V}} \tag{7.154}$$

is made. The energy function (7.153) can be taken as a basis to gain a suitable operator and a corresponding eigenvalue equation. The concept which has to be used is the same which already has been introduced, i. e. the functions $\tilde{\alpha}_{\mathbf{k},\sigma}^{*}(t)$, $\tilde{\alpha}_{\mathbf{k},\sigma}(t)$ have to be replaced by the operators $\alpha_{\mathbf{k},\sigma}^{+}$, $\alpha_{\mathbf{k},\sigma}^{-}$ so that the operator

$$\boxed{\hat{E}_{\text{light}} = \sum_{\mathbf{k},\sigma} \alpha_{\mathbf{k},\sigma}^{+}\alpha_{\mathbf{k},\sigma}^{-}\hbar\omega_k} \tag{7.155}$$

occurs, and the corresponding eigenvalue equation has to be defined by

$$\boxed{\hat{E}_{\text{light}}\Psi_{\text{light},\mathbf{n}} = E_{\text{light}}\Psi_{\text{light},\mathbf{n}}} \quad . \tag{7.156}$$

In this context the total energy of the light field is defined by

$$\boxed{E_{\text{light}} = \sum_{\mathbf{k},\sigma} n_{\mathbf{k},\sigma}\hbar\omega_k} \quad . \tag{7.157}$$

This formalism represents the measurable particle aspect in a direct way, i. e. this formalism makes sense. Due to the fact that photons are bosons, the creation and annihilation operators $\alpha_{\mathbf{k},\sigma}^{+}$, $\alpha_{\mathbf{k},\sigma}^{-}$ fulfill the anti-commutator relation of (7.122) if the identification $\nu \to \mathbf{k}, \sigma$ is made.

If a light field and electrons are considered, all possible energy states of this multi-particle system can be characterized by using the formalism introduced above. This shall be shown. Due to the fact that in this context normally the term *quantum electrodynamics* is used, the reader has to consider:

Quantum Electrodynamics

If electrons and a light field are considered, the introduced operators (7.141) and (7.155) can be used, i. e. this problem can be described by

$$\hat{E}_{\text{electrons,light}} = \sum_{\nu=1}^{\nu_{\text{max}}} \alpha_\nu^+ \alpha_\nu^- E_\nu + \sum_{\mathbf{k},\sigma} \alpha_{\mathbf{k},\sigma}^+ \alpha_{\mathbf{k},\sigma}^- \hbar\omega_k \,, \tag{7.158}$$

where the eigenvalue equation

$$\hat{E}_{\text{electrons,light}} \Psi_{\text{electrons,light},\mathbf{n}} = E_{\text{electrons,light}} \Psi_{\text{electrons,light},\mathbf{n}} \tag{7.159}$$

has to be used. The possible total energies are defined by

$$E_{\text{electrons,light}} = \sum_{\nu=1}^{\nu_{\text{max}}} n_\nu E_\nu + \sum_{\mathbf{k},\sigma} n_{\mathbf{k},\sigma} \hbar\omega_k \,. \tag{7.160}$$

In this context *interaction* means that new electron states are created, where photons are created or annihilated. Therefore, (7.160) is as well valid if interactions can be observed. However, transition probabilties have to be considered which describe the probability of the occurrence of a special new state. As the discussion in section 5.3 showed, an interaction operator (which can be an ordinary interaction function) can be used within the original *Hamiltonian*, in which case such an interaction operator only determines the transition probabilities if the interaction is weak, i. e. if the interaction represents a *perturbation*. Such interaction operators now are needed. Then the question arises whether it is possible to introduce such interaction operators by using a formulation which bases on creation and annihilation operators. This is indeed possible. In the following the problem of an electron-photon coupling shall be discussed.

The energy function of an electron field and an additional light field can be represented by a sum of the energy functions (7.139) and (7.146), i. e. the relation

$$\begin{aligned}
E_{\text{electrons,light}}(t) &= \int_{-\infty}^{+\infty} \Psi^*(\boldsymbol{x},t) \left[\frac{1}{2m_0} \left(\frac{\hbar}{\mathrm{i}} \nabla \right)^2 + V(\boldsymbol{x}) \right] \Psi(\boldsymbol{x},t)\, d\boldsymbol{x} + \\
&\quad \frac{1}{8\pi} \int_{-\infty}^{+\infty} \left(\boldsymbol{E}^2 + \boldsymbol{B}^2 \right)\, d\boldsymbol{x} \\
&= E_{\text{electrons}}(t) + E_{\text{light}}(t) \tag{7.161}
\end{aligned}$$

holds. An electron-photon coupling has to be described by using an additional part which contains the vector potential $\boldsymbol{A}$, i. e.

$$E_{\text{electrons,light,int}}(t) =$$

$$\int_{-\infty}^{+\infty} \Psi^*(\boldsymbol{x},t) \left[\frac{1}{2m_0} \left(\frac{\hbar}{i}\nabla - \frac{e}{c}\boldsymbol{A} \right)^2 + V(\boldsymbol{x}) \right] \Psi(\boldsymbol{x},t)\, d\boldsymbol{x} +$$

$$\frac{1}{8\pi} \int_{-\infty}^{+\infty} \left(\boldsymbol{E}^2 + \boldsymbol{B}^2 \right)\, d\boldsymbol{x}$$

$$= E_{\text{electrons}}(t) + E_{\text{light}}(t) + E_{\text{int}}(t) \tag{7.162}$$

holds. The part which occurs if interactions are observable is defined by

$$E_{\text{int}}(t) = \int_{-\infty}^{+\infty} \Psi^*(\boldsymbol{x},t) \left(-\frac{e\hbar}{im_0 c}\boldsymbol{A}\nabla + \frac{e^2}{2m_0 c^2}\boldsymbol{A}^2 \right) \Psi(\boldsymbol{x},t)\, d\boldsymbol{x}\ , \tag{7.163}$$

where e represents the elementary charge. (7.163) can be taken as a basis to derive an operative representation. In order to derive such a representation, the vector potential (7.147) and the ensemble function (7.138) have to be inserted into (7.163). Then (7.163) has to be replaced by

$$\boxed{\hat{E}_{\text{int}} = \hat{E}_{\text{one-photon processes}} + \hat{E}_{\text{two-photon processes}}\ ,} \tag{7.164}$$

where the operator relations

$$\boxed{\begin{aligned}
\hat{E}_{\text{one-photon processes}} = \sum_{\nu=1}^{\nu_{\max}} \sum_{\tilde{\nu}=1}^{\nu_{\max}} \sum_{\mathbf{k},\sigma} \Big(& M_{\nu,\tilde{\nu},\mathbf{k},\sigma}\, \alpha_\nu^+ \alpha_{\tilde{\nu}}^- \alpha_{\mathbf{k},\sigma}^- + \\
& M_{\nu,\tilde{\nu},-\mathbf{k},\sigma}\, \alpha_\nu^+ \alpha_{\tilde{\nu}}^- \alpha_{\mathbf{k},\sigma}^+ \Big)
\end{aligned}} \tag{7.165}$$

and

$$\boxed{\begin{aligned}
\hat{E}_{\text{two-photon processes}} =& \\
\sum_{\nu=1}^{\nu_{\max}} \sum_{\tilde{\nu}=1}^{\nu_{\max}} \sum_{\mathbf{k},\sigma} \sum_{\check{\mathbf{k}},\check{\sigma}} \Big(& M_{\nu,\tilde{\nu},\mathbf{k},\check{\mathbf{k}},\sigma,\check{\sigma}}\, \alpha_\nu^+ \alpha_{\tilde{\nu}}^- \alpha_{\mathbf{k},\sigma}^- \alpha_{\check{\mathbf{k}},\check{\sigma}}^- + \\
& M_{\nu,\tilde{\nu},\mathbf{k},-\check{\mathbf{k}},\sigma,\check{\sigma}}\, \alpha_\nu^+ \alpha_{\tilde{\nu}}^- \alpha_{\mathbf{k},\sigma}^- \alpha_{\check{\mathbf{k}},\check{\sigma}}^+ + \\
& M_{\nu,\tilde{\nu},-\mathbf{k},\check{\mathbf{k}},\sigma,\check{\sigma}}\, \alpha_\nu^+ \alpha_{\tilde{\nu}}^- \alpha_{\mathbf{k},\sigma}^+ \alpha_{\check{\mathbf{k}},\check{\sigma}}^- + \\
& M_{\nu,\tilde{\nu},-\mathbf{k},-\check{\mathbf{k}},\sigma,\check{\sigma}}\, \alpha_\nu^+ \alpha_{\tilde{\nu}}^- \alpha_{\mathbf{k},\sigma}^+ \alpha_{\check{\mathbf{k}},\check{\sigma}}^+ \Big)
\end{aligned}} \tag{7.166}$$

hold. Within these operator representations matrix elements occur which are defined by

$$\boxed{M_{\nu,\tilde{\nu},\mathbf{k},\sigma} = FIT_1 \int_{-\infty}^{+\infty} \Phi^{*(\nu)}(\boldsymbol{x}) \left[-\frac{e\hbar}{im_0 c}\exp\left(i\mathbf{k}\boldsymbol{x}\right) \boldsymbol{e}_{\mathbf{k},\sigma}\nabla \right] \Phi^{(\tilde{\nu})}(\boldsymbol{x})\, d\boldsymbol{x}}$$

$$\tag{7.167}$$

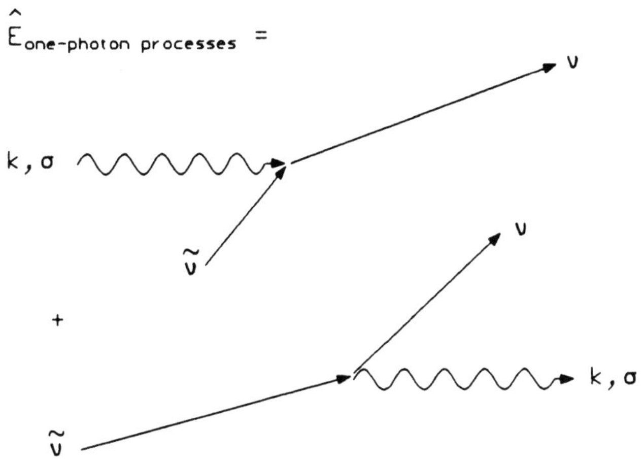

Figure 7.5
A *Feynman* diagram of
one-photon processes

and

$$
\boxed{
\begin{aligned}
M_{\nu,\tilde{\nu},\mathbf{k},\tilde{\mathbf{k}},\sigma,\tilde{\sigma}} &= \\
FIT_2 \int_{-\infty}^{+\infty} \Phi^{*(\nu)}(x) &\left\{ \frac{e^2}{2m_0 c^2} \exp\left[i\left(\mathbf{k} + \tilde{\mathbf{k}}\right) x \right] e_{\mathbf{k},\sigma} e_{\tilde{\mathbf{k}},\tilde{\sigma}} \right\} \Phi^{(\tilde{\nu})}(x)\, dx \; .
\end{aligned}
}
$$

$$(7.168)$$

It is obvious that the operator (7.165) describes the possibility of one-photon processes, i. e. the possibility to create or to annihilate one photon, in which case an electron state changes. By contrast, (7.166) describes the possibility of two-photon processes. Such processes can be visualized by using a so-called *Feynman diagram*. Figure 7.5 shows an example.

In order to explain the meaning of the factors FIT_1, FIT_2, the problem of transition probability again has to be considered.

In chapter 5 a formula was introduced which describes the transition probability $dw_{\kappa \to \mu}/dt$ to reach another physical state μ (see (5.123)) if ensembles of one-particle *Schrödinger* systems are considered. This formula can be written in the form

$$
dw_{\kappa \to \mu}/dt = \frac{2\pi}{\hbar} \int \rho \left| M_{\mu,\kappa}^{(\text{SE,observ})} \right|^2 dE^{(\mu)} \; ,
\tag{7.169}
$$

where the matrix element $M_{\mu,\kappa}^{(\text{SE,observ})}$ has to be defined by

$$
M_{\mu,\kappa}^{(\text{SE,observ})} = M_{\mu,\kappa}^{(\text{SE})} + \sum_{\nu_1=1}^{\nu_{\max}} \frac{M_{\mu,\nu_1}^{(\text{SE})} M_{\nu_1,\kappa}^{(\text{SE})}}{E^{(\kappa)} - E^{(\nu_1)}} K_{\mu,\kappa,\nu_1} + \text{tho} \; .
\tag{7.170}
$$

(K_{μ,κ,ν_1} guarantees that no vanishing denominator occurs. The frequencies $\omega^{(\mu)}$ in (5.123) have been replaced by energies $E^{(\mu)} = \hbar\omega^{(\mu)}$. Due to the fact that ρ shows the dimension $1/(\omega^{(\mu)})^2$, the formula (7.169) shows another form.) Within (7.170) the inner matrix elements

$$M_{\alpha,\beta}^{(\mathrm{SE})} = \left\langle \Phi^{(\alpha)}(\boldsymbol{x}) \left| INT(\boldsymbol{x}) \right| \Phi^{(\beta)}(\boldsymbol{x}) \right\rangle . \tag{7.171}$$

occur which contain the basic interaction operator $INT(\boldsymbol{x})$ as well as solutions $\Phi^{(\alpha)}(\boldsymbol{x})$ of a stationary *Schrödinger* equation. Together with a mediation function ρ, these matrix elements represent the central part of the integral formula. The mediation function ρ is a special density which has to be choosen in a suitable way. Only the correct choice of this mediation function generates a formula which allows to calculate measurable probability densities. This choice represents a heuristic element within the calculation procedure, i. e. ρ represents a heuristic part. However, now multi-particle systems are considered and the operator (7.164) has to be used instead of $INT(\boldsymbol{x})$. Therefore, some modifications are necessary. These modifications shall now be discussed. It shall be assumed that the concept to derive transition probabilities introduced in chapter 5 only has to be extended. The result will then show that this assumption makes sense.

Due to the fact that the conservation law of energy has to be fulfilled, the mediation function has to be replaced by $FIT\,\delta\left(E^{(\kappa)} - E^{(\mu)}\right)$, where FIT represents the necessary heuristic part, and $\delta\left(E^{(\kappa)} - E^{(\mu)}\right)$ is *Dirac's* delta function which directly represents the conservation law of energy. Furthermore, instead of the matrix elements (7.171) the matrix elements

$$\boxed{M_{\alpha,\beta}^{(\mathrm{CA})} = \left\langle \alpha \left| \hat{E}_{\mathrm{int}} \right| \beta \right\rangle} \tag{7.172}$$

have to be used which contain the now relevant operator (7.164) and state functions $|\alpha\rangle$ and $|\beta\rangle$. In this context this state functions only have to fulfill the relations (7.127) and (7.128), and it has to be assumed that an orthonormal system of state functions is given. Due to the fact that $|\alpha\rangle$ has to represent a state which has to be reached and $|\beta\rangle$ has to represent a state before the transition, and due to the fact that such state functions have to represent orthonormal systems, the relation

$$\boxed{\left\langle \alpha \left| \hat{E}_{\mathrm{int}} \right| \beta \right\rangle = \langle \alpha | M_{\alpha,\beta} | \alpha \rangle = M_{\alpha,\beta} \langle \alpha | \alpha \rangle = M_{\alpha,\beta}} \tag{7.173}$$

holds, where $M_{\alpha,\beta}$ represents a special matrix element of the sets of matrix elements (7.167) and (7.168). This means that the matrix elements (7.167) and (7.168) have to determine the probability transition. So far, the parts FIT_1, FIT_2 in (7.167) and (7.168) were not defined. If one assumes that these two parts have to be determined in an heuristic way, the heuristic part FIT of the delta function can be neglected. In this case the result reads

$$\boxed{dw_{\kappa \to \mu}/dt = \frac{2\pi}{\hbar} \int \delta\left(E^{(\kappa)} - E^{(\mu)}\right) \left| M_{\mu,\kappa}^{(\mathrm{CA,observ})} \right|^2 dE^{(\mu)} ,} \tag{7.174}$$

with $M_{\mu,\kappa}^{(\mathrm{SE,observ})}$ being defined by

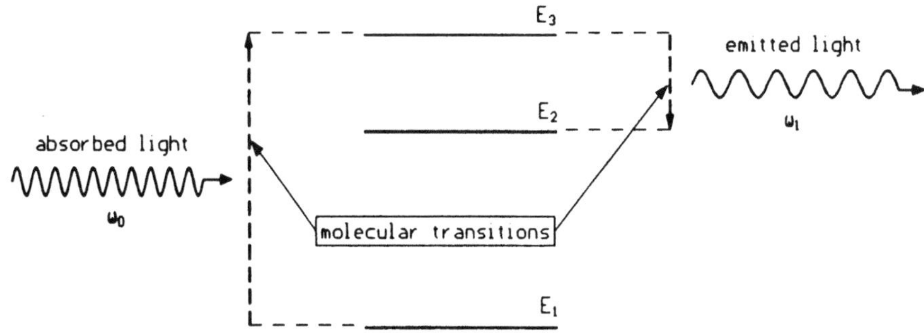

Figure 7.6 Molecules absorb photons so that a new energy state E_3 emerges. This state is not stable so that the state E_2 will be reached. The emitted photons have another energy. An example of such a process is represented by the vibrational *Raman* effect, i. e. molecular vibrations occur.

$$M_{\mu,\kappa}^{(CA,observ)} = M_{\mu,\kappa}^{(CA)} + \sum_{\nu_1=1}^{\nu_{max}} \frac{M_{\mu,\nu_1}^{(CA)} M_{\nu_1,\kappa}^{(CA)}}{E^{(\kappa)} - E^{(\nu_1)}} K_{\mu,\kappa,\nu_1} + tho \ . \tag{7.175}$$

The inner matrix elements of (7.175) have to be defined by (7.172) and (7.173).

As experience shows, (7.174) can be taken as a basis to calculate transition probabilities of multi-particle quantum systems. Therefore, a formula to describe a transition probability is given. A suitable choice of the *fit functions* FIT_1 and FIT_2 allows to calculate the probabilities to find a new particle state. As the reader can see, transitions are possible which show intermediate states, and this is due to experience. If interactions of light and electrons are considered, processes with intermediate states are observable very often. An example is represented by the molecular *Raman* effect (see figure 7.6).

Using the operator (7.164) *Heisenberg* equations which describe electron-photon processes can be calculated. This shall now be shown.

7.5.4 Laser Equations of the Heisenberg Type

In the following the *Heisenberg* equations of electron-photon processes shall be derived. In this context the concept of creation and annihilation operators shall be taken as a basis. As it will be shown, such *Heisenberg* equations are possible, and it will be shown that such equations have the form of the introduced macroscopic laser equations of chapter 6.

In order to describe the interaction of electrons with a light field, the operator

$$\hat{E}_{laser} = \hat{E}_{laser,0} + \hat{E}_{laser,int} \tag{7.176}$$

with

$$\hat{E}_{\text{laser},0} = \sum_{\mathbf{k}} \alpha^+_{\text{F},\mathbf{k}} \alpha^-_{\text{F},\mathbf{k}} \hbar\epsilon_k + \sum_{\mathbf{w}} \alpha^+_{\text{B},\mathbf{w}} \alpha^-_{\text{B},\mathbf{w}} \hbar\omega_w \tag{7.177}$$

and

$$\hat{E}_{\text{laser},\text{int}} = \sum_{\mathbf{k},\mathbf{w}} \left(\hbar M^*_w \alpha^+_{\text{F},\mathbf{k}} \alpha^-_{\text{F},\mathbf{k+w}} \alpha^+_{\text{B},\mathbf{w}} + \hbar M_w \alpha^+_{\text{F},\mathbf{k+w}} \alpha^-_{\text{F},\mathbf{k}} \alpha^-_{\text{B},\mathbf{w}} \right) \tag{7.178}$$

can be used. In order to get the operator (7.176), the sum of (7.158) and (7.164) has to be used, where two-photon processes have to be neglected. If electrons within a crystal are considered, the energy E_ν is of the form $\hbar k^2/2m_{\text{effective}} = \hbar\epsilon_k$, in which case $m_{\text{effective}}$ represents the *effective mass* of the electrons. Such an effective mass allows to neglect the interaction between the electrons and the crystal. The wave vector $\mathbf{k}$ occurs instead of ν and represents the impulse of the electrons. $\mathbf{w}$ represents the wave vector of the photons. Due to the fact that electrons are fermions, the operators $\alpha^\pm_{\text{F},\mathbf{k}}$ are used to describe electron states, and due to the fact that photons are bosons, the operators $\alpha^\pm_{\text{B},\mathbf{w}}$ are used to describe photon states. M_w and M^*_w represent the matrix elements (7.167). Therefore, (7.176) is a special case of the above considered operator $\hat{E}_{\text{electrons},\text{light}} + \hat{E}_{\text{int}}$. As the above considerations showed, $\hat{E}_{\text{laser},0}$ can be used to define an eigenvalue equation which determines all possible states of the considered multi-particle system, and $\hat{E}_{\text{laser},\text{int}}$ can be used to define an equation which determines the transition probabilities. However, in the following it shall be shown that it is possible to derive *Heisenberg* equations by using the operators $\hat{E}_{\text{laser},0}$ and $\hat{E}_{\text{laser},\text{int}}$. In order to show this, it shall be assumed that the introduced calculation procedure (see 7.5.1) holds. Then it will be shown that the result makes sense.

In order to derive the *Heisenberg* equations of a process which is determined by (7.176), the *Heisenberg* operators have to be calculated. Using the introduced transformation (7.114) one obtains the definition

$$\alpha^\pm_{\text{Hb},\text{F},\mathbf{k}}(t) = \exp\left(\frac{\mathrm{i}}{\hbar}\hat{E}_{\text{laser},0}t\right) \alpha^\pm_{\text{F},\mathbf{w}} \exp\left(-\frac{\mathrm{i}}{\hbar}\hat{E}_{\text{laser},0}t\right),$$

$$\alpha^\pm_{\text{Hb},\text{B},\mathbf{k}}(t) = \exp\left(\frac{\mathrm{i}}{\hbar}\hat{E}_{\text{laser},0}t\right) \alpha^\pm_{\text{B},\mathbf{w}} \exp\left(-\frac{\mathrm{i}}{\hbar}\hat{E}_{\text{laser},0}t\right), \tag{7.179}$$

and (due to the fact that $\hat{E}_{\text{laser},0}$ and $\hat{E}_{\text{laser}}$ commute) one obtains

$$\hat{E}_{\text{Hb},\text{laser}} = \exp\left(\frac{\mathrm{i}}{\hbar}\hat{E}_{\text{laser},0}t\right) \hat{E}_{\text{laser}} \exp\left(-\frac{\mathrm{i}}{\hbar}\hat{E}_{\text{laser},0}t\right) = \hat{E}_{\text{laser}}, \tag{7.180}$$

where the operators of the l. h. s. are *Heisenberg* operators. Using the definition (7.117) to determine evolution equations of the *Heisenberg* photon operators $\alpha^\pm_{\text{Hb},\text{B},\mathbf{w}}(t)$ one obtains

$$\frac{d\alpha^\pm_{\text{Hb},\text{B},\mathbf{w}}(t)}{dt} = \frac{\mathrm{i}}{\hbar} \left[\alpha^\pm_{\text{Hb},\text{B},\mathbf{w}}(t), \hat{E}_{\text{Hb},\text{laser}}\right]_- . \tag{7.181}$$

Inserting (7.179) and (7.180) into (7.181) one obtains the explicit formulation of the considered *Heisenberg* equations, i. e. one obtains

$$
\begin{aligned}
\frac{d\alpha^+_{\mathrm{Hb,B,w}}(t)}{dt} &= i\omega_w \alpha^+_{\mathrm{Hb,B,w}}(t) + i\sum_{\mathbf{k}} M_w \alpha^+_{\mathrm{Hb,F,k+w}}(t)\alpha^-_{\mathrm{Hb,F,k}}(t) \,, \\
\frac{d\alpha^-_{\mathrm{Hb,B,w}}(t)}{dt} &= -i\omega_w \alpha^-_{\mathrm{Hb,B,w}}(t) - i\sum_{\mathbf{k}} M^*_w \alpha^+_{\mathrm{Hb,F,k}}(t)\alpha^-_{\mathrm{Hb,F,k+w}}(t) \,.
\end{aligned}
\tag{7.182}
$$

If the reader compares the *Heisenberg* equations (7.182) with the introduced macroscopic laser equations (6.2), it is obvious that these *Heisenberg* equations determine the evolution of the light field of a laser, where the second equation of (7.182) can directly be compared with (6.2), and the first equation of (7.182) has to be compared with the conjugate complex form of (6.2). Instead of the polarizations $\alpha^c_k(t)$ now *Heisenberg*'s operators $\alpha^+_{\mathrm{Hb,F,k}}(t)$, $\alpha^-_{\mathrm{Hb,F,k}}(t)$ occur. Therefore, it has been demonstrated that it is possible to calculate *Heisenberg* equations by using creation and annihilation operators, and it has been demonstrated that *Heisenberg*'s laser equations are similar to the macroscopic laser equations.

In the following some additional comments shall be given.

7.5.5 Additional Comments

In order to calculate equations which are compatibel with experience, additional *fit functions* were necessary. For example, fit functions such as the mediation function ρ and the fit functions FIT_1 and FIT_2 have to be used to derive equations which are able to describe measurable transition probabilities. Another fit function is represented by (7.154) which has to be used to derive a light operator (see (7.155)) which is compatibel with experience. This fact has to be emphasized, i. e. to derive equations which are valid for quantum systems, heuristic elements have to be used.

Another point which has to be emphasized is that the concept of creation and annihilation operators can also be used if macroscopic physical systems are considered. For example, if a physical system shows various attractor states, normally only such attractor states are observable, because intermediate states occur during a short period of time, or such intermediate states can be considered as fluctuating effects. If an ensemble of such physical systems is observed, the measurement shows a distribution of attractor states. Then all possible distributions can be described by using creation and annihilation operators. A very nice example is represented by the human society. If people of such a society are "observed", the result of such a measurement will show that some people are in a *working state* and some people are in a *sleeping state* (in which case the working state sometimes can be identical with the sleeping state). Other states can be characterized by the words *hate* and *love*. Indeed this is an oversimplistic description. Anyhow, if such a description is taken as a basis, all possible states of the human society can be described by using creation and annihilation operators.

The third point I want to stress is that the concept of *Heisenberg* equations can also be used if *Fokker-Planck* systems are considered. However, this I do not want to show.

Therefore, in this chapter it has been shown that it is possible to deal with high-dimensional quantum systems in a totally analytical way. Additionally, it has been shown that the concept of system theory evolved in this book can widely be used, in particular, quantum systems can be described by using the introduced concept of system theory. Thus, the universal aspects of the concept have been demonstrated. Another scheme of description is possible as well. In the following this scheme shall be considered.

8   Information

In chapter 3 the term *information* was introduced which describes the very often used word *information* in an exact way. Strictly speaking, this term represents a measure of the width of a statistical distribution function. In particular, an increase of the width of a distribution function leads to an increase of the value of information (see example (3.2)-(3.5)). Due to the fact that such an increase means that the considered system contains more information (in the ordinary sense of the word), it is obvious that the term *information* can be used to put the often used word *information* in concrete terms. As it was discussed, such a distribution function can describe both stochastic and deterministic systems. For example, if a *Brownian* motion is observed, it makes sense to use a statistical distribution function, in which case such a process represents a stochastic process. By contrast, if a written text is considered, a distribution function can be used to describe the probability to find a special word. Such a system represents a typical deterministic system. Both the *Brownian* motion and the written text contain information (in the ordinary sense of the word). Instead of a written text the human genetic code can be taken as a basis. Then instead of words special molecular configurations occur which represent hereditary information (in the ordinary sense of this word). In order to put the word *information* in concrete terms, the expression

$$\boxed{I_m = -\int_m \rho(\Omega) \ln\left[\rho(\Omega)d\Omega\right] d\Omega \;.}$$

(8.1)

was introduced. (8.1) can be used to describe natural and artifical systems. The level of description is always a compressed one, i. e. a high-dimensional multi-component system can be characterized by using the collective measure *information*. Due to the fact that any statistical system can be described with such a measure, a universal concept is used. Using the quantity *information* does not mean getting new knowledge about a considered system. However, it means having an additional level of consideration (for further information, see [40]).

In the following some of the systems considered in this book shall again be discussed. However, now the level of description shall be the level of information. In this context quantum systems, too, will be considered, i. e. distribution functions of quantum mechanics will be used to determine the information expression (8.1). This is possible, because no special physical background has to be taken into consideration, i. e. any statistical system can be taken into account. This is due to the justification of the term information (see chapter 3), i. e. the given justification does not use special physical conditions, only a space of statistically distributed variables is taken as a basis. And this is due to the fact that the term information

was introduced as a universal expression to put the often used word *information* in
concrete terms. This will lead to a new insight into the discussed problems. It has to
be remarked that then the term *process information* sometimes is used to underline
the dynamical aspects of information.

8.1 Information and Phase Transition

In the following the connection between a phase transition and information shall be
considered. In this context laser systems shall be taken into consideration.

8.1.1 Information of a One-Mode Laser

Analytical and Numerical Facts

If a one-mode laser near a phase transition point is considered, the statistical distribution function is defined by (6.50). Using the information expression (8.1), i. e.
using now the expression

$$I^{(1)}_{m,\text{laser}} = - \int_m \rho(E_w) \ln\left[\rho(E_w)dE_w\right] dE_w \ , \tag{8.2}$$

the information evolution near a phase transition point can numerically be calculated. An illustration of such a calculation shows figure 8.1. The r. h. s. of this
picture shows the region before the phase transition, and the l. h. s. shows the region after the transition. The same picture can be used to describe the evolution of
a magnetization or a polarization in the context of *Landau's* theory, this has to be
remarked additionally. The phase transition point itself ($\lambda^{(2,1)}_w = 0$) is a point near
the maximum of the information curve. This graph characterizes a laser process (or
special magnetization and polarization processes) on the level of information. As it
has been shown in chapter 3, such an information is correlated with a macroscopic
level, i. e. now the relation

$$\begin{aligned}
I^{(1)}_{m,\text{laser}} &= I^{(1)}_{\text{norm,laser}} + \lambda^{(2,1)}_w \langle E^2_w \rangle + \lambda^{(4,1)}_{w,w} \langle E^4_w \rangle \\
&= I^{(1)}_{\text{norm,laser}} + \Delta I^{(1,\text{reduced})}_{m,\text{laser}}
\end{aligned} \tag{8.3}$$

holds (see the general expression (3.16)), with $I^{(1)}_{\text{norm,laser}}$ being a normalization part
which is defined by

$$I^{(1)}_{\text{norm,laser}} = \ln\left(Z^{(1)}_l / dE_w\right) \tag{8.4}$$

($Z^{(1)}_l$ represents the partition function (6.51)). Using the critical hyper-surface equations (6.66) and (6.67) the macroscopic information expression (8.3) can be replaced
by

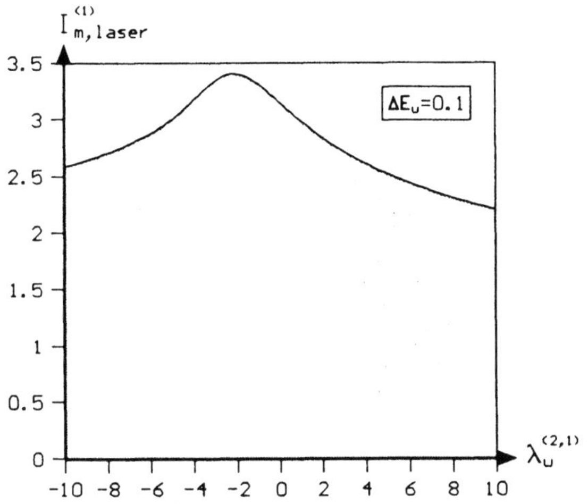

Figure 8.1
Information of a
one-mode laser near
a phase transition
point

$$
I^{(1,\text{crit})}_{m,\text{laser}} = I^{(1,\text{crit})}_{\text{norm,laser}} + \left[\frac{1}{2} - \frac{\langle E^4_w \rangle}{\langle E^2_w \rangle^2} \left(C_1 - \sqrt{C_1^2 - C_2 \frac{\langle E^4_w \rangle}{\langle E^2_w \rangle^2}} \right)^{-2} \right]
$$

$$
+ \left[\frac{\langle E^4_w \rangle}{\langle E^2_w \rangle^2} \left(\sqrt{2} C_1 - \sqrt{2} \sqrt{C_1^2 - C_2 \frac{\langle E^4_w \rangle}{\langle E^2_w \rangle^2}} \right)^{-2} \right] .
$$

$$(8.5)$$

The information expression (8.5) is valid within a narrow surrounding of the critical
point. Figures 8.2 and 8.3 show the correlated reduced process information. The
whole critical region has to be described by using the hyper-surface equations (6.68),
this was discussed in chapter 6. Therefore, information of the one-mode laser is
represented by an ordinary power series, i. e. the relation

$$
I^{(1)}_{m,\text{laser}} = I^{(1)}_{\text{norm,laser}} + \sum_{i=0}^{\infty} \Lambda^{(i)}_{I,1} \left(\frac{\langle E^4_w \rangle}{\langle E^2_w \rangle^2} \right)^i
$$

$$(8.6)$$

holds, wherein the numbers are defined by

$$
\Lambda^{(i=0)}_{I,1} = \Lambda^{(0)}_1, \; \Lambda^{(i>0)}_{I,1} = \Lambda^{(1)}_1 + \Lambda^{(i-1)}_2
$$

$$(8.7)$$

(the numbers $\Lambda^{(i)}_1$ and $\Lambda^{(i)}_2$ are the coefficients of the introduced hyper-surface equations (6.68)).

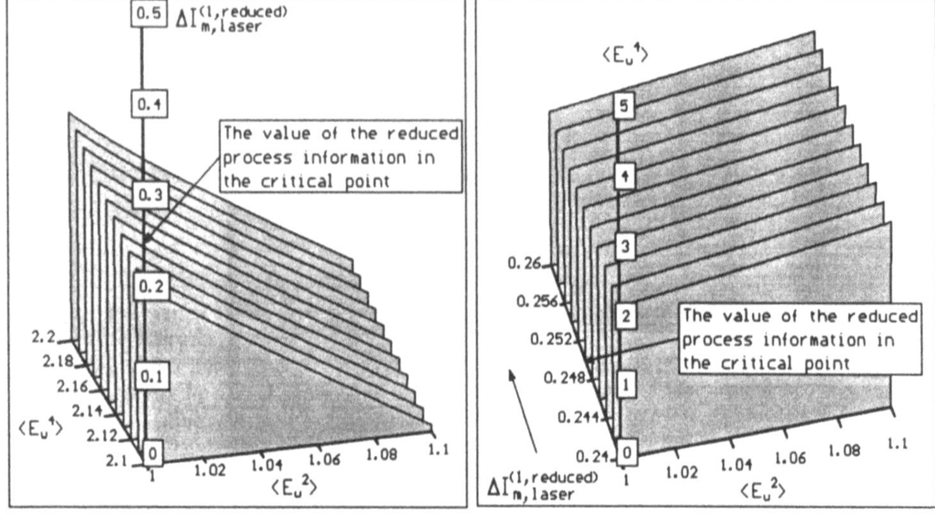

Figure 8.2 The reduced process information of a one-mode laser near a phase transition point as a function of the given correlation functions

The Essence

Therefore, it has to be emphsized that on the level of information the first laser threshold is characterized by an outburst of information (see figure 8.1).

In contrast to the one-mode laser, the multi-mode laser has to be described by using power series functions. This shall now be considered.

8.1.2 The Multi-Mode Laser and Self-Similarity

Analytical Facts

If a multi-mode laser is considered, the information expression (8.2) has to be replaced by

$$I^{(N)}_{m,\text{laser}} = -\int_m \rho(\boldsymbol{E}_w)\ln\left[\rho(\boldsymbol{E}_w)d\boldsymbol{E}_w\right]d\boldsymbol{E}_w\;,\tag{8.8}$$

and the macroscopic information expression is represented by

$$I^{(1)}_{m,\text{laser}} = I^{(N)}_{\text{norm,laser}} + \sum_{w_1}\lambda^{(2,N)}_{w_1}\langle E^2_{w_1}\rangle + \sum_{\substack{w_1,w_2\\w_1<w_2}}\lambda^{(c,N)}_{w_1,w_2}\langle E^2_{w_1}E^2_{w_2}\rangle$$

$$+ \sum_{w_1}\lambda^{(4,N)}_{w_1,w_1}\langle E^4_{w_1}\rangle\;,\tag{8.9}$$

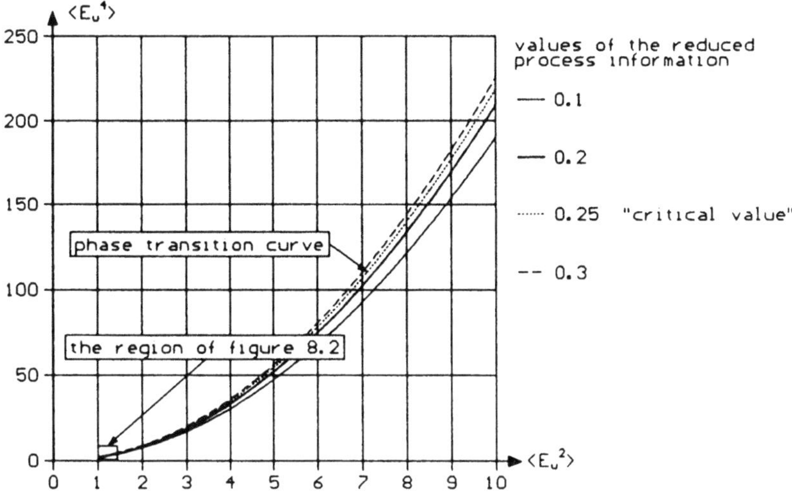

Figure 8.3 The reduced process information of a one-mode laser near a phase transition point. The phase transition curve and the narrow surrounding

with the normalization part being defined by

$$I^{(N)}_{\text{norm,laser}} = \ln\left(Z_l^{(N)}/d\boldsymbol{E}_w\right) .\tag{8.10}$$

(The distribution function of a multi-mode laser is defined by (6.70), the correlated partition function is defined by (6.71), and the general macroscopic information expression is represented by (3.16). $\lambda^{(c,N)}_{w_1,w_2}$ is defined by $\lambda^{(4,N)}_{w_1,w_2} + \lambda^{(4,N)}_{w_2,w_1}$.) Using the hyper-surface equations (6.82) the macroscopic information expression can be replaced, i. e. the expression

$$I^{(N)}_{m,\text{laser}} = I^{(N)}_{\text{norm,laser}} + \left\{ \prod_{\substack{w_3,w_4 \\ w_3 \leq w_4}} \sum_{\epsilon_{w_3,w_4}=0}^{\infty} \Lambda^{(\{\epsilon_{w_3,w_4}\})}_{I,N} \left(\frac{\langle E^2_{w_3} E^2_{w_4}\rangle}{\langle E^2_{w_3}\rangle\langle E^2_{w_4}\rangle}\right)^{\epsilon_{w_3,w_4}} \right\}$$

$$\tag{8.11}$$

holds, in which case the coefficients $\Lambda^{(\{\epsilon_{w_3,w_4}\})}_{I,N}$ are suitable numbers which one can get by inserting the expression (6.82) into the macroscopic information expression (8.9).

The Essence

Information (8.11) is represented by a self-similar power series function (see chapter 4, subsection 4.3.5). This self-similar power series functions represents the connection between the macroscopic level (the mean value level) and information. If the correlation functions are known, a macroscopic determination of information is possible. Due to the fact that such an expression holds near a phase transition point, it has to be noticed that information of a phase transition region can analytically be determined by using a self-similar power series function.

So far, it always was mentioned that the increase of information and the width of a distribution function are correlated. But only one short example was given (see (3.2)-(3.5)). This point shall now be considered in more detail.

8.2 Information and Distribution Width

8.2.1 Analytical and Numerical Facts

An increase of information is correlated with an increase of the width of a considered statistical distribution function. To show this exemplary, the quantum mechanical harmonic oscillator shall be taken as a basis. The indirect state functions (normally called *wave functions*) of the quantum mechanical harmonic oscillator problem were derived in chapter 5 (see (5.95)-(5.98)). Picture 5.5 illustrates these indirect state functions. Using the square of the absolute amount of these indirect state functions special distribution functions $\left|\Phi^{(OSZI,n)}\right|^2$ are given. These distribution functions are shown in figure 8.4. As the reader can see, the width of the functions $\left|\Phi^{(OSZI,n)}\right|^2$ increases if the value of the natural number n increases. Using the information expression (8.1), i. e. using now

$$I_{m,OSZI} = -\int_m \left|\Phi^{(OSZI,n)}\right|^2 \ln\left(\left|\Phi^{(OSZI,n)}\right|^2 dx\right) dx \, , \tag{8.12}$$

the correlated information can numerically be calculated. The result is shown in figure 8.5. As the reader can see, information increases if the width of the distribution functions increases. Due to the fact that the quantum mechanical harmonic oscillator only has discrete stationary states, information shows discrete values, too.

8.2.2 The Essence

This example illustrates the correlation between information and the width of a considered distribution function. However, it has to be remarked that additional patterns normally are overlaid. In the considered example this means that wavy patterns occur. Therefore, it has to be noticed that such patterns have an influence on the value of information, i. e. if distribution functions are compared which nearly

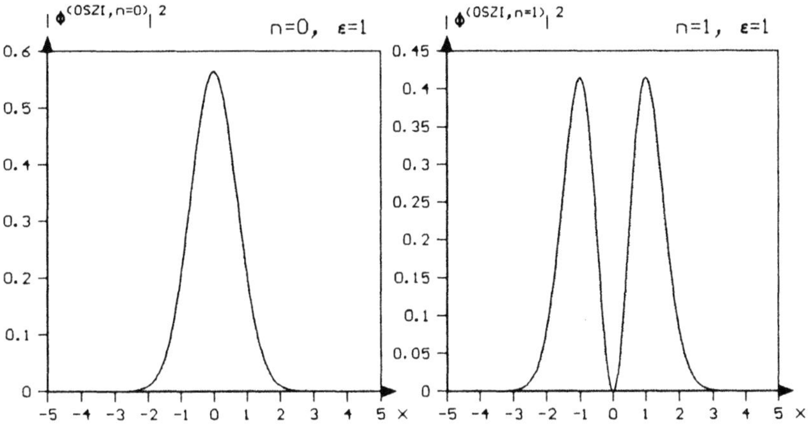

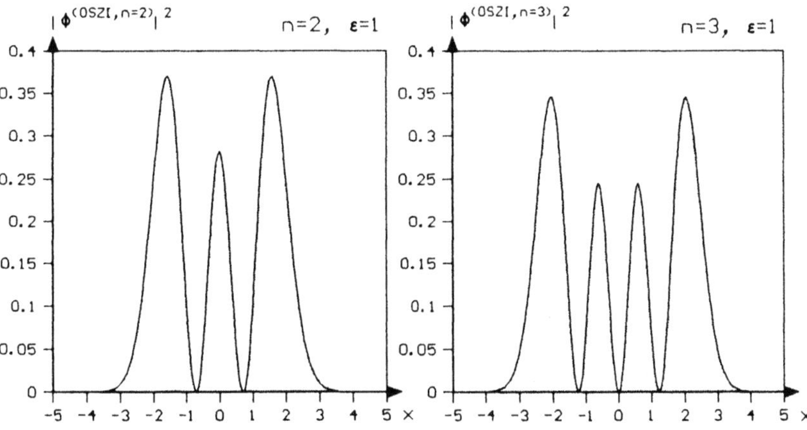

Figure 8.4 The distribution functions $\left|\Phi^{(OSZI,n)}\right|^2$ of the quantum mechanical harmonic oscillator

have the same width, it is possible that the function with the smaller width shows the larger value of information if suitable overlaid patterns are considered. However, in the now considered example this is not the case. Therefore, it has to be emphasized that a transition from one oscillator state to a state of higher energy means to reach a state of higher information if a function $|\Phi|^2$ is taken as a basis. Due to the fact that the ground state is identical with the state which shows the lowest energy, it has to be emphasized that the ground state shows a minimum of information.

Multi-particle quantum system near the zero point of the absolute temperature T show a behavior which can be described by special distribution functions. Such

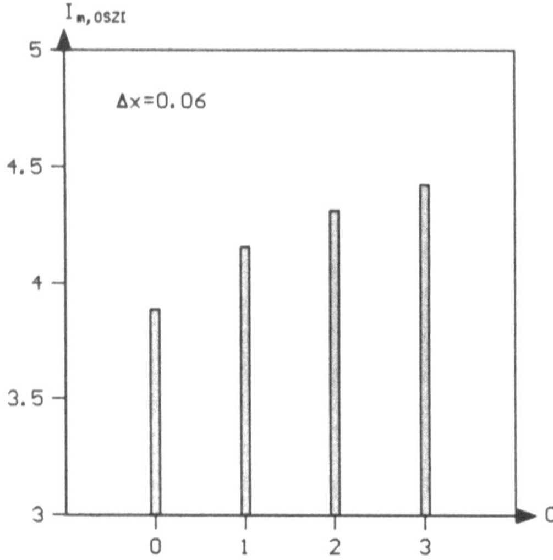

Figure 8.5
Information $I_{m,\mathrm{OSZI}}$ of the quantum mechanical harmonic oscillator

processes, too, can be described in a compressed way by using the term *information*. This shall now be discussed.

8.3   Information and Multi-Particle Systems

8.3.1   Analytical and Numerical Facts

Multi-particle systems are *Fermi-Dirac* systems or *Bose-Einstein* systems, where bosons have a spin which has to be described by an integer, and fermions have a spin which cannot be described by an integer. The distribution function of such multi-particle systems is well-known. In this book this function shall be used in the form

$$\frac{n_i}{N\Delta E_i} = \frac{g_i}{N\Delta E_i}\frac{1}{\exp\left[(E_i - \mu)/kT\right] \pm 1},$$

(8.13)

where the +-sign represents a *Fermi-Dirac* system and the −-sign represents a *Bose-Einstein* system. n_i represents the number of particles in the energy interval ΔE_i, and $n_i/N\Delta E_i$ is the probability to find a particle in such an energy interval. g_i represents a statistical weight function. In the case of non-interacting *Fermi-Dirac* systems the relation $g_i = 2$ holds, because the population of one special energy state at most can be equal two, i. e. if non-interacting *Fermi-Dirac* systems are considered, at most two particles can be found in the same energy state, however, with antiparallel spins. In the case of non-interacting *Bose-Einstein* systems without spin the relation $g_i = 1$ holds. In this context it has to be noticed that non-interacting bosons

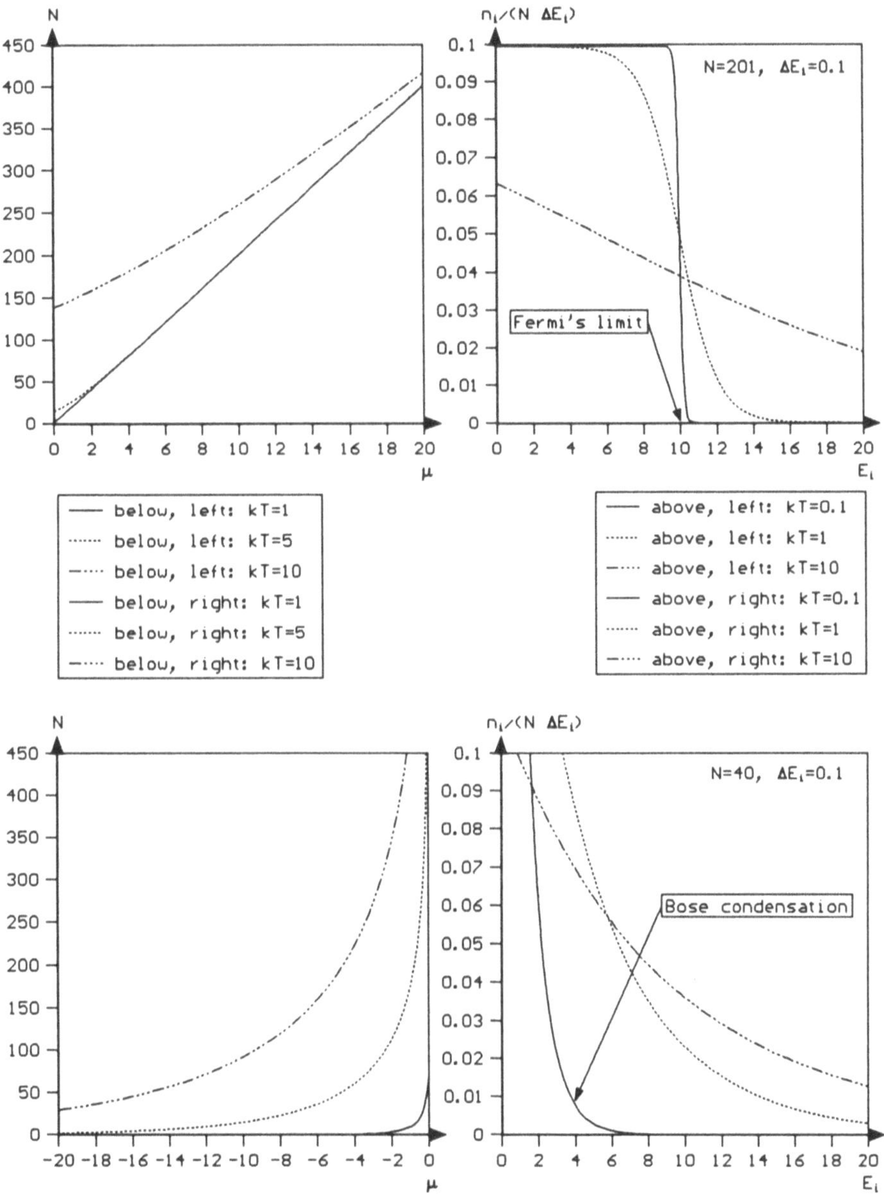

Figure 8.6 *Fermi-Dirac* and *Bose-Einstein* systems. The behavior near the zero point of the absolute temperature

can be found in the same energy state. kT represents the product of *Boltzmann's* constant k and the absolute temperature T. μ is the chemical potential which one

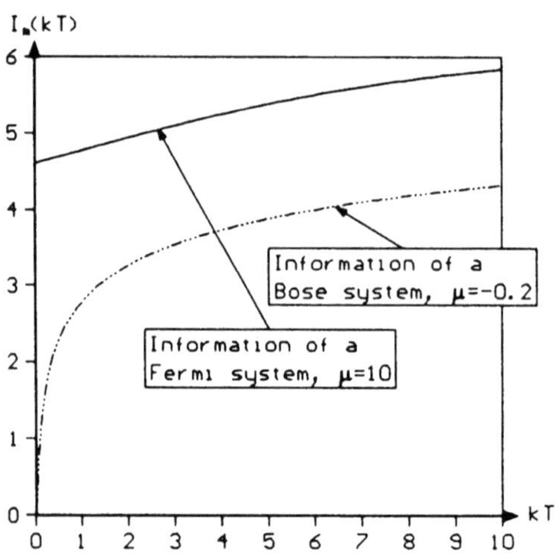

Figure 8.7
Information of *Fermi-Dirac*
and *Bose-Einstein* systems

can get by using the standardization condition

$$\sum_i n_i = N .$$ (8.14)

If the whole number of particles, the number N, is constant, the chemical potential is a variable. Figure 8.6 shows these distribution functions. In this case the information expression (8.1) has to be replaced by the expression

$$I_m(kT) = -\sum_i \frac{n_i}{N\Delta E_i} \ln\left(\frac{n_i}{N\Delta E_i}\Delta E_i\right)\Delta E_i$$
$$= -\sum_i \frac{n_i}{N}\ln\left(\frac{n_i}{N}\right) .$$ (8.15)

Using (8.15), information of *Fermi-Dirac* and *Bose-Einstein* systems can numerically be calculated. Figure 8.7 shows the result.

8.3.2 The Essence

Figure 8.7 shows that near the zero point of the absolute temperature the information of a *Bose-Einstein* vanishes. This behavior represents the *Bose* condensation on the level of information. (*Bose condensation* means that all particles are in the lowest energy state.) By contrast, the information of a *Fermi-Dirac* reaches a special value. This behavior represents the fact that a *Fermi* limit exists. (*Fermi limit* means that all particles can be measured within a small energy band.) It has to be noticed that in this context it was assumed that the lowest energy state can have the value zero. Additionally, a continuous distribution of such energy states was assumed. Therefore, it has to be emphasized that *Fermi-Dirac* systems are systems which reach special values of information, and *Bose-Einstein* systems are systems which show a vanishing information, if the zero point of the absolute temperature is considered.

In particular, information plays a crucial role in human life. Therefore, some additional remarks.

8.4 Information and Human Life

The above considered results show that the physical quantity *information* can be used to characterize the states and the evolution of systems of the inanimated nature. Furthermore, it is possible to use this quantity in the context of the animated nature. Some easy examples already were given. Due to the fact that human behavior can be characterized by a set of quantities such as activities (sleep, thought process, manual processes), this quantity can also be used to characterize the human behavior.

Using the above mentioned set of quantities (i. e. the activities) as a basis, a non-active state such as the sleeping state represents a state of relatively low information, because then the distribution function which describes the probability to observe a special activity has a small width. By contrast, the working state represents a state of high values of information, because many different activities are observable so that a correlated distribution function has a relatively large width. Therefore, it is obvious that the daily routine of people represents a continuous closed curve of low and high information states. Due to the fact that active human states are correlated with a relatively high energy flow, it is obvious that states of high information are states of high energy flow. Furthermore, ground states of quantum systems can be compared with the sleeping state of people. Therefore, it has to be emphasized that ground states of the inanimated and the animated nature are states of relatively low information.

Therefore, it has been shown that the quantity *information* allows to characterize as well the human behavior so that it has been shown that the quantity *information* can be used in a universal way to characterize the behavior of natural and artifical systems.

So far, only non-relativistic systems were considered. In the following some basics of a relativistic system theory shall be considered. The following considerations will then lead to a comprehensive sight of physical dynamics.

9   Basics of Relativistic System Theory

In order to observe a physical system, a reference frame is needed, in which case the physical situation measured by an observer depends on the choosen reference frame.

Special kinds of reference frames are the so-called *inertial reference systems*, which are non-accelerated *Cartesian* reference frames of position coordinates with an additional time reference frame. A position-time point is then defined by

$$r = (x_1, x_2, x_3, t) \ . \tag{9.1}$$

Such a position-time point in the following shall be called *space-time point*. In order to measure the time, measuring instruments have to be installed, in which case one may consider a set of clocks which are placed along the reference frame of the position coordinates. These measuring instruments have to be synchronized. Due to the fact that the light velocity has a finite value (and does not depend on the special initial reference frame), the time within a reference frame of position coordinates has to depend on the considered position point. Figure 9.1 shows this situation, where always shall be assumed that

$$(x_1 = 0, x_2 = 0, x_3 = 0, t = 0) = (\tilde{x}_1 = 0, \tilde{x}_2 = 0, \tilde{x}_3 = 0, \tilde{t} = 0) \tag{9.2}$$

holds. If a special physical situation is observed, and if a special inertial reference system is taken as a basis, the physical situation measured in a relatively moving inertial reference system can be calculated by using the so-called *Lorentz transformations*. If all measurable velocities have small values in comparison with the light velocity of the vacuum ($c = 299792, 458$km/second), the *Lorentz* transformations can approximately be replaced by the so-called *Galilean transformations*. Due to the fact that the form of physical laws must not depend on the special inertial reference system, physical laws have to be *Lorentz covariant*, i. e. if a *Lorentz* transformation is applied, the occuring physical laws have to show the same form as before the transformation (covariance = invariance of the form). Using equations which only remain unchanged with respect to a *Galilean* transformation means then using an approximation which is only valid if low velocities are observable. So far, such approximations were used. (Corresponding systems and quantities shall be called *non-relativistic systems/quantities*. If systems/quantities on the basis of *Lorentz* covariant equations are taken into account, the term *relativistic systems/quantities* shall be used. This term shall be used, too, if a more general class of transformations is taken as a basis.) All these properties are required by *Einstein's special theory of relativity* (see [13, 19]), in which case these properties are well-proved by many experiments.

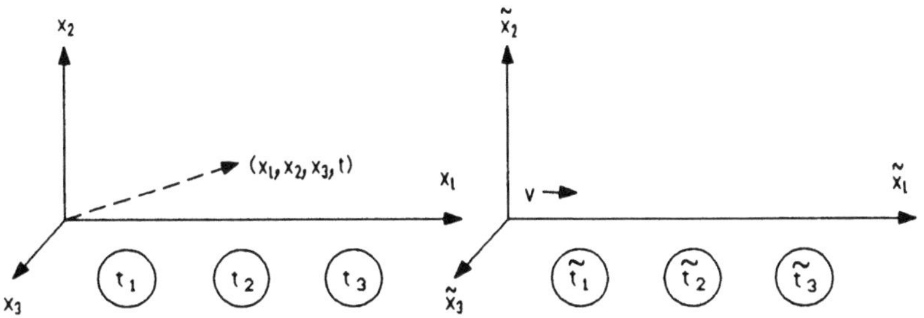

Figure 9.1   Inertial reference systems and *Lorentz* transformations

If a gravitational field exists, inertial reference systems are only possible if relatively small regions of the universe are considered. The total gravitational field corresponds to a curvilinear reference frame, i. e. if gravitational fields exist, a space curvature has to be taken into account. These properties are required by *Einstein's general theory of relativity* (see [13, 68]). These properties are proved as well.

So far, only aspects of non-relativistic system theory were considered. However, if systems of microscopic particles are considered, such particle systems very often show velocities near the light velocity c of the vacuum. (The reader may think of a plasma or of particles in an accelerator.) If such extreme velocities are observable, it has to be taken into account that only equations which are *Lorentz* covariant can represent the measurable behavior in a correct way. Therefore, special kinds of equations have to be used if such relativistic systems are considered. In the following some basics of the then relevant system theory shall be considered. First, the well-known *Lorentz* transformations shall be taken into consideration.

9.1 Lorentz Transformations

If two relatively moving inertial reference systems are considered which move in the same direction (see figure 9.1), a space-time point r has to be transformed by using the *Lorentz* transformations

$$x_1 = \frac{\tilde{x}_1 + v\tilde{t}}{\sqrt{1 - \left(\dfrac{v}{c}\right)^2}} \; , \quad x_2 = \tilde{x}_2 \; , \quad x_3 = \tilde{x}_3 \; , \quad t = \frac{\tilde{t} + \dfrac{v\tilde{x}_1}{c^2}}{\sqrt{1 - \left(\dfrac{v}{c}\right)^2}} \; , \tag{9.3}$$

or the inverse *Lorentz* transformations

$$\tilde{x}_1 = \frac{x_1 - vt}{\sqrt{1 - \left(\dfrac{v}{c}\right)^2}} \ , \quad \tilde{x}_2 = x_2 \ , \quad \tilde{x}_3 = x_3 \ , \quad \tilde{t} = \frac{t - \dfrac{v x_1}{c^2}}{\sqrt{1 - \left(\dfrac{v}{c}\right)^2}} \ , \tag{9.4}$$

respectively. (x_i represents *Cartesian* coordinates, t marks the time point. v is the velocity of the relatively moving inertial reference system.) As it can be seen clearly, position coordinates x_i and the time coordinate t in general are no independent variables. In the sense of this book, these transformations are universal transformations. In the borderline case $v \ll c$ these formulae can be replaced by the *Galilean* transformations

$$x_1 = \tilde{x}_1 + v\tilde{t} \ , \quad x_2 = \tilde{x}_2 \ , \quad x_3 = \tilde{x}_3 \ , \quad t = \tilde{t} \ ,$$
$$\tilde{x}_1 = x_1 - vt \ , \quad \tilde{x}_2 = x_2 \ , \quad \tilde{x}_3 = x_3 \ , \quad \tilde{t} = t \ , \tag{9.5}$$

i. e. in this approximation the position coordinates and the time coordinate are independent variables. So far, this case was assumed in this book.

The validity of (9.3) and (9.4) requires then that the time of observation Δt and the length of an object Δx_1 obey the relation

$$\Delta \tilde{x}_1 = \Delta x_1 \sqrt{1 - \left(\frac{v}{c}\right)^2} \ , \quad \tilde{x}_2 = x_2 \ , \quad \tilde{x}_3 = x_3 \ ,$$
$$\Delta \tilde{t} = \frac{\Delta t}{\sqrt{1 - \left(\dfrac{v}{c}\right)^2}} \ , \tag{9.6}$$

where $\Delta \tilde{t}$ represents the time of observation in a relatively moving inertial reference system, and $\Delta \tilde{x}_1$ represents the observable length in this moving inertial reference system.

Moreover, the observable mass m_0 of an object is a variable if different inertial reference systems are taken as a basis, i. e. the relation

$$m = \frac{m_0}{\sqrt{1 - \left(\dfrac{v}{c}\right)^2}} \tag{9.7}$$

holds. (m represents the observable mass in a relatively moving inertial reference system.) Due to the fact that energy E and mass m are connected by *Einstein's energy-mass relation*

$$E = mc^2 = \sqrt{m_0^2 c^4 + p^2 c^2} \ , \tag{9.8}$$

the energy of a mass depends on the inertial reference system. (p represents the relativistic impulse

$$p = mv \ .) \tag{9.9}$$

Then the kinetic energy has to be defined by

$$E_{\text{kin}} = E - m_0 c^2 = (m - m_0)c^2, \tag{9.10}$$

where the non-relativistic borderline case is defined by

$$E_{\text{kin}}^{(\text{nr})} = \frac{1}{2} m_0 v^2 = \frac{p_0^2}{2m_0} . \tag{9.11}$$

(p_0 represents the non-relativistic impulse

$$p = m_0 v .) \tag{9.12}$$

If basic evolution equations of physical systems are covariant with reference to the *Lorentz* transformations, these equations can be called *Lorentz covariant equations*. As it already was mentioned, such equations are necessary if velocities v are observable which are not restricted by the inequality $v \ll c$. In the following some essential *Lorentz* covariant evolution equations of particle physics shall be considered.

9.2 Lorentz Covariant Evolution Equations of Particle Physics

9.2.1 Macroscopic Equations: The Maxwell Equations

If an electromagnetic field is observed on a macroscopic level, the photon aspect can be neglected. If the *MKSA-system* is taken as a basis, in this case the behavior of such a light field can be described by the differential equations

$$\boxed{\begin{aligned} &\text{curl } \boldsymbol{H} = \boldsymbol{j} + \frac{\partial \boldsymbol{D}}{\partial t} , \quad \text{curl } \boldsymbol{E} = -\frac{\partial \boldsymbol{B}}{\partial t} , \\ &\text{div } \boldsymbol{B} = 0 , \quad \text{div } \boldsymbol{D} = \rho , \end{aligned}} \tag{9.13}$$

or by the integral formulation

$$\boxed{\begin{aligned} &\oint \boldsymbol{H}\, d\boldsymbol{x} = \int_F \left(\boldsymbol{j} + \frac{\partial \boldsymbol{D}}{\partial t} \right) d\boldsymbol{F} , \quad \oint \boldsymbol{E}\, d\boldsymbol{x} = -\int_F \frac{\partial \boldsymbol{D}}{\partial t}\, d\boldsymbol{F} , \\ &\oint \boldsymbol{B}\, d\boldsymbol{F} = 0 , \quad \oint \boldsymbol{D}\, d\boldsymbol{F} = Q , \end{aligned}} \tag{9.14}$$

respectively. ($\boldsymbol{H}$ = magnetic field strength, $\boldsymbol{E}$ = electric field strength, $\boldsymbol{B}$ = magnetic induction, $\boldsymbol{D}$ = electric flux density, $\boldsymbol{j}$ = current density, ρ = charge density, Q = total charge, $d\boldsymbol{F}$ = surface element, $d\boldsymbol{x}$ = line element, curl $\boldsymbol{a} = \nabla \times \boldsymbol{a}$, div $\boldsymbol{a} = \nabla \cdot \boldsymbol{a}$, ∇ = nabla, $\times$ = vector product, . = scalar product.) As it can be seen clearly, these *Maxwell equations* describe also the connection between electromagnetic fields and charges. In particular, the potential function of a charge, the *Coulomb potential*, can be calculated by using these equations. Furthermore, it can be calculated that an electromagnetic field can have the form of a harmonic plane wave. This was used in the context of laser theory (see (6.1)). This system of evolution equations is *Lorentz* covariant. If the transformations (9.3) and (9.4) are applied, this can be shown.

The *Maxwell* equations are evolution equations to describe a special kind of multi-particle systems, namely systems which consist of photons. The level of description is a macroscopic one. If a microscopic level shall be used, other evolution equations are necessary. Such equations shall now be considered.

9.2.2 Microscopic Equations: Klein-Gordon- and Dirac Equation

The Klein-Gordon Equation

A time-dependent *Schrödinger* equation of a one-particle problem, i. e. the evolution equation

$$\left[-\frac{\hbar^2}{2m_0}\Delta_3 + V(\boldsymbol{x}) \right] \Psi(\boldsymbol{x}, t) = i\hbar \frac{\partial}{\partial t}\Psi(\boldsymbol{x}, t) , \tag{9.15}$$

represents nothing but a non-relativistic conservation law of energy, i. e. (9.15) can be replaced by

$$\left[\frac{\boldsymbol{p}_0^2}{2m_0} + V(\boldsymbol{x}) \right] \Phi(\boldsymbol{x}) = E_0\Phi(\boldsymbol{x}) , \tag{9.16}$$

so that the non-relativistic conservation law of energy

$$\frac{\boldsymbol{p}_0^2}{2m_0} + V(\boldsymbol{x}) = E_0 \tag{9.17}$$

has to be taken into account. (E_0 is the non-relativistic total energy, $\boldsymbol{p}_0$ is the non-relativistic impulse vector. $\Phi(\boldsymbol{x})$ represents a stationary state function.) As it can be seen clearly, if the non-relativistic conservation law of energy is taken as a basis, the replacement

$$E_0 \to i\hbar \frac{\partial}{\partial t} , \quad \boldsymbol{p}_0 \to -i\hbar\nabla_3 \tag{9.18}$$

generates the corresponding evolution equation.

Due to the fact that the corresponding relativistic law is defined by (9.8) if interactions are neglected, the relation

$$m_0^2 c^4 + \boldsymbol{p}^2 c^2 = E^2 \tag{9.19}$$

holds if interactions are not considered. (E is the total relativistic energy, p represents the relativistic impulse.) If one assumes the validity of the replacement (9.17) to derive a relativistic evolution equation, too, i. e. if one assumes the replacement

$$E \to i\hbar \frac{\partial}{\partial t} \ , \quad p \to -i\hbar \nabla_3 \ , \tag{9.20}$$

one obtains the evolution equation

$$\boxed{\left(\frac{1}{c^2} \frac{\partial^2}{\partial t^2} - \triangle_3 + \frac{m_0^2 c^2}{\hbar^2} \right) K(x\,,t) = 0 \ ,} \tag{9.21}$$

which is nothing but an ordinary wave equation with an additional rest energy term $\left(m_0^2 c^2 / \hbar^2 \right) K(x,t)$. (9.19) is the so-called *Klein-Gordon equation* if interactions are neglected. Using a relativistic conservation law of energy with an additional interaction function $V(x)$ and a linked electromagnetic field represented by a vector potential A, i. e. using the relation

$$m_0^2 c^4 + (p - eA)^2 c^2 = [E - V(x)]^2 \ , \tag{9.22}$$

one obtains the *Klein-Gordon* equation

$$\boxed{\left\{ - \left[\frac{i}{c} \frac{\partial}{\partial t} - \frac{1}{c\hbar} V(x) \right]^2 + \left(-i\nabla_3 - \frac{e}{\hbar} A \right)^2 + \frac{m_0^2 c^2}{\hbar^2} \right\} K(x,t) = 0} \tag{9.23}$$

which includes interactions. (If the CGS-system is taken as a basis, the term eA has to be replaced by $(e/c)A$.) These equations are *Lorentz* covariant equations which include the relativistic conservation law of energy. $K(x,t)$ represents the now necessary state functions. A suitable formulation of $K(x,t)$ allows to reproduce the underlying conservation law of energy.

Another relativistic quantum mechanical equation is given by *Dirac's* equation. This differential evolution equation shall now be considered.

Dirac's Equation

The time coordinate and the position coordinates have to be treated on an equal basis if *Lorentz* covariant equations shall be derived. Thus, if a differential equation shall be derived which is linear with respect to the time differentiation, combinations of linear position differentiations have to be used. Using the well-known differential *Schrödinger* energy operator $(i\hbar)\partial/\partial t$, a differential evolution equation of the form

$$i\hbar \frac{\partial}{\partial t} D(x,t) = \left(M_1 \frac{\partial}{\partial x_1} + M_2 \frac{\partial}{\partial x_2} + M_3 \frac{\partial}{\partial x_3} + M_4 \right) D(x,t) \tag{9.24}$$

can be taken as a basis. The quantities M_α $(\alpha = 1, 2, 3, 4)$ are matrices which generate combinations of the operators $\partial/\partial x_\beta$ $(\beta = 1, 2, 3)$. Due to the fact that the r. h. s. of (9.24) represents a matrix, the state functions $D(x,t)$ have to be column matrices. Thus, one component $K_i(x,t)$ $(i = 1, 2, \ldots N)$ of the matrix $D(x,t)$ fulfills the relation

$$i\hbar \frac{\partial}{\partial t} K_i(\boldsymbol{x}, t) =$$

$$\sum_{k=1}^{N} \left(M_{1,i,k} \frac{\partial}{\partial x_1} + M_{2,i,k} \frac{\partial}{\partial x_2} + M_{3,i,k} \frac{\partial}{\partial x_3} + M_{4,i,k} \right) K_k(\boldsymbol{x}, t) , \qquad (9.25)$$

wherein $M_{\alpha,i,k}$ are the components of the matrices $\boldsymbol{M}_\alpha$. In order to find a *Lorentz* covariant equation which is connected with the relativistic energy (9.19), it makes sense to require that every component $K_i(\boldsymbol{x}, t)$ has to fulfill the *Klein-Gordon* equation (9.21). In doing so it is possible to derive a matrix relation which allows to determine the unknown matrices $\boldsymbol{M}_\alpha$. This derivation shall now be considered.

Using the operator

$$i\hbar \frac{\partial}{\partial t} = M_{1,i,k} \frac{\partial}{\partial x_1} + M_{2,i,k} \frac{\partial}{\partial x_2} + M_{3,i,k} \frac{\partial}{\partial x_3} + M_{4,i,k} , \qquad (9.26)$$

(9.24) can be replaced by the derivative

$$-\hbar^2 \frac{\partial^2}{\partial t^2} \boldsymbol{D}(\boldsymbol{x}, t) = \sum_{\beta,\gamma=1}^{3} \frac{1}{2} \left(\boldsymbol{M}_\beta \boldsymbol{M}_\gamma + \boldsymbol{M}_\gamma \boldsymbol{M}_\beta \right) \frac{\partial^2}{\partial x_\beta \partial x_\gamma} \boldsymbol{D}(\boldsymbol{x}, t) +$$

$$\sum_{\beta=1}^{3} \left(\boldsymbol{M}_\beta \boldsymbol{M}_4 + \boldsymbol{M}_4 \boldsymbol{M}_\beta \right) \frac{\partial}{\partial x_\beta} \boldsymbol{D}(\boldsymbol{x}, t) +$$

$$\boldsymbol{M}_4 \boldsymbol{M}_4 \boldsymbol{D}(\boldsymbol{x}, t) . \qquad (9.27)$$

Due to the requirement that each component $K_i(\boldsymbol{x}, t)$ shall fulfill the *Klein-Gordon* equation (9.21), equation (9.27) and the *Klein-Gordon* equation can be compared. In doing so one obtains a system of matrix equations, i. e. one obtains the relation

$$\boldsymbol{M}_\beta \boldsymbol{M}_\gamma + \boldsymbol{M}_\gamma \boldsymbol{M}_\beta = -2c^2\hbar^2 \delta_{\beta,\gamma} \mathbf{1}_N ,$$
$$\boldsymbol{M}_\beta \boldsymbol{M}_4 + \boldsymbol{M}_4 \boldsymbol{M}_\beta = 0 ,$$
$$\boldsymbol{M}_4 \boldsymbol{M}_4 = m_0^2 c^4 \mathbf{1}_N , \qquad (9.28)$$

wherein $\mathbf{1}_N$ represents the N-dimensional unit matrix. Introducing the new matrices

$$\boldsymbol{M}_\beta = -ic\hbar \tilde{\boldsymbol{M}}_\beta \ (\beta = 1, 2, 3) ,$$
$$\boldsymbol{M}_4 = m_0 c^2 \tilde{\boldsymbol{M}}_4 , \qquad (9.29)$$

(9.28) has to be replaced by

$$\tilde{\boldsymbol{M}}_\beta \tilde{\boldsymbol{M}}_\gamma + \tilde{\boldsymbol{M}}_\gamma \tilde{\boldsymbol{M}}_\beta = 2\delta_{\beta,\gamma} \mathbf{1}_N ,$$
$$\tilde{\boldsymbol{M}}_\beta \tilde{\boldsymbol{M}}_4 + \tilde{\boldsymbol{M}}_4 \tilde{\boldsymbol{M}}_\beta = 0 ,$$
$$\tilde{\boldsymbol{M}}_4 \tilde{\boldsymbol{M}}_4 = \mathbf{1}_N , \qquad (9.30)$$

and the differential equation

$$-\hbar^2 \frac{\partial^2}{\partial t^2} D(\boldsymbol{x}, t) = -c^2 \hbar^2 \sum_{\beta,\gamma=1}^{3} \frac{1}{2} \left(\tilde{\boldsymbol{M}}_\beta \tilde{\boldsymbol{M}}_\gamma + \tilde{\boldsymbol{M}}_\gamma \tilde{\boldsymbol{M}}_\beta \right) \frac{\partial^2}{\partial x_\beta \partial x_\gamma} D(\boldsymbol{x}, t) - $$

$$i m_0 c^3 \hbar \sum_{\beta=1}^{3} \left(\tilde{\boldsymbol{M}}_\beta \tilde{\boldsymbol{M}}_4 + \tilde{\boldsymbol{M}}_4 \tilde{\boldsymbol{M}}_\beta \right) \frac{\partial}{\partial x_\beta} D(\boldsymbol{x}, t) + $$

$$m_0^2 c^4 \tilde{\boldsymbol{M}}_4 \tilde{\boldsymbol{M}}_4 D(\boldsymbol{x}, t) \tag{9.31}$$

has to be used. The anti-commutator relation (9.30) defines a special matrix algebra. One possibility to fulfill these algebra is given by the choice

$$\tilde{\boldsymbol{M}}_\beta = \begin{pmatrix} 0 & \sigma_\beta \\ \sigma_\beta & 0 \end{pmatrix} \, , \quad \tilde{\boldsymbol{M}}_4 = \begin{pmatrix} \boldsymbol{1}_2 & 0 \\ 0 & -\boldsymbol{1}_2 \end{pmatrix} \, , \tag{9.32}$$

where σ_β represents the so-called *Pauli matrices*

$$\sigma_1 = \begin{pmatrix} 0 & 1 \\ 1 & 0 \end{pmatrix} \, , \quad \sigma_2 = \begin{pmatrix} 0 & -i \\ i & 0 \end{pmatrix} \, , \quad \sigma_3 = \begin{pmatrix} 1 & 0 \\ 0 & -1 \end{pmatrix} \, , \tag{9.33}$$

and where $\boldsymbol{1}_2$ is the two-dimensional unit matrix

$$\boldsymbol{1}_2 = \begin{pmatrix} 1 & 0 \\ 0 & 1 \end{pmatrix} \, . \tag{9.34}$$

The *Pauli* matrices (9.33) occur in non-relativistic quantum mechanics. Mathematically expressed, the set of *Pauli* matrices and $\tilde{M}_4$ form a special representation of the basic relativistic symmetry group. They allow to include the property *spin* into the considerations. Using these Pauli matrices, in particular, the time-dependent Schrödinger equation with an additional vector potential, the equation

$$\left[\frac{1}{2m_0} \left(-i\hbar \nabla_3 - e\boldsymbol{A} \right)^2 + V(\boldsymbol{x}) \right] \Psi(\boldsymbol{x}, t) = i\hbar \frac{\partial}{\partial t} \Psi(\boldsymbol{x}, t) \, , \tag{9.35}$$

can be replaced by

$$\left[\frac{1}{2m_0} \left(-i\hbar \nabla_3 - e\boldsymbol{A} \right)^2 + V(\boldsymbol{x}) - \frac{e\hbar}{2m_0} \boldsymbol{\sigma} \boldsymbol{B} \right] \tilde{\Psi}(\boldsymbol{x}, t) = i\hbar \frac{\partial}{\partial t} \tilde{\Psi}(\boldsymbol{x}, t) \, . \tag{9.36}$$

(9.36) is the so-called *Pauli equation* (for example, see [37]). $\tilde{\Psi}(\boldsymbol{x}, t)$ denotes a two-dimensional function

$$\tilde{\Psi}(\boldsymbol{x}, t) = [\Psi_1(\boldsymbol{x}, t), \Psi_2(\boldsymbol{x}, t)] \tag{9.37}$$

which is called *spinor*. The term $(e\hbar/2m_0)\boldsymbol{\sigma}\boldsymbol{B}$ contains the magnetic induction

$$\boldsymbol{B} = (B_1, B_2, B_3) = (B_{x_1}, B_{x_2}, B_{x_3}) \, , \tag{9.38}$$

where

$$\boldsymbol{\sigma} = (\boldsymbol{\sigma}_1, \boldsymbol{\sigma}_2, \boldsymbol{\sigma}_3) \,,$$
$$\boldsymbol{\sigma}\boldsymbol{B} = \boldsymbol{\sigma}_1 B_1 + \boldsymbol{\sigma}_2 B_2 + \boldsymbol{\sigma}_3 B_3 \tag{9.39}$$

has to be taken into account. This term allows to include the interaction of a magnetic field with the spin of a particle.

The choice (9.32) generates a set of four-dimensional matrices, which fulfills the defined matrix algebra. This can easily be demonstrated. For example, if the choice (9.32) is taken as a basis, the matrix relation

$$\begin{aligned}
\tilde{\boldsymbol{M}}_\beta \tilde{\boldsymbol{M}}_\gamma + \tilde{\boldsymbol{M}}_\gamma \tilde{\boldsymbol{M}}_\beta &= \begin{pmatrix} \boldsymbol{\sigma}_\beta \boldsymbol{\sigma}_\gamma + \boldsymbol{\sigma}_\gamma \boldsymbol{\sigma}_\beta & 0 \\ 0 & \boldsymbol{\sigma}_\beta \boldsymbol{\sigma}_\gamma + \boldsymbol{\sigma}_\gamma \boldsymbol{\sigma}_\beta \end{pmatrix} \\
&= 2\delta_{\beta,\gamma} \mathbf{1}_4 \\
&= \begin{pmatrix} 2\delta_{\beta,\gamma} \mathbf{1}_2 & 0 \\ 0 & 2\delta_{\beta,\gamma} \mathbf{1}_2 \end{pmatrix}
\end{aligned} \tag{9.40}$$

holds so that the matrix equation

$$\boldsymbol{\sigma}_\beta \boldsymbol{\sigma}_\gamma + \boldsymbol{\sigma}_\gamma \boldsymbol{\sigma}_\beta = 2\delta_{\beta,\gamma} \mathbf{1}_2 \tag{9.41}$$

has to be required which is nothing but the well-known *Pauli matrix relation* of non-relativistic quantum mechanics. Using the *Pauli* matrices, introducing

$$\boxed{\tilde{\boldsymbol{M}} = \left(\tilde{\boldsymbol{M}}_1, \tilde{\boldsymbol{M}}_2, \tilde{\boldsymbol{M}}_3 \right)} \tag{9.42}$$

and using

$$\boxed{\hat{\boldsymbol{p}} = -\mathrm{i}\hbar \nabla_3 \,,} \tag{9.43}$$

the differential equation (9.24) can be rewritten. In doing so one obtains the so-called *Dirac equation*

$$\boxed{\mathrm{i}\hbar \frac{\partial}{\partial t} \boldsymbol{D}(\boldsymbol{x}, t) = \left(c\tilde{\boldsymbol{M}}\hat{\boldsymbol{p}} + m_0 c^2 \tilde{\boldsymbol{M}}_4 \right) \boldsymbol{D}(\boldsymbol{x}, t) \,.} \tag{9.44}$$

The column matrix $\boldsymbol{D}(\boldsymbol{x}, t)$ is a *relativistic spinor*. As it can be shown, such a *Dirac* equation is a *Lorentz* covariant equation. In the considered case interactions are neglected.

If an additional interaction with an electromagnetic field has to be considered, the replacement

$$\boldsymbol{p} \to \boldsymbol{p} - e\boldsymbol{A} \,, \quad \tilde{\boldsymbol{p}} \to \tilde{\boldsymbol{p}} - e\boldsymbol{A} \tag{9.45}$$

has to be used. This replacement is the usual procedure to include electromagnetic fields. This replacement was already used. For example, see (7.161)→(7.162) (in this context a CGS-system has to be taken as a basis), (9.19)→(9.22), (7.1)→(9.35) (in this context a one-particle *Schrödinger* equation has to be used). If potential functions $V(\boldsymbol{x})$ have to be included, too, the *Dirac* equation (9.44) has to be replaced by

$$\boxed{\, i\hbar\frac{\partial}{\partial t}\boldsymbol{D}(\boldsymbol{x},t) = \left[c\tilde{\boldsymbol{M}}\left(\hat{\boldsymbol{p}} - e\boldsymbol{A}\right) + V(\boldsymbol{x})\mathbf{1}_4 + m_0c^2\tilde{\boldsymbol{M}}_4\right]\boldsymbol{D}(\boldsymbol{x},t)\,.\,} \quad (9.46)$$

In the following elementary solutions of such a *Dirac* equation will be considered.

Elementary Solutions

In the following *Dirac's* equation shall be taken as a basis, and the problem of solution of such an equation shall be discussed. However, interaction terms shall be neglected so that only elementary solutions will be considered. In this context the solutions of the *Dirac* equation shall be written in the form

$$\boldsymbol{D}^{(\mathrm{fp})}(\boldsymbol{x},t) = \begin{pmatrix} K_{1,1}^{(\mathrm{fp})}(\boldsymbol{x},t) \\ K_{1,2}^{(\mathrm{fp})}(\boldsymbol{x},t) \\ K_{2,1}^{(\mathrm{fp})}(\boldsymbol{x},t) \\ K_{2,2}^{(\mathrm{fp})}(\boldsymbol{x},t) \end{pmatrix}, \quad (9.47)$$

and in the following the definitions

$$\boldsymbol{D}^{(\mathrm{fp})}(\boldsymbol{x}) = \begin{pmatrix} \boldsymbol{D}_1^{(\mathrm{fp})}(\boldsymbol{x}) \\ \boldsymbol{D}_2^{(\mathrm{fp})}(\boldsymbol{x}) \end{pmatrix}, \quad \boldsymbol{D}_l^{(\mathrm{fp})}(\boldsymbol{x}) = \begin{pmatrix} K_{l,1}^{(\mathrm{fp})}(\boldsymbol{x}) \\ K_{l,2}^{(\mathrm{fp})}(\boldsymbol{x}) \end{pmatrix} \quad (l = 1,2) \quad (9.48)$$

and

$$\boldsymbol{D}_0^{(\mathrm{fp})} = \begin{pmatrix} \boldsymbol{D}_{1,0}^{(\mathrm{fp})} \\ \boldsymbol{D}_{2,0}^{(\mathrm{fp})} \end{pmatrix}, \quad \boldsymbol{D}_{l,0}^{(\mathrm{fp})} = \begin{pmatrix} K_{l,1,0}^{(\mathrm{fp})} \\ K_{l,2,0}^{(\mathrm{fp})} \end{pmatrix} \quad (9.49)$$

have to be taken into consideration.

Using the hypothesis

$$\boldsymbol{D}^{(\mathrm{fp})}(\boldsymbol{x},t) = \boldsymbol{D}^{(\mathrm{fp})}(\boldsymbol{x})\exp\left(-\frac{\mathrm{i}}{\hbar}E^{(\mathrm{fp})}t\right) \quad (9.50)$$

the *Dirac* equation (9.44) has to be replaced by

$$E^{(\mathrm{fp})}\boldsymbol{D}^{(\mathrm{fp})}(\boldsymbol{x}) = \left(c\tilde{\boldsymbol{M}}\hat{\boldsymbol{p}} + m_0c^2\tilde{\boldsymbol{M}}_4\right)\boldsymbol{D}^{(\mathrm{fp})}(\boldsymbol{x}) \quad (9.51)$$

or

$$\begin{aligned} E^{(\mathrm{fp})}\boldsymbol{D}_1^{(\mathrm{fp})}(\boldsymbol{x}) &= c\boldsymbol{\sigma}\hat{\boldsymbol{p}}\boldsymbol{D}_2(\boldsymbol{x}) + m_0c^2\mathbf{1}_2\boldsymbol{D}_1^{(\mathrm{fp})}(\boldsymbol{x})\,, \\ E^{(\mathrm{fp})}\boldsymbol{D}_2^{(\mathrm{fp})}(\boldsymbol{x}) &= c\boldsymbol{\sigma}\hat{\boldsymbol{p}}\boldsymbol{D}_1(\boldsymbol{x}) - m_0c^2\mathbf{1}_2\boldsymbol{D}_2^{(\mathrm{fp})}(\boldsymbol{x})\,, \end{aligned} \quad (9.52)$$

respectively. Using the hypothesis

$$\boldsymbol{D}_l^{(\mathrm{fp})}(\boldsymbol{x}) = \boldsymbol{D}_{l,0}^{(\mathrm{fp})}\exp\left(\frac{\mathrm{i}}{\hbar}\boldsymbol{p}\boldsymbol{x}\right) \quad (9.53)$$

this system of coupled equations can be solved. In doing so one obtains the algebraic system

$$\left(E^{(\mathrm{fp})} - m_0 c^2\right) \mathbf{1}_2 \boldsymbol{D}_{1,0}^{(\mathrm{fp})} - c\boldsymbol{\sigma}\boldsymbol{p}\boldsymbol{D}_{2,0} = 0 \;,$$

$$-c\boldsymbol{\sigma}\boldsymbol{p}\boldsymbol{D}_{1,0} + \left(E^{(\mathrm{fp})} + m_0 c^2\right) \mathbf{1}_2 \boldsymbol{D}_{2,0}^{(\mathrm{fp})} = 0 \;,$$

$$(9.54)$$

which is a linear, homogeneous, algebraic system of equations. Such a system has non-trivial solutions if the determinant of the coefficients vanishes, i. e. if

$$\begin{vmatrix} \left(E^{(\mathrm{fp})} - m_0 c^2\right) \mathbf{1}_2 & -c\boldsymbol{\sigma}\boldsymbol{p} \\ -c\boldsymbol{\sigma}\boldsymbol{p} & \left(E^{(\mathrm{fp})} + m_0 c^2\right) \mathbf{1}_2 \end{vmatrix} = 0 \qquad (9.55)$$

holds. Due to the fact that the explicit form of this determinant is given by

$$\left(E^{(\mathrm{fp})^2} - m_0^2 c^4\right) \mathbf{1}_2 - c^2 \left(\boldsymbol{\sigma}\boldsymbol{p}\right)^2 = \left(E^{(\mathrm{fp})^2} - m_0^2 c^4\right) \mathbf{1}_2 - \boldsymbol{p}^2 c^2 \mathbf{1}_2$$
$$= 0 \;, \qquad (9.56)$$

the energy eigenvalues are defined by

$$E^{(\mathrm{fp})^2} = m_0^2 c^4 + \boldsymbol{p}^2 c^2 \qquad (9.57)$$

or

$$E^{(\mathrm{fp})} = \xi E \;, \quad \xi = \pm 1 \;, \quad E = \sqrt{m_0^2 c^4 + \boldsymbol{p}^2 c^2} \;, \qquad (9.58)$$

respectively. This result is identical with the relativistic conservation law of energy (see (9.19)) if interactions are neglected. Due to the fact that the derivation of *Dirac's* equation makes use of the *Klein-Gordon* equation which includes the energy expression (9.19), this is the result which one has to expect. A schematic representation of the possible energy eigenvalues is shown in figure 9.2.

The solution of the problem of this relativistic problem is then given by

$$\boldsymbol{D}^{(\mathrm{fp})}(\boldsymbol{x}, t) = \boldsymbol{D}_0^{(\mathrm{fp})} \exp\left[\frac{\mathrm{i}}{\hbar}\left(\boldsymbol{p}\boldsymbol{x} - \xi E t\right)\right] \;, \qquad (9.59)$$

in which case $\boldsymbol{D}_0^{(\mathrm{fp})}$ can be determined by using (9.54) and an additional condition such as

$$\int \boldsymbol{D}^{*(\mathrm{fp})}(\boldsymbol{x}, t) \boldsymbol{D}^{(\mathrm{fp})}(\boldsymbol{x}, t) \, d\boldsymbol{x} = \delta\left(\boldsymbol{p} - \tilde{\boldsymbol{p}}\right) \delta_{\xi, \tilde{\xi}} \;. \qquad (9.60)$$

Thus, *Dirac* spinors and correlated energy eigenvalues of a problem without interaction were calculated. Other examples can be calculated as well, however, this shall not be done. Instead of such a discussion, the meaning of the introduced relativistic evolution equations, *Dirac* spinors and *Klein-Gordon* state functions shall be discussed. The occurrence of negative energy eigenvalues shall be discussed later.

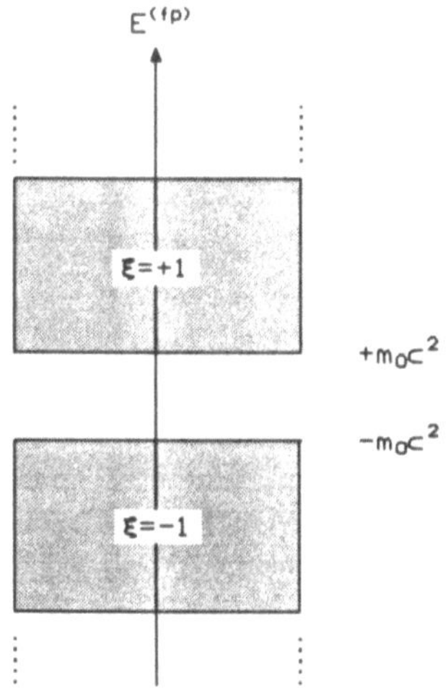

Figure 9.2
The spectrum of the possible energy eigenvalues
of *Dirac's* equation. The case without
interactions

9.2.3 Wave Functions and Spinors

So far, *Lorentz* covariant evolution equations, namely the *Klein-Gordon-* and the
Dirac equation, were considered. As it was demonstrated, solutions of such relativistic evolution equations can systematically be calculated. Due to the fact that
the introduced equations include the measurable relativistic conservation law of
energy of a particle with the rest mass m_0, these equations are usable relativistic
particle equations. Thus, the calculated solutions (9.59) represent the solutions of
a relativistic free particle. A non-relativistic borderline case of the *Dirac* equation
(9.46) is represented by the *Pauli* equation (9.36). This shall not be shown. However,
this has to be noted, because this means that *Dirac's* equation includes, in contrast
to the *Klein-Gordon* equation, the interaction of an electromagnetic field with a
spin of a particle. (This does not mean that a spin is a relativistic effect. Such an
interpretation is a mistake, because a *Pauli* equation can be derived by using only
non-relativistic basics. This has to be emphasized.) Furthermore, this means that
the *Dirac* equation directly represents a relativistic extension of the time-dependent
Schrödinger equation. In contrast to an ordinary time-dependent *Schrödinger* equation, the measurement basis of such a relativistic evolution equation is given by an
inertial reference system in which the time coordinate and the position coordinates
are not independent any more.

Now it has to be explained which physical meaning the *Klein-Gordon* state functions and the *Dirac* spinors have. This shall now be discussed. The physical meaning
of *Schrödinger's* indirect state functions has to be explained, too, because so far only
the connection between *Schrödinger's* indirect state functions (so-called *wave functions*) and measurement values was discussed (see chapter 5 and subsection 7.4).
Thus, in the following *Schrödinger* wave functions, *Klein-Gordon* state functions
and *Dirac* spinors shall be considered.

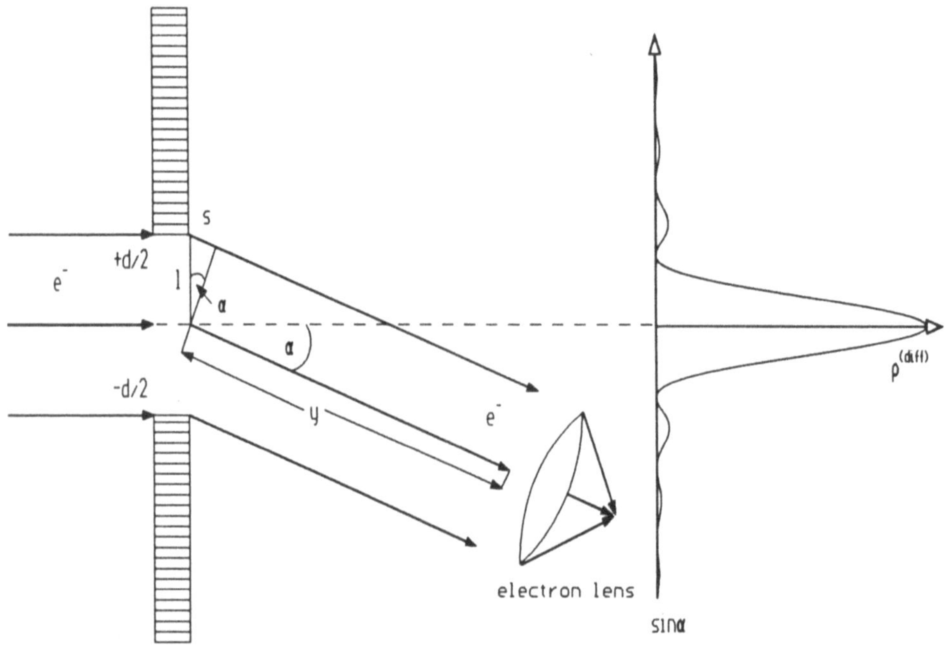

Figure 9.3   Electron diffraction if a slot experiment is considered

Diffraction of Particles

Observing a single particle or a stream of particles which move towards a suitable
grating (for example, a crystal) or a single small slot a diffraction is observable.
Figure 9.3 shows a schematic representation of such a behavior if a particle stream
and a single slot is considered. e^- represents electrons, d is the width of the single
slot, $\rho^{(\mathrm{diff})}$ is the measured probability density to find an electron in the direction
$\sin \alpha$. Such a measurement result is typical for the diffraction of particles (and wave
phenomena such as the phenomenon of diffraction of electromagnetic waves or the
phenomenon of diffraction of water waves) if one single slot is taken as a basis. The
measurable result of such a diffraction process can be described by

$$\rho^{(\mathrm{diff})} \sim \left[\frac{\sin\left(\dfrac{\pi d}{w}\sin\alpha\right)}{\dfrac{\pi d}{w}\sin\alpha} \right]^2 \underset{k=2\pi/w}{=} \left[\frac{\sin\left(k\dfrac{d}{2}\sin\alpha\right)}{k\dfrac{d}{2}\sin\alpha} \right]^2 , \tag{9.61}$$

where w has the meaning of a wavelength. In figure 9.3 this measurement result is
taken down. If the interaction function $V(\boldsymbol{x})$ of the multi-particle equation (7.1) is

neglected, such a measurement result can be calculated by using a solution of this time-dependent multi-particle *Schrödinger* equation. This shall now be shown.

If interactions are neglected, and if approximately an infinite set of particles is taken as a basis, a solution of (7.1) is given by

$$\Psi[\{\boldsymbol{x}(\nu)\}, t] = \int_{-\infty}^{+\infty} C(\nu) \exp\left\{i\left[\mathbf{k}\boldsymbol{x}(\nu) - \omega_k t\right]\right\} d\nu , \tag{9.62}$$

where a suitable choice of $C(\nu)$ allows to find a special representation of the wave function $\Psi[\{\boldsymbol{x}(\nu)\}, t]$. $\{\boldsymbol{x}(\nu)\}$ denotes an infinite set of position vectors. (If (7.1) is taken as a basis, a solution of the kind $\Psi(\{\boldsymbol{x}_i\}, t) = \sum_\nu C_\nu \exp[i(\mathbf{k}\boldsymbol{x}_\nu - \omega_k t)]$ has to be used, wherein $\boldsymbol{x}_\nu$ denotes the position vectors of the single particles. $\{\boldsymbol{x}_\nu\}$ denotes a finite set of position vectors. If approximately an infinite set of particles shall be taken as a basis, (9.62) has to be used instead of this sum.) In particular, a special choice of $C(\nu)$ allows to describe the experiment shown in figure 9.3 and to reproduce the result (9.61) if the special geometry and *Huygens' principle* (which can be called *Huygens-Fresnel principle*, too) is taken as a basis. Using *Huygens'* principle and considering the special geometry one obtains

$$\begin{aligned}
\exp\left\{i\left[\mathbf{k}\boldsymbol{x}(\nu) - \omega_k t\right]\right\} &= \exp\left[i\left\{\mathbf{k}\left[\boldsymbol{y} + \boldsymbol{s}(\nu)\right] - \omega_k t\right\}\right] \\
&= \exp\left\{i\left[ky + ks(\nu) - \omega_k t\right]\right\} \\
&= \exp\left[i\left(ky - \omega_k t\right)\right]\exp\left[ikl(\nu)\sin\alpha\right] ,
\end{aligned} \tag{9.63}$$

and the superposition (9.62) has to be replaced by

$$\begin{aligned}
\Psi^{(\mathrm{diff})}(k, y, \alpha, t) &= \exp\left[i\left(ky - \omega_k t\right)\right]\int_{\nu_1}^{\nu_2} C(\nu)\exp\left[ikl(\nu)\sin\alpha\right] d\nu \\
&= \exp\left[i\left(ky - \omega_k t\right)\right]\int_{-d/2}^{+d/2} C\exp\left(ikl\sin\alpha\right) dl .
\end{aligned} \tag{9.64}$$

If C is a constant, the expression

$$\Psi^{(\mathrm{diff})}(k, y, \alpha = 0, t) = C\exp\left[i\left(ky - \omega_k t\right)\right] d \tag{9.65}$$

holds. In this context $\mathbf{k}$ is the wave vector with the amount k, and y, $s(\nu)$ are the amounts of the vectors $\boldsymbol{y}$, $\boldsymbol{s}(\nu)$, where $\boldsymbol{x}(\nu) = \boldsymbol{y} + \boldsymbol{s}(\nu)$ holds. (9.63) and (9.64) contain *Huygens'* principle in an implicit way. (*Huygens' principle* means that a diffraction phenomenon can be described by using an infinite number of elementary waves, where in every direction $\mathbf{k}$ a superposition has to be carried out. This principle of construction is the basis of (9.63) and (9.64).) Inserting (9.65) into (9.64) the constant C can be eliminated so that one obtains the diffraction function

$$\Psi^{(\mathrm{diff})}(k, y, \alpha, t) = \frac{1}{d}\Psi^{(\mathrm{diff})}(k, y, \alpha = 0, t) \int_{-d/2}^{+d/2} \exp\left(ikl \sin \alpha\right) dl$$

$$= \frac{1}{idk \sin \alpha}\Psi^{(\mathrm{diff})}(k, y, \alpha = 0, t)\left[\exp\left(ikl \sin \alpha\right)\right]_{-d/2}^{+d/2}$$

$$= \Psi^{(\mathrm{diff})}(k, y, \alpha = 0, t)\left[\frac{\sin\left(k\dfrac{d}{2}\sin \alpha\right)}{k\dfrac{d}{2}\sin \alpha}\right], \qquad (9.66)$$

where the square of the amount of (9.66) obviously represents the measured signal (9.61). Therefore, the interpretation

$$\left|\Psi^{(\mathrm{diff})}(k, y, \alpha, t)\right|^2 = \rho^{(\mathrm{diff})} \qquad (9.67)$$

has to be required, i. e. the square of the amount of the used wave function is identical with the measurable probability density of the diffraction process. Due to the fact that the r. h. s. of (9.67) can be written as a product of a complex and a conjugate complex function, the relation

$$\Psi^{*(\mathrm{diff})}(k, y, \alpha, t)\Psi^{(\mathrm{diff})}(k, y, \alpha, t) = \rho^{(\mathrm{diff})} \qquad (9.68)$$

holds, too.

Probability Density, Schrödinger Wave Function, Klein-Gordon State Function and Dirac Spinor

Schrödinger wave functions, *Klein-Gordon* state functions and *Dirac* spinors describe a quantum mechanical system in an indirect way. If products of complex and conjugate complex *Schrödinger* wave functions or *Dirac* spinors are considered, i. e. if the functions

$$\boxed{\Psi^*(\boldsymbol{x}, t)\Psi(\boldsymbol{x}, t) := \rho_{\mathrm{S}}(\boldsymbol{x}, t) , \;\; \boldsymbol{D}^*(\boldsymbol{x}, t)\boldsymbol{D}(\boldsymbol{x}, t) := \rho_{\mathrm{D}}(\boldsymbol{x}, t)} \qquad (9.69)$$

are considered, special functions are given. As the above considered example shows, at least under certain conditions such functions can be interpreted as probability densities. Such an interpretation is then compatible with the fact that $\rho_{\mathrm{S}}(\boldsymbol{x}, t)$ and $\rho_{\mathrm{D}}(\boldsymbol{x}, t)$ are real and always positive functions. Due to the fact that products of complex and conjugate complex *Klein-Gordon* state functions, the functions

$$K^*(\boldsymbol{x}, t)K(\boldsymbol{x}, t) := \rho_{\mathrm{KG}}(\boldsymbol{x}, t) , \qquad (9.70)$$

can have negative values, in this case such an interpretation cannot be taken as a basis.

Using indirect state functions, functions can be formed which contain further information. These functions shall be called *background functions* and shall be introduced. However, first some facts about the problem of *metric* and the problem of *Einstein's field equation of gravitation* shall be presented. Then the meaning of such background functions can be explained.

9.3 Metric

If non-inertial reference frames are taken as a basis, forces of inertia are measurable. As it can experimentally be shown, the inert mass and the gravitational mass have to be treated in the same way. (For example, this shows the experiment of *Eötvös*.) Therefore, gravitational fields and fields of inertia have to be treated in the same way, and it has to be required that gravitational fields/ fields of inertia and special non-inertial reference systems are mutually conditional. In this book this principle shall be called *1. Einstein principle*.

Furthermore, it has to be noticed that equations which are valid for relatively high velocities have to be form-invariant with respect to *Lorentz* transformations which correspond to inertial systems. Due to the fact that also non-inertial systems are usable, the hypothesis makes sense that general equations have to be form-invariant with respect to any transformations. This principle shall be called *2. Einstein principle*.

Using *Einstein's* principles a general theory of gravitation can be introduced. The basis of this theory is *Einstein's* field equation of gravitation which connects the distribution of mass (expressed by an *energy-impulse tensor*) with the special reference frame (expressed by *Ricci's tensor* which includes the *fundamental metric tensor*). With this field equation a suitable reference frame can be calculated, i. e. a special *metric* can be determined. If a gravitational field is considered, the determined metric corresponds to a curvilinear reference frame, i. e. a *space curvature* has to be taken into account. In this context the movement of a mass point is given by an equation which determines the geodetic lines, i. e. the lines which represent the shortest possible connections between two points of the space considered. Both *Einstein's* field equation and the equation of the geodetic lines are form-invariant with respect to any transformations.

This scheme of description is a very successful one, i. e. the validity of this scheme can be proved by many experiments (for example, this scheme requires the rotation of the perihelion of planet orbits, and this can experimentally be shown), and well-proved borderline cases such as *Newton's equation of motion* are included.

In the following these facts shall be considered in more detail. First, some mathematical properties shall be presented.

9.3.1 Fundamental Metric Tensor, Co- and Contravariance

The Fundamental Metric Tensor

If a general reference frame is considered, the square of the distance between two points is given by

$$(ds)^2 = \sum_{\mu,\nu} g_{\mu,\nu} dx^\mu dx^\nu \, ,$$

(9.71)

in which case the differentials dx^μ normally correspond to a curvilinear reference frame. The elements $g_{\mu,\nu}$ represent a tensor called *fundamental metric tensor*, which determines the considered space, i. e. the special metric. ds is invariant with respect to any transformations.

In the following a four-dimensional position-time space will be relevant so that $\alpha, \beta, \nu, \mu, \kappa = 1, 2, 3, 4$ has to be used.

If an inertial reference system can be taken as a basis, and if the coordinates are denoted by x^1, x^2, x^3, $x^4 = ict$, the components of the fundamental metric tensor are defined by

$$g_{\mu,\nu}^{(\text{inertial})} = \delta_{\mu,\nu} \; , \tag{9.72}$$

and $(ds)^2$ shows the form

$$\left(ds^{(\text{inertial})} \right)^2 = \left(dx^1 \right)^2 + \left(dx^2 \right)^2 + \left(dx^3 \right)^2 + (icdt)^2$$

$$= -c^2 \left[1 - \left(\frac{v}{c} \right)^2 \right] dt^2 \; . \tag{9.73}$$

These two relations define an *Euclidean position-time space*. If the coordinates x^1, x^2, x^3, ct are taken as a basis, such coordinates represent a so-called *Minkowski universe*. A *Minkowski* universe represents a *pseudo-Euclidean* position-time space. In such a case the fundamental metric tensor has to be defined by $\pm\delta_{\mu,\nu}$. *Lorentz* transformations are then rotations within such a *Minkowski* universe.

Co- and Contravariant Quantities

If a transformation is considered, the differentials of the new reference frame can be calculated by using the rule

$$\widetilde{dx^\mu} = \sum_\nu \frac{\partial \widetilde{x^\mu}}{\partial x^\nu} dx^\nu \; . \tag{9.74}$$

Every transformed vector (vector = tensor of first order) which can be calculated by using the rule (9.74) is called a *contravariant vector*. Thus, a contravariant is defined by

$$\widetilde{A^\mu} = \sum_\nu \frac{\partial \widetilde{x^\mu}}{\partial x^\nu} A^\nu \; . \tag{9.75}$$

In the same way tensors of higher order are defined. For example, the rule

$$\widetilde{A^{\mu,\nu}} = \sum_{\alpha,\beta} \frac{\partial \widetilde{x^\mu}}{\partial x^\alpha} \frac{\partial \widetilde{x^\nu}}{\partial x^\beta} A^{\alpha,\beta} \; , \tag{9.76}$$

has to be used to define a contravariant tensor of second order. In this sense the differentials dx^μ represent a contravariant vector. The property *contravariance* shall be denoted by using upper indices. Lower indices characterize *covariant quantities*, in which case a covariant vector is defined by

$$\widetilde{B}_\mu = \sum_\nu \frac{\partial x^\nu}{\partial \widetilde{x}^\mu} B_\nu \, , \tag{9.77}$$

and a covariant tensor of second order is defined by

$$\widetilde{B_{\mu,\nu}} = \sum_{\alpha,\beta} \frac{\partial x^\alpha}{\partial \widetilde{x}^\mu} \frac{\partial x^\beta}{\partial \widetilde{x}^\nu} B_{\alpha,\beta} \, . \tag{9.78}$$

In this sense a fundamental metric tensor $g_{\mu,\nu}$ represents a covariant tensor of second order. Co- and contravariant tensors of higer order are defined in the same way.

The product of a covariant and of a contravariant vector defines a *scalar SC*, where a scalar is a one-dimensional quantity which is invariant with respect to any transformations, i. e. the relation

$$\widetilde{SC} = \sum_\mu \widetilde{A^\mu} \widetilde{B}_\mu = \sum_{\alpha,\beta} \frac{\partial \widetilde{x}^\mu}{\partial x^\alpha} \frac{\partial x^\beta}{\partial \widetilde{x}^\mu} A^\alpha B_\beta = \sum_{\alpha,\beta} \delta_\alpha^\beta A^\alpha B_\beta = \sum_\alpha A^\alpha B_\alpha = SC \tag{9.79}$$

holds. δ_α^β represents *Kronecker's* delta, in which case

$$\delta_\alpha^\beta = \sum_\mu g_{\alpha,\mu} g^{\mu,\beta} \tag{9.80}$$

holds. $g^{\mu,\beta}$ represents the inverse fundamental metric tensor. Such an inverse tensor represents a contravariant tensor of second order.

If general reference frames have to be used, this mathematical properties have to be taken as a basis (for more information, for example, see [68]).

The shortest connection between two points of a general reference frame is a so-called *geodetic line*. Such a line can be defined by using a differential equation of the *Euler-Lagrange* type. This equation shall now be considered.

9.3.2   Geodetic Lines

The Equation of Geodetic Lines

The shortest possible connection between two points of a reference frame is identical with a curve which has an extreme value. In order to find an equation which is able to determine such a curve, the requirement

$$\delta \int_{P_1}^{P_2} ds = 0 \tag{9.81}$$

can be taken as a basis. (9.81) is a variational integral. Using (9.71) the relation (9.81) has to be replaced by

$$\delta \int_{P_1}^{P_2} \sqrt{\sum_{\mu,\nu} g_{\mu,\nu} dx^\mu dx^\nu} = \delta \int_{P_1}^{P_2} \sqrt{\sum_{\mu,\nu} g_{\mu,\nu} \frac{dx^\mu}{ds} \frac{dx^\nu}{ds}} ds$$

$$= 0 \, . \tag{9.82}$$

(9.82) is a functional integral of the form

$$\delta \int_{P_1}^{P_2} G\left(x^\mu, \frac{dx^\mu}{ds}\right) ds = 0 \,. \tag{9.83}$$

A neccessary condition that $G\left(x^\mu, \frac{dx^\mu}{ds}\right)$ fulfills such a functional integral is given by the *Euler-Lagrange equation*, i. e.

$$\frac{d}{ds}\left[\frac{\partial G\left(x^\mu, \frac{dx^\mu}{ds}\right)}{\partial\left(\frac{dx^\alpha}{ds}\right)}\right] - \frac{\partial G\left(x^\mu, \frac{dx^\mu}{ds}\right)}{\partial x^\alpha} = 0 \tag{9.84}$$

holds. In the considered case the function $G\left(x^\mu, \frac{dx^\mu}{ds}\right)$ is of the form

$$G\left(x^\mu, \frac{dx^\mu}{ds}\right) = \sqrt{\sum_{\mu,\nu} g_{\mu,\nu} \frac{dx^\mu}{ds} \frac{dx^\nu}{ds}} \,. \tag{9.85}$$

If $G\left(x^\mu, \frac{dx^\mu}{ds}\right)$ is identical with the *Lagrangian* function L, and if dt instead of ds is used, the integral of (9.83) represents the deterministic action of a mechanical system. In this case (9.83) is identical with the *Hamiltonian principle of deterministic mechanics*, and the corresponding *Euler-Lagrange* equation is the basic deterministic equation of the system. This has to be noted.

Inserting (9.85) into (9.84) one obtains the result

$$\frac{d^2 x^\kappa}{ds^2} + \sum_{\mu,\nu} \Gamma^\kappa_{\mu,\nu} \frac{dx^\mu}{ds} \frac{dx^\nu}{ds} = 0 \,, \tag{9.86}$$

where $\Gamma^\kappa_{\mu,\nu}$ denotes so-called *Christoffel symbols*. (These *Christoffel* symbols represent no tensors!) Such *Christoffel* symbols are defined by

$$\Gamma^\kappa_{\mu,\nu} = \sum_\alpha \left[g^{\kappa,\alpha} \frac{1}{2} \left(\frac{\partial g_{\nu,\alpha}}{\partial x^\mu} + \frac{\partial g_{\alpha,\mu}}{\partial x^\nu} - \frac{\partial g_{\mu,\nu}}{\partial x^\alpha} \right) \right] \,. \tag{9.87}$$

(9.86) defines the geodetic lines of a general reference frame. Within relatively small space elements an inertial reference system can be taken as a basis. In this case (9.86) defines a straight line. If the whole space is considered, normally a space curvature has to be taken into account so that the geodetic lines are curves. The equation (9.86) is form-invariant with respect to any transformations.

Newton's Equation of Motion

The equation of geodetic lines, the equation (9.86), includes *Newton's equation of motion*. In order to show this, the hypothesis

$$g_{\mu,\nu} := g_{\mu,\nu}^{(\text{inertial})} + \gamma_{\mu,\nu} = \delta_{\mu,\nu} + \gamma_{\mu,\nu} \tag{9.88}$$

shall be taken as a basis, where $\gamma_{\mu,\nu}$ represents a perturbation. In this case the *Christoffel* symbols (9.87) can be rewritten. In doing so one obtains

$$\Gamma_{\mu,\nu}^{(\text{Newton})\kappa} = \sum_{\alpha} \left[\delta^{\kappa,\alpha} \frac{1}{2} \left(\frac{\partial \gamma_{\nu,\alpha}}{\partial x^\mu} + \frac{\partial \gamma_{\alpha,\mu}}{\partial x^\nu} - \frac{\partial \gamma_{\mu,\nu}}{\partial x^\alpha} \right) \right] , \tag{9.89}$$

where it was used that $\gamma_{\mu,\nu}$ represents a small perturbation, i. e. terms of higher order can be neglected. Assuming an inertial reference system the line element ds is defined by (9.73), where in the borderline case $v \ll c$ the relation

$$ds^2 = -c^2 dt^2 \tag{9.90}$$

holds. Using (9.90) and neglecting all terms with v/c (this is due to the requirement $v \ll c$) the equation of geodetic lines (9.86) can be rewritten. In doing so one obtains

$$-\frac{1}{c^2} \frac{d^2 x^\epsilon}{dt^2} - \frac{1}{c^2} \Gamma_{4,4}^\epsilon \frac{dx^4}{dt} \frac{dx^4}{dt} = -\frac{1}{c^2} \frac{d^2 x^\epsilon}{dt^2} + \Gamma_{4,4}^\epsilon = 0 \ (\epsilon = 1, 2, 3) ,$$
$$-\frac{1}{c^2} \frac{d^2 x^4}{dt^2} = 0 . \tag{9.91}$$

Inserting the *Christoffel* symbol $\Gamma_{4,4}^\epsilon$ into (9.91) one obtains the equation

$$\frac{d^2 x^\epsilon}{dt^2} = \frac{c^2}{2} \left(\frac{\partial \gamma_{4,\epsilon}}{\partial x^4} + \frac{\partial \gamma_{\epsilon,4}}{\partial x^4} - \frac{\partial \gamma_{4,4}}{\partial x^\epsilon} \right) , \tag{9.92}$$

which can be replaced by

$$\boxed{\frac{d^2 x^\epsilon}{dt^2} = -\frac{c^2}{2} \frac{\partial \gamma_{4,4}}{\partial x^\epsilon} = -\frac{\partial \phi}{\partial x^\epsilon} , \quad \phi = \frac{c^2}{2} \gamma_{4,4}} \tag{9.93}$$

if a time-independent fundamental metric tensor is taken into account. Obviously, (9.93) represents *Newton's* equation of motion with a potential function ϕ.

Thus, it was shown that the equation of geodetic lines (9.86) includes *Newton's* equation of motion. In order to gain this equation, it was necessary to use an inertial reference system and to require a borderline case $v \ll c$. However, a potential function only occurs if the fundamental metric tensor has a deviation term $\gamma_{\mu,\nu}$ which represents the deviation from the *Lorentz metric* (in the borderline case $v \ll c$), i. e. the use of an inertial reference system (in the borderline case $v \ll c$) is only possible within special parts of the derivation procedure. Otherwise no potential function would occur within *Newton's* equation.

Interpretation

Due to *Einstein's* principles the idea makes sense that the equation of geodetic lines (9.86) represents a generalized form of *Newton's* equation of motion, where a fundamental metric tensor replaces an ordinary potential function. If this interpretation is the correct one, the example shows, only in special cases an ordinary potential function can be used, normally a tensor has to be taken into account. Furthermore, if this interpretation is correct, the example shows that a potential (represented by a tensor or an ordinary potential function), which corresponds to a field, and a curvilinear reference frame are mutually conditional, in particular, only in an approximative way an inertial reference system and a potential function can be used at the same time. Due to the fact that many experiments show that the above given interpretation is the correct one, such an interpretation can be taken as a basis.

A potential and a distribution of mass are mutually conditional. So it has to be possible to derive an equation which describes the connection between a fundamental metric tensor and such a distribution. This shall now be discussed.

9.3.3   Einstein's Field Equation of Gravitation

The Basic Equation

Einstein's field equation of gravitation determines the connection between an energy-impulse tensor and the fundamental metric tensor. This field equation fulfills *Einstein's* principles, in particular, this field equation is form-invariant with respect to any transformations. This field equation is well-proved and can be written in the form

$$\boxed{R_{\mu,\nu} = -K\tilde{T}_{\mu,\nu} \, ,} \tag{9.94}$$

where

$$\boxed{\tilde{T}_{\mu,\nu} = T_{\mu,\nu} - \frac{1}{2K}g_{\mu,\nu}R} \tag{9.95}$$

holds. The components $R_{\mu,\nu}$ represent *Ricci's* tensor and are defined by

$$R_{\mu,\nu} = \sum_{\alpha,\beta} \delta_\beta^\alpha R_{\mu,\nu,\alpha}^\beta$$

$$= -\sum_\alpha \frac{\partial \Gamma_{\mu,\alpha}^\alpha}{\partial x^\nu} + \sum_\alpha \frac{\partial \Gamma_{\mu,\nu}^\alpha}{\partial x^\alpha} + \sum_{\alpha,\beta} \Gamma_{\mu,\nu}^\beta \Gamma_{\beta,\alpha}^\alpha - \sum_{\alpha,\beta} \Gamma_{\mu,\alpha}^\beta \Gamma_{\beta,\nu}^\alpha \tag{9.96}$$

($\Gamma_{\mu,\nu}^\kappa \dots$ *Christoffel* symbols). $R_{\mu,\nu,\alpha}^\beta$ represents *Riemann's tensor of curvature*. This tensor is an anti-symmetrical tensor of fourth order. In contrast to *Riemann's* tensor, the Ricci tensor is a symmetrical tensor of second order. R is the *scalar of curvature* and is defined by

$$R = \sum_{\alpha,\beta} g^{\alpha,\beta} R_{\alpha,\beta} \, , \tag{9.97}$$

K is *Einstein's constant of gravitation*

$$K = \frac{8\pi}{c^4}G \tag{9.98}$$

(*G ... Newton's constant of gravitation*), and $T_{\mu,\nu}$ represents the energy-impulse tensor. Due to the fact that this energy-impulse tensor contains the energy density, the impulse density, the density of the energy flow and the density of the impulse flow, and due to the fact that the *Christoffel* symbols contain the fundamental metric tensor, the field equation (9.94) connects a dynamical system, which includes the energy of mass (mass = ponderable mass, matter = ponderable mass + fields), with the fundamental metric tensor which determines the gravitation. *Einstein's* field equation of gravitation can be used to calculate cosmological problems.

In order to elucidate the meaning of an energy-impulse tensor, some basic facts shall now be considered.

The Energy-Impulse Tensor

The energy-impulse tensor is a tensor of second order, where this tensor can be written in the form

$$(T_{\mu,\nu}) = \begin{pmatrix} T_{1,1} & T_{1,2} & T_{1,3} & T_{1,4} \\ T_{2,1} & T_{2,2} & T_{2,3} & T_{2,4} \\ T_{3,1} & T_{3,2} & T_{3,3} & T_{3,4} \\ T_{4,1} & T_{4,2} & T_{4,3} & T_{4,4} \end{pmatrix} . \tag{9.99}$$

If an *Euclidean* position-time space is taken as a basis, a *Lorentz* covariant energy-impulse tensor can be used which is defined by

$$\left(T_{\mu,\nu}^{(\text{inertial})}\right) = \begin{pmatrix} S_{1,1}^{(p)} & S_{1,2}^{(p)} & S_{1,3}^{(p)} & ic\rho_1^{(p)} \\ S_{2,1}^{(p)} & S_{2,2}^{(p)} & S_{2,3}^{(p)} & ic\rho_2^{(p)} \\ S_{3,1}^{(p)} & S_{3,2}^{(p)} & S_{3,3}^{(p)} & ic\rho_3^{(p)} \\ S_1^{(E)}/ic & S_2^{(E)}/ic & S_3^{(E)}/ic & \rho^{(E)} \end{pmatrix} . \tag{9.100}$$

$\rho^{(E)}$ represents the energy density, $\rho_k^{(p)}$ ($k = 1,2,3$) represents impulse densities, $S_k^{(E)}$ ($k = 1,2,3$) represents the density components of the energy flow, and $S_{k,l}^{(p)}$ ($k,l = 1,2,3$) denotes the density components of the impulse flow. If an *Euclidean* position-time space is taken as a basis, an energy-impulse tensor allows to calculate basic conservation laws by using the rule

$$\sum_\nu \frac{\partial T_{\mu,\nu}}{\partial x^\nu} = 0 . \tag{9.101}$$

Strictly speaking, (9.101) represents basic equations of continuity, namely the equations

$$\sum_l \frac{\partial S_{k,l}^{(p)}}{\partial x^l} + ic\frac{\partial \rho_k^{(p)}}{\partial ict} = 0 , \tag{9.102}$$

$$\sum_l \frac{1}{ic} \frac{\partial S_l^{(E)}}{\partial x^l} + \frac{\partial \rho^{(E)}}{\partial ict} = 0 \,. \tag{9.103}$$

(9.102) describes the conservation of impulse and (9.103) the conservation of energy. In this context it has to be emphasized that in the *Euclidean* case

$$T_{\mu,\nu} = T_\nu^\mu = T^{\mu,\nu} \tag{9.104}$$

holds, i. e. in the *Euclidean* case there is no difference between co- and contravariance. This is due to the fact that in such a case the fundamental metric tensor can be replaced by *Kronecker's* delta. For example, in such a case the transformation rule

$$T_\nu^\mu = \sum_\alpha g^{\mu,\alpha} T_{\alpha,\nu} \overset{Euclidean}{=} \sum_\alpha \delta^{\mu,\alpha} T_{\alpha,\nu} = T_{\mu,\nu} \tag{9.105}$$

holds. Thus, the identity (9.104) is valid.

Furthermore, it has to be emphasized that in a general case the used energy-impulse tensor has to be form-invariant with respect to any transformations. In the *Euclidean* case only *Lorentz* covariance has to be taken as a basis.

Evolution of Mass

In a general case, (9.100) has to be replaced by

$$\boxed{\sum_\mu T_{\nu;\mu}^\mu = 0 \,,} \tag{9.106}$$

where the sign $;\mu$ denotes the *absolute differentiation*. The general equation of continuity (9.101) represents a borderline case of (9.106) which arises if an *Euclidean* position-time space is taken as a basis. (9.106) determines the evolution of mass. This evolution equation a suitable energy-impulse tensor has to fulfill. Thus, not only *Einstein's* field equation has to be used, in order to determine the metric of a mass system, (9.106) has to be considered, too.

Absolute Differentiation

(9.106) contains an absolute differentiation. If a contravariant vector is considered, such an absolute differentiation is defined by

$$A_{;\nu}^\mu = \frac{\partial A^\mu}{\partial x^\nu} + \sum_\alpha \Gamma_{\alpha,\nu}^\mu A^\alpha \,, \tag{9.107}$$

and if a covariant vector is considered, such an absolute differentiation is defined by

$$A_{\mu;\nu} = \frac{\partial A_\mu}{\partial x^\nu} - \sum_\alpha \Gamma_{\mu,\nu}^\alpha A_\alpha \tag{9.108}$$

so that

$$\left(\sum_\mu A^\mu A_\mu \right)_{;\nu} = 0 \tag{9.109}$$

holds. If a general tensor is considered, the rule

$$A^{\mu\ldots\nu}_{\alpha\ldots\beta;\tau} = \frac{\partial A^{\mu\ldots\nu}_{\alpha\ldots\beta}}{\partial x^\tau} - \sum_\kappa \Gamma^\kappa_{\alpha,\tau} A^{\mu\ldots\nu}_{\kappa\ldots\beta} - \cdots - \sum_\kappa \Gamma^\kappa_{\beta,\tau} A^{\mu\ldots\nu}_{\alpha\ldots\kappa}$$

$$+ \sum_\kappa \Gamma^\mu_{\kappa,\tau} A^{\kappa\ldots\nu}_{\alpha\ldots\beta} + \cdots + \sum_\kappa \Gamma^\nu_{\kappa,\tau} A^{\mu\ldots\kappa}_{\alpha\ldots\beta} \tag{9.110}$$

has to be used. Thus, the absolute differentiation $T^\mu_{\nu;\mu}$ has to be defined by

$$T^\mu_{\nu;\mu} = \frac{\partial T^\mu_\nu}{\partial x^\mu} - \sum_\kappa \Gamma^\kappa_{\nu,\mu} T^\mu_\kappa + \sum_\kappa \Gamma^\mu_{\kappa,\mu} T^\kappa_\nu \; . \tag{9.111}$$

Such an absolute differentiation has to replace an ordinary differentiation. The reason of this replacement is that in the considered general case the equations have to be form-invariant with respect to any transformations, i. e. such equations have to contain tensors which are defined in 9.3.1. Due to the fact that an ordinary differentiation such as $\partial T^\mu_\nu / \partial x^\mu$ does not represent a tensor, additional terms have to be introduced which generate form-invariance. If the special absolute differentiation $T^\mu_{\nu;\mu}$ is taken into account, this means that additional terms $\Gamma^\kappa_{\nu,\mu} T^\mu_\kappa$, $\Gamma^\mu_{\kappa,\mu} T^\kappa_\nu$ have to be taken into consideration. These remarks shall be enough. In the following a short derivation of *Einstein's* field equation shall be presented.

The Foundation of Einstein's Field Equation of Gravitation

In order to derive the field equation (9.94), it can be taken as a basis that in an *Euclidean* position-time space the connection of a distribution of mass with a potential function is given by the *Poisson equation*

$$\boxed{\Delta\phi = -\frac{c^2}{2} K\, Mc^2 = \frac{c^2}{2}\Delta\gamma_{4,4} = \frac{c^2}{2}\Delta g_{4,4} \;,} \tag{9.112}$$

wherein M represents the density of mass, and wherein Δ is the *Laplacian*. K is *Einstein's* constant of gravitation which is defined by (9.98). $\gamma_{4,4}$ represents a component of the deviation term of the metric tensor if an *Euclidean* position-time space with an additional potential function ϕ shall be considered (see (9.88)ff.). $g_{4,4}$ is the corresponding metric tensor component. This *Poisson* equation is an essential equation in *Newton's* theory. In particular, (9.112) allows to calculate potential functions ϕ if a density of mass is given.

Due to the fact that the term Mc^2 represents nothing but the energy density of mass (comparison with (9.8)!), and due to the fact that the component $T_{4,4}$ of the energy-impulse tensor represents such an energy density, the *Poisson* equation (9.112) can be replaced by

$$-KT_{4,4} = \triangle g_{4,4} \, , \tag{9.113}$$

where $g_{4,4}$ is a component of the fundamental metric tensor (9.88).

(9.113) shows that in the *Euclidean* case, instead of a potential function and a density of mass, a component of the fundamental metric tensor and a component of the energy-impulse tensor can be used. Therefore, the idea is obvious that in a general case a connection between the general fundamental metric tensor and the general energy-impulse tensor has to be used. Mathematically expressed, this means the hypothesis

$$-KT_{\mu,\nu} = F_{\mu,\nu}\left(g_{\alpha,\beta}\right) \tag{9.114}$$

has to be taken as a basis. $F_{\mu,\nu}\left(g_{\alpha,\beta}\right)$ has to represent a form-invariant differential tensor. Using special requirements this differential tensor can be determined, and the field equation (9.94) can be calculated. However, this shall not be shown in this book. Instead of such a discussion, again the problem of indirect state functions shall be considered, and it shall be shown that functions can be introduced which contain further information about the basic physical system. During the following considerations the facts presented above have to be borne in mind.

9.4 Background Functions

9.4.1 The Definition

Solutions of relativistic and non-relativistic quantum mechanical evolution equations describe a physical system in an indirect way, i. e. such functions are no measurable quantities. If a kinetic energy function is given (for example, in the *Schrödinger* case this means that the energy $E - V(\boldsymbol{x})$ has to be given), such indirect state functions can be calculated. Thus, such a quantum mechanical evolution equation describes the connection between kinetic energies and non-measurable indirect state functions. If the notation of chapter 5 is used, i. e. if the kinetic energy is denoted by $KIN(\boldsymbol{x}, t)$, this means that

$$\boxed{KIN(\boldsymbol{x}, t) = W^{-1}(\boldsymbol{x}, t)\hat{O}_{KIN}(\boldsymbol{x})W(\boldsymbol{x}, t)} \tag{9.115}$$

holds. Such a kinetic energy function characterizes the whole space considered. Such a function can be measured on a macroscopic level. $\hat{O}_{KIN}(\boldsymbol{x})$ denotes an operator which depends on position coordinates and generates the kinetic energy, and $W(\boldsymbol{x}, t)$ or $W^{-1}(\boldsymbol{x}, t)$ denote the considered functions $\Psi(\boldsymbol{x}, t)$, $K(\boldsymbol{x}, t)$ and $\boldsymbol{D}(\boldsymbol{x}, t)$ or the inverse functions, respectively. For example, if a time-dependent one-particle *Schrödinger* equation is taken as a basis, $\hat{O}_{KIN}(\boldsymbol{x})$ has to be identified with $-\frac{\hbar^2}{2m_0}\Delta_3 = E - V(\boldsymbol{x}) = i\hbar\frac{\partial}{\partial t} - V(\boldsymbol{x})$, and $W(\boldsymbol{x}, t)$ has to be identified with $\Psi(\boldsymbol{x}, t)$. Due to the fact that spinors can be considered, $KIN(\boldsymbol{x}, t)$ can be a matrix.

Without any restrictions $W(\boldsymbol{x}, t)$ can be used in the form

$$\boxed{W(\boldsymbol{x},t) = \int W(\boldsymbol{x},\boldsymbol{v},t)\, d\boldsymbol{v}\, ,} \tag{9.116}$$

where $\boldsymbol{v}$ can be a velocity vector. Therefore, (9.115) can be replaced by

$$
\begin{aligned}
KIN(\boldsymbol{x},t) &= \left[\int W(\boldsymbol{x},\boldsymbol{v},t)\, d\boldsymbol{v}\right]^{-1} \hat{O}_{KIN}(\boldsymbol{x}) \int W(\boldsymbol{x},\boldsymbol{v},t)\, d\boldsymbol{v} \\
&= \left[\int W(\boldsymbol{x},\boldsymbol{v},t)\, d\boldsymbol{v}\right]^{-1} \int \hat{O}_{KIN}(\boldsymbol{x}) W(\boldsymbol{x},\boldsymbol{v},t)\, d\boldsymbol{v} \\
&= \left[\int W(\boldsymbol{x},\boldsymbol{v},t)\, d\boldsymbol{v}\right]^{-1} \int w W(\boldsymbol{x},\boldsymbol{v},t)\, d\boldsymbol{v}\, ,
\end{aligned}
\tag{9.117}
$$

where w is a function of the considered variables. Introducing the new function

$$\boxed{\tilde{W}(\boldsymbol{x},\boldsymbol{v},t) = \left[\int W(\boldsymbol{x},\boldsymbol{v},t)\, d\boldsymbol{v}\right]^{-1} W(\boldsymbol{x},\boldsymbol{v},t)\, ,} \tag{9.118}$$

(9.117) can be replaced by

$$\boxed{KIN(\boldsymbol{x},t) = \int w\tilde{W}(\boldsymbol{x},\boldsymbol{v},t)\, d\boldsymbol{v}\, .} \tag{9.119}$$

(9.119) replaces one of the considered evolution equations. For example, (9.119) replaces the *Schrödinger* equation (9.15). (9.119) represents a decomposition of a kinetic energy into a set of elementary functions $\tilde{W}(\boldsymbol{x},\boldsymbol{v},t)$. This superposition includes functions (a vector potential, a scalar potential function V) which in last consequence determine possible mass states, where an *Euclidean* position-time space is taken as a basis. Such functions shall be called *background functions*. If a *Schrödinger* case is considered, such background functions are non-relativistic approximations.

The possibility of such a superposition is the normal case. This shall now be discussed.

9.4.2   The Principle of Superposition of Elementary Functions

The possibility to decompose a measurable signal into a set of elementary functions which contain information about the underlying physical system in a more or less direct way is the normal case. For example, if an electromagnetic field signal is observed, such a signal can be decomposed by *Fourier's* method, in which case the elementary functions are real trigonometrical functions or complex *Euler functions* (comparison with formula (6.1)), respectively. Another example is represented by a fluid wave. In such a case the observable signal can be decomposed into a set of basic waves. Furthermore, the theoretical experience shows that self-organizing systems can be described by using a set of elementary functions. For example, pattern formations of *Taylor-* and *Bénard instabilities* of fluid dynamics can be decomposed into a set of elementary functions (see [5]). Furthermore, mean value representations of the kind

$$\langle f_1(\alpha)\rangle\,(\beta) = \int f_1(\alpha)f_2(\alpha,\beta)\,d\alpha \tag{9.120}$$

are representations on the basis of special elementary functions. For example, such mean value representations occur in the theory of *Fokker-Planck* equations. Other examples of such superpositions are given by path integral representations which are formulated by using eigenfunctions of *Schrödinger's* equation (see (7.80)).

These examples represent a principle which shall be called *principle of superposition of elementary functions*. It can be interpreted as a special case of the introduced *principle of coupling of elementary systems* (see 7.3.7). The decomposition (9.119) represents an example which shows that this principle also holds if quantum systems are considered. Due to the fact that the special kind of decomposition represents special properties of the underlying physical system (for example, if a fluid is considered, the *underlying physical system* is represented by coupled molecules, where the whole system of molecules can be described by the *Navier-Stokes equation*), such background functions represent properties of the underlying physical system.

In order to understand what the term *underlying physical system* in the context of background functions means, the connection with *Einstein's* field equation shall be considered.

9.4.3 Self-Consistency Equations

As it was discussed, *Einstein's* field equation of gravitation (9.94) connects the energy-impulse tensor (which contains energy- and impulse densities of mass) with the fundamental metric tensor (which represents the reference frame, where a reference frame and a potential of a gravitational field are mutually conditional). As it was discussed, if relatively small space regions are considered, instead of a metric tensor, an ordinary potential function and an additional *Euclidean* reference frame can be taken as a basis, where in this approximation reference frame and potential function can separately be considered. In such an approximation *Einstein's* field equation of gravitation can be replaced by an equation which bases on an *Euclidean* reference frame and connects the potential function of the problem with the energy of the considered mass. This energy of mass consists of two parts, namely the energy of the rest mass and the additional energy of the dynamic mass which corresponds to the kinetic behavior of the mass (comparison with (9.7)+(9.8)). In the general case and in the approximative case *Einstein's* field equation determines a *self-consistent state of mass and field*, which shall be called an *eigenstate of matter*. Thus, *Einstein's* field equation shall be called *self-consistency equation*. If an additional gravitational field is considered, which is generated by an accelerating mass, the eigenstate will be disturbed, i. e. the considered mass will be attracted. Using the self-consistency equation of *Einstein*, i. e. the field equation (9.94), the effect of the additional mass can be included by using a modified fundamental metric tensor. Then the l. h. s. and the r. h. s. of *Einstein's* field equation are not equal anymore, i. e. the self-concistency is disturbed. However, *Einstein's* field equation of gravitation requires the equivalence of both sides. Therefore, the field equation determines another mass state of the considered mass, i. e. the mass will be accelerated so that an additional dynamic mass occurs. On the mathematical level of *Einstein's* equation this means that then both sides of the equation are equal so that self-consistency is guaranteed. On the experimental level this means that an accelerated mass will be observed. In the same way the self-consistency of both masses can be determined by using *Einstein's* field equation. Such processes can be considered as self-organization processes, too.

Using an equation of *Einstein's* type, electromagnetic problems can be included, too. The possibility to include electromagnetic problems can easily be seen. For example, it has to be borne in mind that the energy-impulse tensor (9.100) is a tensor used in classical electrodynamics. Then the density of the electromagnetic field energy occurs. Furthermore, it has to be borne in mind that the derivation of *Newton's* equation of motion, see (9.88)-(9.93), does not require a gravitational field. A metric tensor which includes an electric potential can be used, too. In both cases the well-proved *Newton* equation of motion can be derived. Then, in particular, it has to be required that charged masses can be considered as well. Furthermore, it has to be required that the notion *metric* can be used, too, if electromagnetic problems are considered. In such a case an equation of *Einstein's* type describes the self-consistency of charged masses and fields.

As it was discussed, if the introduced quantum mechanical evolution equations (which characterize a charged mass) are considered, a formulation of the kind (9.115) or – if background functions are considered – of the kind (9.119) is possible. Such a formulation shows that a quantum mechanical evolution equation connects the observable kinetic energy with an ordinary potential function, where an *Euclidean* reference frame has to be taken as a basis. Additionally, indirect state functions have to be taken into account. If background functions are used (see (9.119)), such background functions include both a potential function and the corresponding indirect state function. The indirect state functions (and thus the background functions) depend on the movement of the considered charged mass, i. e. they depend on the region of the potential function which is relevant for the movement, where a characteristic quantity for a special movement is given by the total energy of the considered movement. If a special movement is considered, the corresponding kinetic energy of the charged mass – measured on a macroscopic level within the region of the particle movement – is determined by the decomposition (9.119) or the operative relation (9.115), respectively. Thus, an equation of the kind (9.119) (or (9.115), respectively) connects the kinetic energy of a special movement of a charged mass with the relevant potential values. This means that a formulation (9.119) (or (9.115), respectively) determines the state of a charged mass and an additional field in a self-consistent way. If the above introduced notation is used, this means that such an equation represents a *self-consistency equation in the context of quantum system theory*. Therefore, such an equation can be compared with an equation of *Einstein's* type. However, instead of a fundamental metric tensor (or a borderline case) and an energy-impulse tensor (or a borderline case), a macroscopic energy function and a decomposition on the basis of background functions occur.

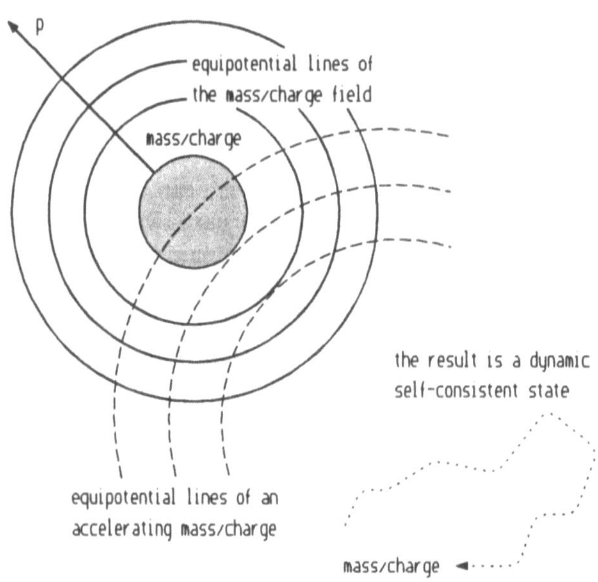

Figure 9.4
Self-consistency of a matter state. The mass can be charged. If the mass and the corresponding potential represent an eigenstate, only a movement with a constant impulse p is possible. If an additional potential has to be taken into account (for example, if an additional mass/charge causes interaction forces), the eigenstate will be disturbed and a non-constant impulse which describes a space-dependent kinetic energy occurs. Both cases are self-consistent states which can be represented by a suitable self-consistency equation

(If the operative equation (9.115) is taken as a basis, indirect state functions and an additional potential function replace the background functions.)

Thus, the background functions represent the self-consistent field- (or potential-) background of the mass-field system. These functions are special mediation functions. Due to the fact that these functions cannot be choosen in an arbitrary way, such background functions have to contain information about the possible states of the mass-field system, i. e. they characterize the underlying physical system. Figure 9.4 shows a schematic representation of an example of a self-consistent state.

As the above considerations show, macroscopic states of matter can be described by using the notion *metric*. However, the basis of every macroscopic state of matter are quantum systems. Thus, a connection between a *macroscopic metric* and quantum systems has to exist, i. e. it has to be possible to use a suitable kind of metric in the context of quantum system theory. Therefore, macroscopic as well as microscopic physical systems can be submitted under the notion *metric*.

Thus, a connection between the theory of micro-systems (molecules, atoms, elementary particles) and *Einstein's* theory of gravitation has been worked out, and it has been worked out that the notions *metric* and *energy-impulse tensor* are *universal notions*. Furthermore, then it is obvious that the notion *metric* represents a notion under which all systems considered in this book can be submitted. This possibility shall be discussed in more detail.

9.5 The Riemann Universe

9.5.1 The Definition

As the above considerations showed, observable events can be described by using the following procedure:

1. A suitable kind of metric has to be introduced. This metric represents both the necessary measurement devices and the observed events. Such a metric represents a special position-time space. In order to describe all possibilities, in this context the term *Riemann universe* shall be used.

2. A special metric and special physical quantities such as the energy of mass have to be mutually conditional. Therefore, equations have to be introduced which connect metrices with such physical quantities. Equations which describe such a connection are self-consistency equations. Examples are *Einstein's* field equation or the decomposition (9.119).

Then the dynamics of all physical systems considered in this book can be interpreted as a metric dynamics within the *Riemann* universe. Every special metric represents then a possible physical situation. If one physical situation is considered, another physical situation can be gained by using a special reference frame transition. This possibility shall now be discussed.

9.5.2 Reference Frame Transformations and Riemann Universe

If a physical event is considered, this event can be changed by suitable forces. For example, a charge can be accelerated. In the accelerated case a magnetic field is observable. However, the same result can be achieved by changing the situation of the observer, i. e. by changing the used reference frame. In the above example this means using a relatively moving inertial reference system as a measurement basis. However, not only in the *Euclidean* case such effects are observable, if a curvilinear metric has to be taken as a basis, such reference frame transformations, too, generate new physical situations. For example, due to *Einstein's* principles a local gravitational field can be changed by using an accelerated reference frame. Other examples were discussed in subsection 7.3.9, where it was discussed that path integrals of different physical situations can be gained by using special reference frame transformations. A visualization of a special reference frame transformation shows figure 9.5. As it can be seen, a special geometrical body is visualized. Both pictures show the same body, however, in two different reference frames. For an unbiased observer these two pictures show two different physical situations. On the higher level of a *Riemann* universe all these examples are represented by special metric transformations.

As the example (9.58)+(9.59) showed, negative energy eigenvalues arise if quantum mechanical evolution equations are considered. Such eigenvalues cannot be un-

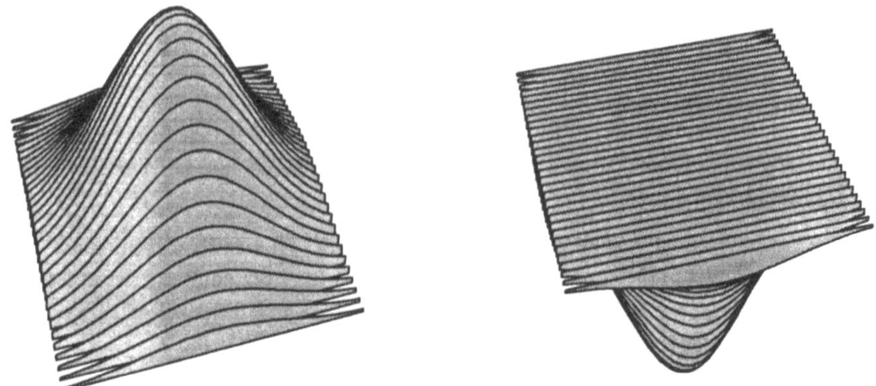

Figure 9.5   Reference frame transformation and physical situation. An ordinary two-dimensional Gaussian function. Two different reference frames. The physical situation which an observer can measure depends on the choosen reference frame

derstood if a *Riemann* universe is taken as a basis. Therefore, another interpretation is necessary. A possible interpretation shall now be given.

9.6   The Underlying Universe

If background functions are considered, corresponding negative energy eigenvalues are possible. This showed the example (9.58)+(9.59). Such negative energies cannot describe a measurable mass m_0 with an impulse p. However, the hypothesis makes sense that background functions with corresponding negative energy eigenvalues characterize a defect of an *underlying universe*, where the defect arises because a measurable particle with rest mass m_0 comes into being. This means that an observable mass m_0 and a defect are mutually conditional. In the sense of this interpretation, a *Riemann* universe and defects of an underlying universe are mutually conditional. This interpretation is then similar to *Dirac's* interpretation.

This remarks shall be enough. In the last chapter the basic ideas of this book and the problem of universality in statistical physics shall be presented in a compressed way.

10 Universality in Statistical Physics and Synergetics

In the following the basic ideas considered in this book shall be presented in a compressed way. First, the highest level of consideration used in this book shall be presented in a compressed way.

10.1 The Highest Level of Consideration

10.1.1 The Riemann Universe

All possible events of the universe represent more or less dynamic states of mass (which can be charged) and fields. As it was discussed, observable states can be interpreted as states within a *Riemann universe*. Under such a notion physical systems can be submitted, i. e. a universal description of different physical systems is possible. Such a *Riemann* universe represents possible *metrices*, where such metrices represent the necessary *reference frames*. As it was discussed, such reference frames in the general case are *curvilinear reference frames*, where necessary reference frames and the *energy of mass/electromagnetic fields* are mutually conditional. Due to the fact that energy of mass/electromagnetic fields and potentials are mutually conditional, curvilinear reference frames and potentials are mutually conditional, too.

10.1.2 Self-Consistency Equations

Such a connection can be described by using special *self-consistency equations*. The considered self-consistency equations represent self-consistent states of matter. A relatively general self-consistency equation is given by *Einstein's field equation of gravitation*, which connects energy of mass with a corresponding potential (which characterizes the basic metric). As it was discussed, basic relativistic and non-relativistic evolution equations of quantum mechanics (*Schrödinger equation, Klein-Gordon equation, Dirac equation*) are valid if an *Euclidean reference frame* can be taken as a basis. As it was shown, such evolution equations also represent self-consistency equations which connect energy of charged masses with a corresponding potential (which represents a special metric). However, such evolution equations have to be considered as special borderline cases, in which case a reference frame and a potential can separately be treated. Furthermore, such equations represent borderline cases where an ordinary potential function can be used to characterize the basic potential.

10.1.3 The Euclidean Space

These examples showed that when a special reference frame such as an inertial
reference system can be taken as a basis, instead of the notions *metric* and *Riemann
universe*, other notions can be introduced to describe the dynamics of a physical
system. In this book normally systems were considered which have to be described by
taking an *Euclidean position-time space* as a basis. If velocities near the light velocity
occur, inertial reference systems were used, and in the borderline case *velocity* $\ll$
light velocity special non-relativistic reference frames, namely *Cartesian reference
frames* with additional time coordinates, were used.

   In the following the physical systems used in this book shall be considered in a
compressed way.

10.2 The Physical Systems Considered in this Book

In particular, *thermodynamic systems, laser systems* and *high-dimensional quantum
systems* were the physical systems considered in this book. In the last chapter aspects
of cosmological systems were taken into account. Very short, biological (sociological)
systems were considered, too. However, such systems can also be classified by using
the term *physical system*.

10.2.1 Multi-Component Systems

Such physical systems are typical multi-component systems, i. e. they represent
ensembles of single components. These components are material components (molecules, atoms, elementary particles) or non-material components such as amplitudes
of elementary functions (for example, mode-amplitudes of a multi-mode laser).

   Such non-material parameters very often are *order parameters*. Such order parameters can be used to describe the essential behavior of the physical system.

   Due to the fact that especially multi-component systems were considered in this
book, the term *system theory* was normally used.

10.2.2 Statistical Description

If such multi-component systems are taken as a basis, normally the behavior has to
be described by using statistical methods. Such statistical methods of description
were taken as a basis.

10.2.3 Synergetic Systems

Such multi-component systems show a collective behavior. In particular, if critical points are reached, a spontaneous emergence of more or less complicated patterns is observable. This effect is called *self-organization*, and the systems which are able to generate pattern in a self-organizing way are called *synergetic systems*. A paradigm of synergetics is represented by a laser system. Such systems were discussed.

10.2.4 Micro- and Macro-Levels

If multi-component systems are considered, various levels of consideration are possible. If one level of consideration is called *microscopic level*, a corresponding mean value level represents then a relatively *macroscopic level* of description. For example, the level of the statistical behavior of the mode-amplitudes of a laser can be considered as a microscopic level, and the level of the measurable intensity correlations can be considered as a macroscopic level.

A real physical system shows a cascade of different levels. For example, a human society can be considered as a multi-component system. The society represents then a macroscopic level, and the people themselves are the single components which represent a relatively microscopic level. Due to the fact that every person consists of organs, further microscopic levels can be defined. In last consequence a cascade of dependent levels can be introduced which includes also systems of elementary particles.

The connection between microscopic and relatively macroscopic levels of multi-component systems was widely discussed in this book.

If such physical systems are considered, the property *universality* is an essential property. In the following this shall be explained in a compressed way.

10.3 Universality

Considering multi-component systems, it was worked out that many kinds of *universality* have to be taken into account, in which case the term *universality* was used when laws, principles, structures, mathematical and physical properties were considered which are able to characterize many classes of physical systems.

10.3.1 Extreme Principles

As it was discussed, *universal extreme principles* such as the *Hamiltonian principle* and the *maximum information entropy principle* (MIEP) can be used as starting points of the discussion, where the *Hamiltonian* principle is a first principle used in deterministic physics, and where the MIEP represents a first principle of statistical physics.

10.3.2 Covariance

If a special physical system is considered, the corresponding evolution equations
have to be form-invariant (covariant) with respect to special transformations. In
the sense of this book, such transformations are *universal transformations*. In this
book, in particular, equations were taken as a basis which are form-invariant with
respect to *Galilean transformations* and to *Lorentz transformations*. However, the
general covariance (form-invariance with respect to arbitrary transformations) was
discussed, too.

10.3.3 Patterns, Self-Similarity

If multi-component systems are considered, a wealth of characteristic patterns is
observable. For example, *periodic motions* and *chaotic states* are observable if oscillatory systems such as laser systems are considered. Other characteristic patterns
of multi-component systems can be characterized by using the term *self-similarity*,
where self-similarity especially is observable if biological systems are considered. For
example, ferns represent self-similar objects. Therefore, such patterns are *universal
patterns*.

10.3.4 Self-Organization

Complicated structures very often emerge in a self-organizing way. The *principle of
self-organization* is a universal principle of nature, in particular, this principle has
to be taken into account if the animated nature is considered.

10.3.5 The Slaving Principle

The emergence of complicated patterns can be considered as a consequence of the
so-called *slaving principle*. This principle is a universal principle which holds near
critical points, where critical points mark regions which are characterized by dramatic changes of basic physical quantities.

10.3.6 Phase Transitions

If *control parameters* are changed in such a way that critical points are reached, *phase
transitions* are observable. Such phase transitions represent a universal phenomenon
of the animated and inanimated nature.

Far from thermodynamic equilibrium and near the thermodynamic equilibrium
such phase transitions are observable. For example, the occurrence of magnetism,
the change of horses' gait form trot to gallop and the occurrence of laser activity
are special kinds of phase transitions.

10.3.7   Hyper-Surface Equations

On a mathematical level such phenomena can be described in a universal way. Such mathematical relations were discussed in this book. In particular, *hyper-surface equations* were introduced which allow to determine statistical distribution function parameters by using only macroscopic measurement quantities. Such hyper-surface equations are totally analytical functions which show the property *self-similarity*. Such self-similar hyper-surface equations can be used in the context of non-linear thermodynamics, in laser physics, in quantum system theory and brain research. In the context of quantum system theory these equations allow to determine path integrals of high-dimensional quantum systems by starting form a macroscopic measurement level.

10.3.8   The Statistical Basic Function

The statistical behavior of many classes of different physical systems can be calculated by using only one statistical function, the *statistical basic function*. This statistical basic function includes partition functions of thermodynamics and laser physics as well as the central parts of path integrals of quantum system theory and path integrals of the theory of *Fokker-Planck systems*. Therefore, such a statistical basic function is a *universal function* in the context of statistical system theory.

10.3.9   Path Integrals, a General Calculation Procedure

A general calculation procedure to calculate high-dimensional partition functions and path integrals in a systematic way was introduced, where the starting point of the calculation is represented by the statistical basic function. This concept is a *universal concept*. This concept can be extended to solve many problems of statistical system theory and quantum system theory.

10.3.10   The Principle of Coupling of Elementary Systems

As it was discussed, this calculation procedure bases on the *principle of coupling of elementary systems*. This principle includes the *principle of superposition of elementary functions* and is a *universal principle*, too. For example, due to this principle electromagnetic phenomena can be decomposed into a set of elementary wave functions by *Fourier's method*. Furthermore, the possibility to replace quantum mechanical evolution equations by a macroscopic energy function and a set of *background functions* can be considered as a consequence of this principle.

10.3.11 Statistical Evolution Equations

As it was discussed, evolution equations of kinds used in statistical physics can be derived by starting from only one basic evolution equation. A successive determination
procedure led then to the mentioned evolution equations. In particular, equations
of the *Fokker-Planck* type and of the *Schrödinger* type were derived, where it was
shown that in the context of such a derivation the quantum mechanical scheme of
description automatically occurs. In this context the close relation between equations of the *Fokker-Planck* type and the *Schrödinger* type was discussed. This close
relation can be used to determine solutions of one type of equation if solutions of
the other type of equation are known. Thus, the mentioned derivation procedure
can be named a *universal calculation procedure*, and the basic evolution equation
can be named a *universal evolution equation*.

10.3.12 Information

In order to have a *universal access* to all considered physical systems, the quantity
information can be taken as a basis. This was discussed in this book. Using this
term the behavior of different classes of physical systems can be characterized in a
universal way. Due to the fact that the mathematical definition of the term *information* allows to put the often used word *information* in concrete terms, a universal
method of description on the basis of information theory was introduced.

10.3.13 Reference Frame Transformations

As it was discussed, a special physical situation can be changed by changing the
situation of the observed object or by changing the situation of the observer, i. e. by
changing the used reference frame. This was discussed in the context of *Einstein's
theory of gravitation*, in the context of inertial reference system transformations and
in the context of path integral transformations in non-relativistic physics. Thus, the
universality of the possibility to achieve any physical situation by using special
reference frames was discussed.

10.3.14 Riemann Universe, Metric

In the general case a reference frame transformation means to generate a new physical situation within a *Riemann universe*, where the special physical situation can
be described by using a suitable kind of *metric*. This was discussed, too. Thus,
Riemann universe and *metric* are universal notions.

   This compressed discussion of the property *universality* shall be enough. In the
following some concluding remarks shall be given.

10.4   The Comprehensive Structure of this Book

In this book systems of the micro-world, of the direct observable surrounding and – in the last chapter – cosmological systems were considered. In particular, the universal aspects were worked out. Universal calculation procedures to calculate high-dimensional multi-components systems of system theory and quantum system theory were introduced. Thus, the structure of this book can be characterized by using the terms *comprehensive* or *holistic*.

With these short remarks I want to close this book. I hope that this book may contribute to a more unified sight of modern theoretical physics.

Volker Achim Weberruß

Bibliography

[1] Abraham, P./ McCullum, W. C.: J. Electroenc. Clin. Neurophysiol., 43, 533 (1977)

[2] Ed.: Abramowitz, M./ Stegun, I.: *Handbook of Mathematical Functions*

[3] Albeverio, S. /Combe, Ph./et al. (Ed.): *Lecture Notes in Physics – Feynman Path Integrals*, Proceedings, Marseille 1978 (Springer-Verlag, Berlin, Heidelberg, New York 1979)

[4] Albeverio, S. /Hoegh-Krohn, R.: *Oscillatory Integrals and the Method of Stationary Phase in Infinitely Many Dimensions*, Inventiones Mathem. 40, 59-106 (1977)

[5] Bestehorn, M.: Dissertation: *Verallgemeinerte Ginzburg-Landau-Gleichungen für die Musterbildung in Konvektions-Instabilitäten* (Universität Stuttgart 1988)

[6] Bronstein, I. N./ Semendjajew, K.A.: *Taschenbuch der Mathematik*, 19. Auflage (Teubner Verlagsgesellschaft, Leipzig 1979)

[7] Clegg, J. C.: *Variationsrechnung* (B. G. Teubner, Stuttgart 1968)

[8] Courant, R./ Hilbert, D.: *Methoden der mathematischen Physik*, 2 Bände, 2. Auflage (Springer-Verlag, Berlin, Heidelberg, New York, Tokyo 1968)

[9] Creutzfeldt, O. D.: *Cortex Cerebri* (Springer-Verlag, Berlin, Heidelberg, New York, Tokyo 1983)

[10] Creutzfeldt, O. D./ Houchin, J.: In A. Remond (Ed.): Handbook of Electroenc. Clin. Neurophysiol., part 2, Amsterdam: Elsevier, pp. 5-55 (1974)

[11] Duering, E./ Otero, D./ Plastino, A./ Proto, A.N.: Phys. Rev. A, Vol. 32, Numb. 2 (1985)

[12] Eckmiller, R./ v. d. Malsburg, C. (eds): *Neural Computers* (Springer-Verlag, Berlin 1988, reprinted 1989 and 1990)

[13] Einstein, A.: *Über die spezielle und die allgemeine Relativitätstheorie*, Wissenschaftliche Taschenbücher, Band 59, 21. Auflage (Vieweg Verlag, 1981)

[14] Endl, K./ Luh, W.: *Analysis* II (5. Auflage), III (4. Auflage) (Akademische Verlagsgesellschaft, Wiesbaden 1981)

[15] *Fachlexikon ABC Physik*, Band 1 (Verlag Harri Deutsch, Zürich, Frankfurt/Main 1974)

[16] *Fachlexikon ABC Physik*, Band 2 (Verlag Harri Deutsch, Zürich, Frankfurt/Main 1974)

[17] Feynman, R. P.: *Space-Time Approach to Non-Relativistic Quantum Mechanics*, Rev. Mod. Phys. 20, 367-387 (1948)

[18] Frank, W.: *Thermodynamik und Statistik*, Vorlesungsmanuskript, Vorlesung gehalten an der Universität Stuttgart 1970/71

[19] French, A. P.: *Die spezielle Relativitätstheorie*, M. I. T. Einführungskurs Physik (Vieweg Verlag, 1982)

[20] Fu, K. S.: *Digital Pattern Recognition*, 2nd edition (Springer-Verlag, Berlin, Heidelberg, New, York, Tokyo 1980)

[21] Fuchs, A./ Haken, H.: In Z. Phys. B – Condensed Matter 71, 519-520 (1988)

[22] Fuchs, A.: *Synergetische Systeme zur Mustererkennung und zur phänomenologischen Modellierung raum-zeitlich aufgelöster gemessener EEGs* (Dissertation, Universität Stuttgart 1990)

[23] Gradshteyn, I. S./ Ryzhik, I. M.: *Tables of Integrals, Series and Products* (Academic Press, New York, London, Toronto, Sydney, San Francisco 1980)

[24] Grosche, Ch./ Steiner, F.: *A Table of Feynman Path Integrals* (to be published, Juni 1992)

[25] Grossberg, S.: Proc. Natl. Acad. Sci. 59, 368 (1968)

[26] Grossberg, S.: In *Synergetics of the brain*. Basar, E./ Flohr, H./ Haken, H./ Mandell, H. (Eds.) (Springer-Verlag, Berlin, Heidelberg, New, York, Tokyo 1983)

[27] Großmann, S.: *Funktionalanalysis* I, 3. Auflage (Akademische Verlagsgesellschaft, Wiesbaden 1975)

[28] Haken, H.: Z. Phys. 181, 96 (1964)

[29] Haken, H.: *Quantenfeldtheorie des Festkörpers* (B. G. Teubner, Stuttgart 1973)

[30] Haken, H.: *Synergetik*, 2. Auflage (Springer-Verlag, Berlin, Heidelberg, New York, Tokyo 1983)

[31] Haken, H.: *Advanced Synergetics* (Springer-Verlag, Berlin, Heidelberg, New York, Tokyo 1983)

[32] Haken, H.: *Laser Theory*, reprint edition (Springer-Verlag, Berlin, Heidelberg, New York, Tokyo 1983)

[33] Haken, H.: In Z. Phys. B – Condensed Matter 61, 335-338 (1985)

[34] Haken, H.: In Z. Phys. B – Condensed Matter 61, 329-334 (1985)

[35] Haken, H.: In Z. Phys. B – Condensed Matter 62, 255-259 (1986)

[36] Haken, H.: In Z. Phys. B – Condensed Matter 63, 487-491 (1986)

[37] Haken, H.: *Atom- und Quantenphysik*, dritte Auflage (Springer-Verlag, Berlin, Heidelberg, New York 1987)

[38] Haken, H.: In Z. Phys. B – Condensed Matter 71, 521-526 (1988)

[39] Haken, H. (Ed.): *Computational Systems, Natural and Artificial* (Springer-Verlag, Berlin, Heidelberg, New, York, Tokyo 1988)

[40] Haken, H.: *Information and Self-Organization* (Springer-Verlag, Berlin, Heidelberg, New York, Tokyo 1988)

[41] Haken, H.: *Licht und Materie*, 2 Bände, 2. Aufl. (Wissenschaftsverlag, Mannheim, Wien, Zürich 1989)

[42] Haken, H.: *Synergetic Computers and Cognition* (Springer-Verlag, Berlin, Heidelberg, New York 1991)

[43] Haken, H./ Wolf, H. C.: *Molekülphysik und Quantenchemie* (Springer-Verlag, Berlin, Heidelberg, New York, Tokyo 1992)

[44] Haken, H.: Z. Phys. B. – Condensed Matter 729

[45] Harris, E. G.: *Quantenfeldtheorie* (R. Oldenbourg Verlag, München, Wien, John Wiley & Sons Gmbh Frankfurt 1975)

[46] Janz, D.: *Die Epilepsien* (Thieme, Stuttgart 1969)

[47] Jaynes, E. T.: Phys. Rev. 106, 4, 620 (1957)

[48] Jaynes, E. T.: Phys. Rev. 108, 171 (1957)

[49] Katz, B.: *Nerv, Muskel und Synapse* (deutsche Übersetzung), 5. Auflage (Thieme, Stuttgart 1987)

[50] Kelso, J. A. S.: *Phasetransitions and critical behavior in human bimanual coordination*, Am. J. Physiol.: Reg. Integ. Comp., 15 (1984)

[51] Kelso, J. A. S./ Southard, D. L./ Goodman, D.: *On the Nature of Human Interlimb Coordination*, Science, 203, 1029 (1979)

[52] Kittel, C.: *Einführung in die Festkörperphysik*, 6. Auflage (R. Oldenbourg Verlag, München, Wien 1983)

[53] Kleinert, H.: *Path Integrals in Quantum Physics, Statistics and Polymer Physics* (World Scientific, 1990)

[54] Klimontovic, Yu. L.: Physica A 142, 390 (1987) und Z. Phys. B – Condensed Matter 65, 125 (1987)

[55] Klimontovic, Yu. L.: *Kleine naturwissenschaftliche Bibliothek, R. Physik: Laser und nichtlineare Optik* (Teubner-Verlag 1971)

[56] Kociszewski, A.: In J. Phys. A: Math. Gen. 18 (1985) L 337-L 339

[57] Kohonen, T.: *Associative Memory* (Springer-Verlag, Berlin, Heidelberg, New, York, Tokyo 1977)

[58] Kohonen, T.: *Self-Organization and Associative Memory*, Springer Ser. Inf. Sci., Vol. 8 (Springer-Verlag, Berlin, Heidelberg, New, York, Tokyo 1983)

[59] Laugwitz, D.: *Ingenieurmathematik*, 5 Bände (Bibliographisches Institut Mannheim, Wien, Zürich 1973)

[60] Lutzenberger, Th. E./ Rockstroh, B./ Birbaumer, N.: *Das EEG, Psychophysiologie und Methodik von Spontan-EEG und ereigniskorrelierten Potentialen* (Springer-Verlag, Berlin, Heidelberg, New York, Tokyo 1985)

[61] Otero, D./ Plastino, A./ Proto, A. N./ Mizrahi, S.: Phys. Rev. A, Vol. 33, Numb. 15 (1986)

[62] Pelster, Axel: *Untersuchung der Eigenschaften spezieller Ljapunov-Funktionen in der Umgebung von Instabilitäten in Nichtgleichgewichtssystemen* (Diplomarbeit, Universität Stuttgart 1992)

[63] Pick, Spaeth, Paus und andere: *Experimentalphysik, Vorlesungsmanuskript, Vorlesung gehalten an der Universität Stuttgart*, erste Fassung des Manuskripts 1969/70

[64] Risken, H.: In Z. Phys. 186, 85 (1965)

[65] Risken, H.: Z. Phys. 186, 85 (1965)

[66] Roepstorff, G.: *Pfadintegrale in der Quantenphysik* (Vieweg Verlag, 1990)

[67] Rösler, F.: *Hirnelektrische Korrelate Kognitiver Prozesse* (Springer-Verlag, Berlin, Heidelberg, New York, Tokyo 1982)

[68] Ruder, H.: *Allgemeine Relativitätstheorie*, Vorlesungsmanuskript (7400 Tübingen, Auf der Morgenstelle 12C, 1984)

[69] Schindel, M.: *Information und Informationsgewinn bei Nichtgleichgewichts-Phasenübergängen* (Diplomarbeit 1987)

[70] Schmidt, R. F. (Hrsg.)/ Thews, G. (Hrsg.): *Physiologie des Menschen*, 21. Auflage (Springer-Verlag, Berlin, Heidelberg, New York, Tokyo 1983)

[71] Schöner, G.: Dissertation: *Das Versklavungsprinzip für stochastische Differentialgleichungen und Anwendungen der Stochastik auf synergetische Systeme* (Universität Stuttgart 1985)

[72] Simon, O.: *Das Elektroenzephalogramm* (Urban & Schwarzenberg, München 1977)

[73] Sommerfeld, A.: *Thermodynamik und Statistik* (Verlag Harri Deutsch, Thun, Frankfurt/Main 1977)

[74] Sommerfeld, A./ Bethe, A.: *Elektronentheorie der Metalle* (Berlin, Göttingen, Heidelberg 1967)

[75] Tass, P.: Dissertation (im entstehen, 1991)

[76] Taylor, G. I.: Philos. Trans. R. Soc. London, A 223, 289 (1923)

[77] Tuszynski, I. A.: In Z. Phys. B – Condensed Matter 65, 375-379 (1987)

[78] Universität Stuttgart, Sonderforschungsbereich 329: *Physikalische und chemische Grundlagen der Molekularelektronik*, Arbeits- und Ergebnisbericht 1986-1988

[79] Universität Stuttgart, Sonderforschungsbereich 329: *Physikalische und chemische Grundlagen der Molekularelektronik*, Arbeits- und Ergebnisbericht 1989-1991

[80] Weberruß, V. A.: *Die Erkennung von Mustern bei Nichtgleichgewichts-Phasenübergängen mit Hilfe des Maximum-Information-Entropie-Prinzips* (Diplomarbeit 1987)

[81] Weberruß, V. A.: *Eine neue Methode zur Berechnung der Lagrangeschen Parameter des Maximum-Information-Entropie-Prinzips, eine Anwendung auf den Laser und der Übergang zu Feynmanschen Wegintegralen* (Dissertation, Universität Stuttgart, 1992)

[82] Witschel, W.: J. Phys. A: Math. Gen. 8, 143 (1975)

[83] Witschel, W.: Chemical Physics 50, 265-271 (1980)

[84] Witschel, W./ Bohmann, J.: J. Phys. A: Math. Gen. 13, 2735 – 2750 (1980)

[85] Witschel, W.: Z. Naturforsch. 36a, 481-488 (1981)

Index